Introduction to Proofs and Proof Strategies

Emphasizing the creative nature of mathematics, this conversational textbook guides students through the process of discovering a proof. The material revolves around possible strategies to approaching a problem without classifying "types of proof" or providing proof templates. Instead, it helps students develop the thinking skills needed to tackle mathematics when there is no clear algorithm or recipe to follow. Beginning by discussing familiar and fundamental topics from a more theoretical perspective, the book moves on to inequalities, induction, relations, cardinality and elementary number theory. The final supplementary chapters allow students to apply these strategies to the topics they will learn in future courses. With its focus on "doing mathematics" through 200 worked examples, over 370 problems, illustrations, discussions, and minimal prerequisites, this book will be indispensable to first- and second-year students in mathematics, statistics and computer science. Instructor resources include solutions to select problems.

Shay Fuchs is Associate Professor (Teaching Stream) in the Department of Mathematical and Computational Sciences at the University of Toronto, Mississauga, Canada, and a member of the Mathematical Association of America. He has been a mathematics educator for more than 25 years. His course based on this text has been taken by more than 1500 students and used by dozens of his colleagues in the past 3 years.

CAMBRIDGE MATHEMATICAL TEXTBOOKS

Cambridge Mathematical Textbooks is a program of undergraduate and beginning graduate-level textbooks for core courses, new courses, and interdisciplinary courses in pure and applied mathematics. These texts provide motivation with plenty of exercises of varying difficulty, interesting examples, modern applications, and unique approaches to the material.

A complete list of books in the series can be found at
www.cambridge.org/mathematics

Recent titles include the following:
Chance, Strategy, and Choice: An Introduction to the Mathematics of Games and Elections, S. B. Smith
Set Theory: A First Course, D. W. Cunningham
Chaotic Dynamics: Fractals, Tilings, and Substitutions, G. R. Goodson
A Second Course in Linear Algebra, S. R. Garcia & R. A. Horn
Introduction to Experimental Mathematics, S. Eilers & R. Johansen
Exploring Mathematics: An Engaging Introduction to Proof, J. Meier & D. Smith
A First Course in Analysis, J. B. Conway
Introduction to Probability, D. F. Anderson, T. Seppäläinen & B. Valkó
Linear Algebra, E. S. Meckes & M. W. Meckes
A Short Course in Differential Topology, B. I. Dundas
Abstract Algebra with Applications, A. Terras
Complex Analysis, D. E. Marshall
Abstract Algebra: A Comprehensive Introduction, J. W. Lawrence & F. A. Zorzitto
An Invitation to Combinatorics, S. Shahriari
Modern Mathematical Logic, Joseph Mileti
Introduction to Proofs and Proof Strategies, Shay Fuchs

Introduction to Proofs and Proof Strategies

Shay Fuchs
University of Toronto

Shaftesbury Road, Cambridge CB2 8EA, United Kingdom

One Liberty Plaza, 20th Floor, New York, NY 10006, USA

477 Williamstown Road, Port Melbourne, VIC 3207, Australia

314–321, 3rd Floor, Plot 3, Splendor Forum, Jasola District Centre,
New Delhi – 110025, India

103 Penang Road, #05–06/07, Visioncrest Commercial, Singapore 238467

Cambridge University Press is part of Cambridge University Press & Assessment,
a department of the University of Cambridge.

We share the University's mission to contribute to society through the pursuit of
education, learning and research at the highest international levels of excellence.

www.cambridge.org
Information on this title: www.cambridge.org/highereducation/isbn/9781009096287

DOI: 10.1017/9781009089005

First published 2023

A catalogue record for this publication is available from the British Library.

A Cataloging-in-Publication data record for this book is available from the Library of Congress.

ISBN 978-1-009-09628-7 Paperback

Additional resources for this publication at www.cambridge.org/fuchs

"Every student in the sciences should be exposed to the basic language of modern mathematics, and standard courses such as calculus or linear algebra do not play this role. The ideal textbook for such a course should not attempt to be encyclopedic and should not assume special prerequisites. It should cover a carefully chosen selection of topics efficiently, engagingly, thoroughly, without being overbearing. Fuchs' text fits this description admirably. The level is right, the math is rock solid, the writing is very pleasant. The book talks to the reader, without ever sounding patronizing. A vast selection of problems, many including solutions, will be splendidly helpful both in a classroom setting and for self-study."

Paolo Aluffi, *Florida State University*

"This well-written text strikes a good balance between conciseness and clarity. Students are led from looking more deeply into familiar topics, such as the quadratic formula, to an understanding of the nature, structure, and methods of proof. The examples and problems are a strong point. I look forward to teaching from it."

Eric Gottlieb, *Rhodes College*

"Fuchs' text is an excellent addition to the 'transitions to proof' literature. I will use it when I next teach such a course. Except for the excellent 'Additional Topics' sections, the content is standard, but the spiraling presentation and helpful narrative around proofs are what truly elevate this text. Fuchs has made every attempt to connect the structure and rigor of mathematics with the intuition of the student. For example, the notion of function arises in three different chapters, with two increasingly rigorous 'provisional definitions,' before a complete definition is given within a wider discussion of relations. I anticipate this approach resonating with students. Fuchs' Chapter 3, which introduces logic and proof strategies, is the most usable presentation of the material I have seen or used. The practice of mathematics and mathematical thinking is communicated well, while opportunities for confusion and obfuscation via a blizzard of symbols are minimized."

Ryan Grady, *Montana State University*

"This book is a must-have resource for an undergraduate mathematics student or interested reader to learn the fundamental topics in how to prove things. The text is thorough and of top quality, yet it is conversational and easy to absorb. Maybe the most important quality, it offers advice about how to approach problems, making it perfect for an introduction to proofs class."

Andrew McEachern, *York University, Canada*

"This is a great choice of textbook for any course introducing undergraduates to mathematical proofs. What makes this book stand out are the early chapters, as well as the 'Additional Topics,' both with accompanying exercises. The book begins by gently introducing proof-based thinking by posing well-motivated prompts and exercises concerning familiar arithmetic of real numbers and the integers. It then introduces fields as a playground to practice working with axioms and drawing (sometimes surprising) conclusions from them. The book proceeds with introducing formal logic, mathematical induction, set theory, and relations on sets. The book's design nicely enables framing classes around a choice sampling among the abundant exercises. The book's 'Additional Topics' can serve to engage those students with a brimming imagination and who are already familiar with basic notions of proofs."

David Ayala, *Montana State University*

"Fuchs' *Introduction to Proofs and Proof Strategies* is an excellent textbook choice for an undergraduate proof-writing course. The author takes a friendly and conversational approach, giving many worked examples throughout each section. Furthermore, each section is replete with exercises for the reader, along with fully worked solutions at chapter's end. This is exactly the 'get your hands dirty' approach students and readers will benefit greatly from!"

Frank Patane, *Samford University*

"The book *Introduction to Proofs and Proof Strategies* by Shay Fuchs takes the problem-solving approach to the forefront by accompanying the reader in the construction and deconstruction of proofs through numerous examples and challenging exercises. The fundamental principles of mathematics are introduced in a creative and innovative way, making learning an enjoyable journey."

Roberto Bruni, *Università di Pisa*

"This textbook is easy to read and designed to enhance students' problem-solving skills in their first year of university. The book really stands out due to the variety and quality of exercises at the end of each chapter. The latter chapters dive into more advanced topics for interested students."

Marina Tvalavadze, *University of Toronto Mississauga*

CONTENTS

SYMBOLS AND NOTATION

Δ	discriminant of a quadratic equation
$\square$	end of proof
$\mathbb{C}$	the field of complex numbers
$\mathbb{N}$	the set of natural numbers
$\mathbb{Q}$	the set of rational numbers
$\mathbb{R}$	the set of real numbers
$\mathbb{Z}$	the set of integers
$\lvert x \rvert$	the absolute value of x
$[a, b]$	a closed interval
(a, b)	an open interval
∞ or $+\infty$	positive infinity
$-\infty$	negative infinity
$\sum$	summation (Sigma) notation
$\prod$	product (Pi) notation
$n!$	n factorial
$b \mid a$	b divides a (or a is divisible by b)
$b \nmid a$	b does *not* divide a
$\gcd(a, b)$	the greatest common divisor of a and b
$\operatorname{lcm}(a, b)$	the least common multiple of a and b
$[x]$	the equivalence class of x
$a \equiv b \pmod{n}$	a and b are congruent modulo n
$\binom{n}{k}$	the binomial coefficient n *choose* k
$\lim a_n$	the limit of the sequence (a_n)
$N_r(c)$	an r-neighbourhood of c
$N_r^*(c)$	a punctured r-neighbourhood of c
$\{\cdots\}$	a set
$y \in C$	y is an element of the set C
$y \notin C$	y is *not* an element of the set C
$C \subseteq D$ or $D \supseteq C$	C is a subset of D

$C \not\subseteq D$ or $D \not\supseteq C$	C is a *not* subset of D
$\emptyset$	the empty set
$A \cap B$	the intersection of A and B
$A \cup B$	the union of A and B
$A \setminus B$	the difference between A and B
A^c	the complement of the set A
$A \times B$	the Cartesian product of A and B
$S/\sim$	the quotient set (or space)
$\bigcap_{\alpha \in J} A_\alpha$	the intersection of sets A_α for $\alpha \in J$
$\bigcup_{\alpha \in J} A_\alpha$	the union of sets A_α for $\alpha \in J$
$P(X)$	the power set of X
$\|A\|$	the number of elements in the finite set A
$\|A\| = \|B\|$	the sets A and B have the same cardinality
$\|A\| \leq \|B\|$	A has cardinality less than or equal to the cardinality of B
$\|A\| \geq \|B\|$	A has cardinality greater than or equal to the cardinality of B
$\forall$	for all/for every (the universal quantifier)
$\exists$	there is/there exists (the existential quantifier)
$\neg$	not (negation)
$\wedge$	and (conjunction)
$\vee$	or (disjunction)
$\Rightarrow$	if-then (implication)
$\Leftrightarrow$	if-and-only-if (equivalence)
$f: A \to B$	f is a function with domain A and codomain B
$f(C)$	the image of the set C under the function f
f^{-1}	the inverse function of f
$f^{-1}(D)$	the pre-image of the set D under the function f
$\lim_{x \to c} f(x)$	the limit of the function f as x approaches c
$\lim_{x \to c^-} f(x), \lim_{x \to c^+} f(x)$	the left- and right-hand limits of f as x approaches c
$f'(c)$	the derivative of the function f at c
$Ker(T), Im(T)$	the kernel and image of a linear map T
i	the complex imaginary unit satisfying $i^2 = -1$
$Re(z), Im(z)$	the real and imaginary parts of the complex number z
$\overline{z}$	the conjugate of the complex number z
$\underline{u}, \underline{v}, \underline{w}$	vectors in $\mathbb{R}^n$
$\vec{AB}$	a geometric vector from A to B
$\|\|\underline{v}\|\|$	the length of a vector in $\mathbb{R}^n$

PREFACE

The main purpose of this book is to help students develop problem-solving and mathematical thinking skills required for the advanced study of mathematics and related fields. Quite often, high school mathematics focuses on applying techniques, carrying out computations and following prescribed algorithms, which are insufficient for being successful in post-secondary mathematics and mathematics-related disciplines. This book presents mathematics as a creative field, which involves experimentation and the development of new ideas as part of the journey.

In terms of content, the first seven chapters cover fundamental topics in mathematics, such as sets, functions, relations and more. The remaining four chapters cover additional topics that can be introduced as a preview for upcoming courses or as a place for formalizing ideas presented more intuitively in previous courses. Chapters 8–11 are independent of each other, and so the instructor may choose to cover any of them, in any order, or none of them. Our recommendation is that a one-semester course cover Chapters 1–7, and then, as time permits, one or more chapters from the additional topics (Chapters 8–11).

A few features distinguish this book from similar books on the subject.

- The book offers a *unique and informal treatment of logic* in Chapter 3. This allows students to experiment with mathematical thinking and writing from the beginning, and to develop problem-solving skills in a more natural and motivated way. The topic of logic is covered lightly, so that students become familiar with the logic symbols and related notions, such as negation and proof by contrapositive.
- Exercises are embedded throughout the text, requiring students to pause and think as they read it. Solutions are available at the end of each chapter.
- Every chapter ends with a Problems section, containing more advanced exercises without solutions. Overall, there are more than 350 problems in the book, most of which are *non-technical*, and require students to be creative and combine previous ideas in a novel way. Working on these problems is an integral part of the learning process, and it is crucial that students attempt them on a regular basis. It is through working on new problems

that students develop the required thinking and creative skills. Solutions to about half of these problems are provided to instructors in a solution manual, which is available at www.cambridge.org/fuchs.
- Quite often, there is a discussion on *how to approach* a problem, and the process of discovering a valid proof.
- The book touches on familiar topics, such as the quadratic formula, functions and prime numbers, but from a *more advanced and theoretical point of view*. This makes the presentation relevant and naturally bridges students' previous knowledge with a deeper and more advanced treatment of these topics.

The book is aimed at first-year students, and there are no required prerequisites other than basic high school algebra. It will require, however, persistence, maturity, and readiness perhaps to abandon old habits in studying mathematics.

PART I
Core Material

1 Numbers, Quadratics and Inequalities

We begin our journey by taking a closer look at some familiar notions, such as quadratic equations and inequalities. And, rather than using mechanical computations and algorithms, we focus on more fundamental questions. Where does the quadratic formula come from and how can we prove it? What are the rules that can be used with inequalities, and how can we justify them? These questions will lead us to look at a few proofs and mathematical arguments. We highlight some of the main features of a mathematical proof, and discuss the process of constructing mathematical proofs.

We also review informally the types of numbers often used in mathematics, and introduce relevant terminology.

1.1 The Quadratic Formula

The general formula for solving an equation of the form $ax^2 + bx + c = 0$,

$$x = \frac{-b \pm \sqrt{b^2 - 4ac}}{2a},$$

was most likely presented to you in high school, and you learned how to use it for solving quadratic equations in various settings. However, if you have not seen a proof, or some sort of explanation, it would be hard to see where this formula is coming from, and why it works. In fact, the proof of this formula is quite straightforward, and requires only certain algebraic manipulations. We therefore start by properly stating a theorem on quadratic equations, and then present a proof using the method the *completing the square*.

Theorem 1.1.1 (The Quadratic Formula) *Let a, b, c be real numbers, with $a \neq 0$. The equation $ax^2 + bx + c = 0$ has*

1. *no real solutions if $b^2 - 4ac < 0$,*
2. *a unique solution if $b^2 - 4ac = 0$, given by $x = -\frac{b}{2a}$,*
3. *two distinct solutions if $b^2 - 4ac > 0$, given by*

$$x = \frac{-b + \sqrt{b^2 - 4ac}}{2a} \quad \textit{and} \quad x = \frac{-b - \sqrt{b^2 - 4ac}}{2a}.$$

Remarks.

- The quantity $b^2 - 4ac$ is called the *discriminant* of the quadratic equation, and is often denoted by Δ, the capital Greek letter *Delta*. The theorem implies that the number of real solutions depends on whether $\Delta < 0$, $\Delta = 0$ or $\Delta > 0$.
- We used the terms *real numbers* and *real solutions* in the statement of the theorem. For now, let us think of real numbers as representing points on an infinite number line. A real number may be a whole number, positive or negative, a fraction, etc. We will discuss later in more detail the notion of a real number and the real number system.

Proof First, let us multiply both sides of the equation by $4a$. As $a \neq 0$, this leads to the following equivalent equation:

$$4a^2x^2 + 4abx + 4ac = 0.$$

Next, we add and subtract the term b^2 to the left-hand side:

$$4a^2x^2 + 4abx + b^2 - b^2 + 4ac = 0.$$

We now observe that the expression $4a^2x^2 + 4abx + b^2$ or, equivalently, $(2ax)^2 + 2 \cdot 2ax \cdot b + b^2$, is a perfect square. Replacing these three terms by $(2ax + b)^2$ and moving the remaining terms to the right-hand side leads to

$$(2ax + b)^2 = b^2 - 4ac.$$

The resulting equation is simpler than the original one, as the unknown x appears only once. It will be easier now to *solve for* x and obtain the quadratic formula. Nevertheless, we must be careful. Solving for x will involve using square roots, which cannot be applied to negative numbers. We therefore consider three possible cases.

1. If $b^2 - 4ac < 0$, then the equation has no real solutions, as $(2ax + b)^2 \geq 0$ for all real numbers x.
2. If $b^2 - 4ac = 0$, then the equation becomes $(2ax + b)^2 = 0$. This implies that $2ax + b = 0$, from which it follows that $x = -\frac{b}{2a}$. Consequently, the equation has a unique solution in this case.
3. If $b^2 - 4ac > 0$, then the equation has two real solutions, given by

   $$2ax + b = \pm\sqrt{b^2 - 4ac}.$$

   We can rearrange this equality to obtain the familiar quadratic formula

   $$x = \frac{-b \pm \sqrt{b^2 - 4ac}}{2a},$$

   as needed. □

Exercise 1.1.2

Suppose that a, b and c are real numbers with $a > 0$ and $c < 0$. How many solutions does the quadratic equation $ax^2 + bx + c = 0$ have?

This was our first proof. Take a close look at it! What features can you identify in that proof? Here are a few important remarks.

Remarks.

- The proof included quite a few words and sentences, in natural language, and not just mathematical symbols such as equations, numbers and formulas. This will happen with most mathematical proofs. A mathematical argument should be made out of complete sentences, which may contain words, symbols, or a combination of both. The words are meant to help the reader follow the logical flow of the argument, explain the main steps, and connect the various parts of the proof. Words such as *if-then*, *and*, and *or* often appear in mathematical arguments and should be used properly. In Chapter 3 we discuss in detail the meaning of these words in mathematics, and how to use them in mathematical proofs.
- At the end of the proof, we placed the symbol □. This is a common way to denote the end of a mathematical proof or, more generally, the end of an argument. In other books you might see the symbol ■ or the acronym *Q.E.D.* used instead. The latter comes from the Latin phrase *Quod Erat Demonstrandum*, meaning "that which was to be shown." In this book we will use our square □.
- A mathematical argument is normally based on facts that have been previously validated, or agreed upon. For example, in the proof of Theorem 1.1.1, we used the identity $(x + y)^2 = x^2 + 2xy + y^2$, which is valid for every two real numbers x and y. Should we have also proved this formula? Well, we could, but we assumed that it was well established prior to proving the theorem, and so there was no need to re-explain or prove it again. This sort of judgment needs to be done each time a mathematical argument is presented to an audience, and you will need to ask yourself which facts should be well known to the reader? What other theorems or claims may I refer to in my proof? What are the main steps, or ideas, in the argument? What is the main tool (or tools) used in my proof? Should I mention them explicitly? With time and practice, you will develop your own style of writing mathematical proofs. The feedback you will get from your teachers and classmates will help you improve your writing and polish your arguments.

There are several reasons why proofs are important. First, a proof *validates* the truth of a general statement. Once a theorem is proved, it remains true forever (unless an error is found). For instance, Theorem 1.1.1 implies that a quadratic equation can never have three distinct solutions, no matter how hard you try to find one, or how much time you spend searching. This is the strength of a proof. We can say now, without a doubt, that every given quadratic equation must have zero, one or two real solutions, and there are no other options or exceptions.

Second, a proof often gives us an insight as to *why* the theorem is valid, and may suggest strategies for proving other related statements. For instance, can we prove a similar theorem on cubic equations? Quite often, discovering a proof for a given statement serves as a step in proving other related or more general results.

1.2 Working with Inequalities – Setting the Stage

In your high school years, you must have spent a substantial amount of time on equations. You had to rearrange, simplify, and solve equations regularly. However, working with inequalities can be more challenging, and one has to be much more careful with arguments and computations involving inequalities.

Example 1.2.1 Consider the equation $\frac{1}{x} = x$. Solving it is quite straightforward. We multiply both sides by x to get the equation $x^2 = 1$, which has solutions $x = 1$ and $x = -1$.

On the other hand, how would one solve the inequality $\frac{1}{x} > x$? Here, we cannot multiply both sides by x as we did previously, since the inequality sign would need to be reversed if $x < 0$. Instead, we consider two cases.

- If $x > 0$, then multiplying by x gives $1 > x^2$, and the *positive* xs satisfying this inequality are those between 0 and 1. Namely, we conclude that $0 < x < 1$.
- If $x < 0$, we get $1 < x^2$ (the inequality sign is reversed), and the *negative* xs satisfying this condition are those which are less than -1. That is, $x < -1$.

In summary, the set of real xs for which $\frac{1}{x} > x$ are the numbers between 0 and 1, and those that are smaller than -1. We can write:

$$\frac{1}{x} > x \quad \text{if and only if} \quad x < -1 \text{ or } 0 < x < 1.$$

In mathematics, the words "*if and only if*" indicate a two-sided implication. If x solves the inequality, then it must satisfy the condition "$x < -1$ or $0 < x < 1$," and if this condition is satisfied, then x solves the inequality.

This example shows some of the complications that may arise while working with inequalities, and how important it is to be able to manipulate them properly.

We begin by listing a few basic properties involving inequalities, which we temporarily refer to as *Basic Facts*.

Basic Facts. *Suppose that x, y, z are real numbers.*

1. *Exactly one of the following must occur: $x < y$, $y < x$ or $x = y$.*
2. *If $x < y$ and $y < z$ then $x < z$.*
3. *If $x < y$ then $x + z < y + z$.*
4. *If $x < y$ and $z > 0$ then $xz < yz$.*
5. *If $z > 0$, there is exactly one positive number $\sqrt{z}$, whose square is z.*

Note that the condition $a < b$ has the same meaning as $b > a$. Moreover, we allow ourselves to use symbols such as $\leq$ and $\geq$ to mean "*less than or equal to*," and "*greater than or equal to*," respectively.

For now, we accept the Basic Facts without proof. As we will discuss later, certain theories in mathematics are built on some foundational assumptions, often called *axioms*, which we accept without proof.

There are more basic properties involving inequalities. We have decided not to include them as they can be derived as consequences from the above basic facts.

Proposition 1.2.2 *For all real numbers x, y, w, z, the following hold true.*

1. *If $x < y$ and $z < 0$ then $xz > yz$.*
2. *If $x < y$ and $z < w$ then $x + z < y + w$.*
3. *If $0 < x < y$ and $0 < z < w$ then $xz < yw$.*

Proof 1. Using Basic Fact 3, we add $-z$ to the inequality $z < 0$, to get

$$z + (-z) < 0 + (-z) \quad \text{which simplifies to} \quad 0 < -z.$$

Now, we use Basic Fact 4 and multiply both sides of $x < y$ by $-z$, which gives us $(-z)x < (-z)y$ or, equivalently, $-zx < -zy$.

Finally, we add zx and zy to both sides (using Basic Fact 3 again), and get $zy < zx$, or $xz > yz$, as needed.

2. We first use Basic Fact 3 twice. Adding z to both sides of $x < y$ gives $x + z < y + z$. Adding y to both sides of $z < w$ gives $y + z < y + w$.

   From Basic Fact 2 it follows that $x + z < y + z$ and $y + z < y + w$ imply $x + z < y + w$, as needed.
3. See the exercise below. □

Note again how our proofs contained words, and that complete sentences were used. If we remove all words from, say, the proof of Part 1 above, we would get something like

$$z < 0 \quad z + (-z) < 0 + (-z) \quad 0 < -z \quad x < y \quad -zx < -zy \quad xz > yz,$$

which cannot be considered a proof (even though it contains the key steps). Without words and complete sentences, it would be very hard for the reader to follow the argument, and the logic that was used. The reader may conclude that the argument is incomplete, unclear, or even flawed.

If we replace all the inequality signs $<$ and $>$ in Proposition 1.2.2 by $\leq$ and $\geq$, respectively, we obtain another valid proposition, that can be proved using similar arguments.

Exercise 1.2.3
Prove Part 3 of Proposition 1.2.2.

Exercise 1.2.4
Use the Basic Facts and Proposition 1.2.2 to prove the following.

- $0 < 1$. (Hint: Show that $1 < 0$ is impossible.)
- For every non-zero real number x, we have $x^2 > 0$.

Our next proposition involves squaring and square-rooting inequalities.

Proposition 1.2.5 *Let a and b be two real numbers.*

1. *If $0 < a < b$ then $a^2 < b^2$ and $\sqrt{a} < \sqrt{b}$.*
2. *Similarly, if $0 \leq a \leq b$, then $a^2 \leq b^2$ and $\sqrt{a} \leq \sqrt{b}$.*

Exercise 1.2.6
Show that the assumption that a and b are positive is crucial. That is, show that if a and b are two real numbers, and $a < b$, then $a^2 < b^2$ might be false.

Proof of Proposition 1.2.5 We prove Part 1 only. The proof of Part 2, which is almost identical, is left to the reader.

Suppose that $0 < a < b$. As $a < b$ and $a > 0$, we can use Basic Fact 3 with $x = z = a$ and $y = b$ to get $a^2 < ab$. Similarly, as $b > 0$, we can use Basic Fact 3 again to get $ab < b^2$. Now, from $a^2 < ab$ and $ab < b^2$ we get, from Basic Fact 1, that $a^2 < b^2$.

To prove the second inequality, we rearrange the inequality $a < b$ and use the difference of squares formula $x^2 - y^2 = (x + y)(x - y)$:

$$a < b \quad \Rightarrow \quad b - a > 0 \quad \Rightarrow \quad (\sqrt{b} + \sqrt{a})(\sqrt{b} - \sqrt{a}) > 0.$$

Note that Basic Fact 5 has been used here implicitly. The symbol $\Rightarrow$ means "*implies that,*" and will be discussed in detail in Chapter 3. Finally, we multiply both sides, using Basic Fact 4 with $z = \frac{1}{\sqrt{b}+\sqrt{a}}$, to get $\sqrt{b} - \sqrt{a} > 0$, or $\sqrt{a} < \sqrt{b}$, as needed. □

1.3 The Arithmetic-Geometric Mean and the Triangle Inequalities

In this section, we present two fundamental and important inequalities in mathematics: the Arithmetic-Geometric Mean Inequality and the Triangle Inequality. Both are central to many areas of mathematics and have numerous applications in physics, statistics and other sciences. Moreover, the approach we use to prove these inequalities is quite general, and can be used to prove other useful statements.

The Arithmetic-Geometric Mean Inequality

We begin with the following definition.

Definition 1.3.1 The *arithmetic mean* of two real numbers x and y is $\frac{x+y}{2}$. If $x, y \geq 0$, then their *geometric mean* is $\sqrt{x \cdot y}$.

You may be already familiar with the arithmetic mean, often called the *average* of two numbers. The geometric mean is another type of *average* which shows up frequently in various applications. Let us look at a few examples.

Example 1.3.2 The arithmetic mean of 2 and 8 is $\frac{2+8}{2} = 5$, and their geometric mean is $\sqrt{2 \cdot 8} = 4$.

The arithmetic mean of 5 and 45 is $\frac{50}{2} = 25$ and their geometric mean is $\sqrt{225} = 15$.

Note how in both cases, the arithmetic mean is greater than the geometric mean. As we will shortly see, this is not a coincidence.

The arithmetic mean of -10 and 7 is -1.5, and their geometric mean is undefined.

Example 1.3.3 A bank offers a savings account that pays interest once a year as follows. The rate for the first year is 10%, and for the second year it is 20%. For instance, if the initial investment is \$250, then after one year, this amount grows to $\$250 \cdot 1.1 = \$275$, and after two years to $\$275 \cdot 1.2 = \$330$. In general, if the initial investment is x, then after two years, it grows to $x \cdot 1.1 \cdot 1.2 = 1.32x$.

What would be a sensible way to define an "*average rate*" for the first two-year period?

Solution

We might want to look for a hypothetical fixed rate r that would lead to the same final amount. Thus, we want r to satisfy the condition $x \cdot r \cdot r = x \cdot 1.1 \cdot 1.2$ (for every value of x). We get

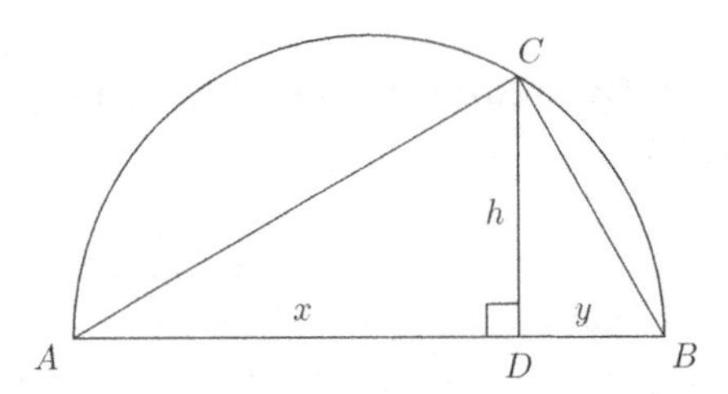

Figure 1.1 A right triangle inscribed in a (half) circle.

$$r^2 = 1.1 \cdot 1.2 \qquad \Rightarrow \qquad r = \sqrt{1.1 \cdot 1.2} = \sqrt{1.32} \approx 1.1489.$$

We conclude that the average interest rate is about 14.89% (and *not* 15%). Note that 1.1489 is the *geometric mean* of 1.1 and 1.2.

Example 1.3.4 In Figure 1.1, AB is a diameter of a circle, and CD is perpendicular to AB. If x, y and h are the lengths of AD, BD and CD, respectively, how can we express h in terms of x and y?

Solution

One way to proceed, is to observe that triangles ADC, CDB and ACB are all right triangles (remember that inscribed angles in a circle, subtended by a diameter, are right angles). Therefore, we can apply the Pythagorean Theorem to get

$$AC^2 = AD^2 + DC^2, \quad CB^2 = CD^2 + DB^2 \quad \text{and} \quad AB^2 = AC^2 + CB^2.$$

We now use the first two equalities to replace AC^2 and CB^2 by $AD^2 + DC^2$ and $CD^2 + DB^2$ in the third equality:

$$AB^2 = (AD^2 + DC^2) + (CD^2 + DB^2) = AD^2 + 2CD^2 + DB^2.$$

Expressing all quantities in terms of x, y and h leads to

$$(x+y)^2 = x^2+2h^2+y^2 \quad \Rightarrow \quad x^2+2xy+y^2 = x^2+2h^2+y^2 \quad \Rightarrow \quad h^2 = xy,$$

which gives $h = \sqrt{xy}$. In other words, h is the *geometric mean* of x and y.

Exercise 1.3.5

Use similar triangles instead of the Pythagorean Theorem, to provide an alternative solution to Example 1.3.4.

We are now ready to present the Arithmetic-Geometric Mean Inequality.

Theorem 1.3.6 (The Arithmetic-Geometric Mean Inequality) *For every two real numbers x and y, we have $x \cdot y \leq \left(\frac{x+y}{2}\right)^2$, and equality holds if and only if $x = y$.*

If, in addition, $x \geq 0$ *and* $y \geq 0$*, then* $\sqrt{x \cdot y} \leq \frac{x+y}{2}$.

The last statement says that when $x, y \geq 0$, the geometric mean is less than or equal to the arithmetic mean of x and y.

Let us first make sure that we fully understand the statement, and what needs to be proved. Keeping in mind that "*if and only if*" means a double-sided implication, we see that there are three statements included in the first sentence.

1. For all real numbers x and y, $x \cdot y \leq \left(\frac{x+y}{2}\right)^2$.
2. If $x = y$, then $x \cdot y = \left(\frac{x+y}{2}\right)^2$.
3. If $x \cdot y = \left(\frac{x+y}{2}\right)^2$, then $x = y$.

The second sentence in Theorem 1.3.6 also needs to be proved, but this will follow by applying square roots to both sides of $x \cdot y \leq \left(\frac{x+y}{2}\right)^2$.

Let us start by focusing on Part 1. How would one prove such an inequality for all x and y? We cannot substitute numbers for x and y, as our argument must be completely general. But, as a start, we can try to rewrite the given inequality, with the hope of simplifying it to an inequality that would be easier to prove. We call this kind of work "rough work," since we are not writing an actual proof yet, but only doing preliminary experimentation to try and *discover a proof.*

Rough Work

$$x \cdot y \leq \left(\frac{x+y}{2}\right)^2 \quad \Rightarrow \quad x \cdot y \leq \frac{x^2 + 2xy + y^2}{4} \quad \Rightarrow \quad 4xy \leq x^2 + 2xy + y^2$$

$$\Rightarrow \quad 0 \leq x^2 - 2xy + y^2 \quad \Rightarrow \quad 0 \leq (x-y)^2.$$

Can this rough work be considered a proof? No. First, there are no words and full sentences explaining the argument. More importantly, a proof cannot begin with the statement that needs to be proved. Remember, we cannot assume the validity of the inequality $x \cdot y \leq \left(\frac{x+y}{2}\right)^2$. Our task is to provide a proof that validates the inequality. We may only use facts that are known to be true, such as elementary high school algebra.

However, we did achieve something. Using algebraic manipulations, we were able to obtain a simpler inequality, namely $0 \leq (x-y)^2$, which holds true for all x and y, by Exercise 1.2.4. We might be able to use it as our starting point, and work backwards in the rough work. If we manage to reverse all the steps, we will end up with the desired inequality, and that would be our proof.

We are now ready to prove the theorem.

Proof of Theorem 1.3.6 1. For every two real numbers x and y, we have $0 \leq (x-y)^2$, as $a^2 \geq 0$ for every real number a. We expand and add $4xy$ to both sides, and get

$$0 \leq (x-y)^2 \quad \Rightarrow \quad 0 \leq x^2-2xy+y^2 \quad \Rightarrow \quad 4xy \leq x^2+2xy+y^2.$$

We divide by 4, and notice that the right-hand side is a perfect square:

$$xy \leq \frac{x^2+2xy+y^2}{4} \quad \Rightarrow \quad xy \leq \left(\frac{x+y}{2}\right)^2.$$

We conclude that the last inequality is valid for all real numbers x and y, as needed.

2. To prove this part, we do not need rough work. We can simply replace y by x and verify that we get an equality. Indeed, if $x = y$, then the left-hand side becomes $xy = x^2$, and the right hand side becomes

$$\left(\frac{x+y}{2}\right)^2 = \left(\frac{2x}{2}\right)^2 = x^2.$$

We have proved that when $x = y$ we have an equality, as needed.

3. Finally, we assume that $xy = \left(\frac{x+y}{2}\right)^2$, and prove that $x = y$. This can be done by simplifying the former equality, with the hope that, at some point, it will become clear that x and y must be equal to each other. The steps we follow resemble the rough work:

$$x \cdot y = \left(\frac{x+y}{2}\right)^2 \quad \Rightarrow \quad x \cdot y = \frac{x^2+2xy+y^2}{4} \quad \Rightarrow \quad 4xy = x^2+2xy+y^2$$

$$\Rightarrow \quad 0 = x^2-2xy+y^2 \quad \Rightarrow \quad 0 = (x-y)^2.$$

The only number that squares to zero is 0, and so $x - y = 0$, from which we conclude that $x = y$, as needed.

To prove the second sentence in the theorem, suppose that $x, y \geq 0$. We already know that the inequality

$$xy \leq \left(\frac{x+y}{2}\right)^2$$

holds true. Using Proposition 1.2.5, we apply square roots to both sides, and get

$$\sqrt{xy} \leq \frac{x+y}{2},$$

which concludes our proof. □

Here is an example in which the Arithmetic-Geometric Mean Inequality is used.

Example 1.3.7 Suppose that a and b are real numbers satisfying $2a^3 + 5b^3 = 200$. What is the largest possible value of $a \cdot b$?

Solution

Let us first make sure we understand the question. There are many pairs of real numbers satisfying the condition $2a^3 + 5b^3 = 200$.

For instance, here are three such pairs:

$$a = 0,\ b = \sqrt[3]{40}, \qquad a = \sqrt[3]{80},\ b = 2, \qquad a = 5,\ b = -\sqrt[3]{10}.$$

For each pair, we can calculate the product $a \cdot b$. For the above pairs, we get the following products:

$$0, \qquad 2 \cdot \sqrt[3]{80}, \qquad -5 \cdot \sqrt[3]{10}.$$

Our task is to find the largest possible product that can be achieved. Although calculus can be used to answer the question, we offer an alternative approach using Theorem 1.3.6.

Let a and b be two real numbers for which $2a^3 + 5b^3 = 200$. Then, by the Arithmetic-Geometric Mean Inequality, with $x = 2a^3$ and $y = 5b^3$, we get

$$(2a^3) \cdot (5b^3) \leq \left(\frac{2a^3 + 5b^3}{2}\right)^2.$$

Using the fact that $2a^3 + 5b^3 = 200$, we get

$$10a^3b^3 \leq \left(\frac{200}{2}\right)^2 \quad \Rightarrow \quad 10(ab)^3 \leq 100^2 \quad \Rightarrow \quad (ab)^3 \leq 1000,$$

from which it follows that $ab \leq 10$. This means that the product ab can never be more than 10.

We now show that the product 10 can actually be achieved for some specific values of a and b. Theorem 1.3.6 states that equality is obtained if and only if $x = y$. In our case, this means that $2a^3 = 5b^3$, and combining the last equality with $2a^3 + 5b^3 = 200$, we get

$$2a^3 + 2a^3 = 200 \quad \Rightarrow \quad a^3 = 50 \quad \Rightarrow \quad a = \sqrt[3]{50}$$

and

$$5a^3 + 5a^3 = 200 \quad \Rightarrow \quad b^3 = 20 \quad \Rightarrow \quad b = \sqrt[3]{20}.$$

Indeed, if $a = \sqrt[3]{50}$ and $b = \sqrt[3]{20}$, then the equality $2a^3 + 5b^3 = 200$ is satisfied, and $ab = 10$.

We conclude that the largest possible value of ab is 10.

The Triangle Inequality

The next inequality we discuss involves absolute values, and so we begin with the following definition.

Definition 1.3.8 The *absolute value* of a real number x, denoted as $|x|$, is defined as

$$|x| = \begin{cases} x & \text{if } x \geq 0 \\ -x & \text{if } x < 0. \end{cases}$$

Exercise 1.3.9
Find the absolute values of the following numbers:

$$0, \qquad \frac{4}{7}, \qquad -\sqrt{2}, \qquad -\pi.$$

Geometrically, we can interpret absolute values as measuring *the distance of a number from the origin* (see Figure 1.2). For instance, $|-4| = 4$ as -4 is four units away from the origin.

The following proposition lists some basic properties of absolute values.

Proposition 1.3.10 *For all real numbers x and y, we have the following:*

$$\sqrt{x^2} = |x|, \qquad |x|^2 = x^2, \qquad -|x| \leq x \leq |x| \qquad \textit{and} \qquad |x \cdot y| = |x| \cdot |y|.$$

Proof Proofs involving absolute values are often done by looking at cases. This is not surprising, given that the absolute value of a real number x is defined through cases: $x \geq 0$ and $x < 0$.

Let us prove the first equality. If $x \geq 0$, then $\sqrt{x^2} = x$, as x is the only non-negative real number that, when squared, gives x^2. If $x < 0$, then $-x > 0$ and $(-x) \cdot (-x) = x^2$. That is, $-x$ is the only positive number that squares to x^2. Thus, $\sqrt{x^2} = -x$ when $x < 0$. To summarize, we have

$$\sqrt{x^2} = \begin{cases} x & \text{if } x \geq 0 \\ -x & \text{if } x < 0, \end{cases}$$

and indeed we see that $\sqrt{x^2} = |x|$ as needed.

The proof of the second equality also follows directly from the definition of absolute values. If $x \geq 0$, then $|x| = x$, and so $|x|^2 = x^2$. If $x < 0$, then $|x| = -x$, and we also get $|x|^2 = (-x)^2 = x^2$.

We leave the remaining properties as an exercise. □

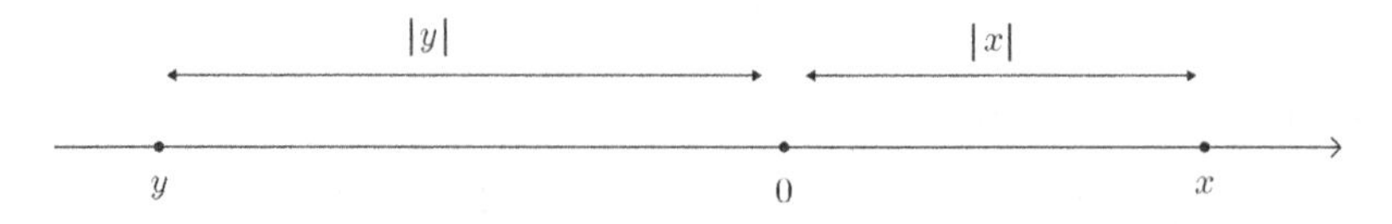

Figure 1.2 Absolute values on the number line.

Exercise 1.3.11
Complete the proof of Proposition 1.3.10.

The equality $|x \cdot y| = |x| \cdot |y|$ shows that products are preserved by absolute values. Namely, the absolute value of a product of two numbers equals the product of their absolute values. This property does not hold for sums. However, there is still a relation between the absolute value of a sum, and the sum of absolute values. This is the well-known Triangle Inequality, stated below.

Theorem 1.3.12 (The Triangle Inequality) *For every two real numbers x and y, we have*

$$|x+y| \leq |x| + |y|.$$

In other words, the absolute value of a sum is always less than or equal to the sum of the absolute values.

Exercise 1.3.13
Find two numbers x and y for which $|x+y| \neq |x| + |y|$.

You might wonder why this inequality is called the *Triangle Inequality*, as there do not seem to be any triangles involved here. The version stated above is the one-dimensional version of the Triangle Inequality, concerning numbers on the number line, but the Triangle Inequality can be generalized to higher dimensions. For instance, in two dimensions the Triangle Inequality implies that the sum of the lengths of any two sides of a triangle must be greater than the length of the remaining side. For now, we will restrict ourselves to the one-dimensional version.

How can we go about proving the Triangle Inequality? There are a few possibilities. For example, we could proceed by cases, according to the sign of x, y and $x+y$. We prefer, however, to follow an approach similar to the one we used for the Arithmetic-Geometric Mean Inequality, and begin with some rough work.

Rough Work

$$|x+y| \leq |x|+|y| \quad \Rightarrow \quad |x+y|^2 \leq (|x|+|y|)^2 \quad \Rightarrow \quad (x+y)^2 \leq |x|^2+2|x||y|+|y|^2$$

$$\Rightarrow \quad x^2+2xy+y^2 \leq x^2+2|xy|+y^2 \quad \Rightarrow \quad xy \leq |xy|.$$

Note how we relied on Propositions 1.2.5 and 1.3.10, and that the last inequality is valid for all real numbers x, y. This will be the starting point of the proof.

Proof of Theorem 1.3.12 For every real number a we have $a \leq |a|$. Using basic algebra and Proposition 1.3.10 we get

$$xy \leq |xy| \Rightarrow x^2+2xy+y^2 \leq x^2+2|xy|+y^2 \Rightarrow x^2+2xy+y^2 \leq |x|^2+2|x||y|+|y|^2$$

$$\Rightarrow \quad (x+y)^2 \leq (|x|+|y|)^2.$$

By square-rooting both sides (Proposition 1.2.5) and using the fact that $\sqrt{a^2} = |a|$ for all real numbers a, we get

$$|x+y| \leq |x|+|y|,$$

as needed. □

The Triangle Inequality is frequently used in many areas of mathematics and science. Here is an example in which we use this inequality to bound an expression.

Example 1.3.14 Find a real number M such that $|x^6 - 3x^3 + 5x| \leq M$ whenever $|x| \leq 2$.

Solution

The key to finding such an M is to relate the absolute value of $x^6 - 3x^3 + 5x$ to the absolute value of x. This can be done by applying the Triangle Inequality:

$$\begin{aligned} |x^6 - 3x^3 + 5x| &\leq |x^6 - 3x^3| + |5x| = |x^6 + (-3x^3)| + |5x| \\ &\leq |x^6| + |-3x^3| + |5x| = |x|^6 + 3|x|^3 + 5|x|. \end{aligned}$$

Now, using the assumption that $|x| \leq 2$, we get

$$\leq 2^6 + 3 \cdot 2^3 + 5 \cdot 2 = 64 + 24 + 10 = 98.$$

Therefore, we can take $M = 98$. Other numbers may work too. For instance, we can take M to be any number larger than 98.

1.4 Types of Numbers

In this section we review, informally, the basic types of numbers used in mathematics, and introduce relevant terminology. The first type of numbers one normally encounters as a child are the *natural numbers* $1, 2, 3, 4, 5, \ldots$. These are also often called the *positive integers*, or the *counting numbers*, as we use them to count objects or people on a regular basis. Some consider 0 to be a natural number as well. However, in this book, we employ the convention that 0 is *not* a natural number.

What can we do with natural numbers? We can add and multiply them to produce new natural numbers: $3 + 5 = 8$, $7 \cdot 3 = 21$, etc. We can take powers of natural numbers: $2^3 = 8$, $5^4 = 625$, etc. We can also subtract and divide natural numbers, for instance: $13-9 = 4$ and $42/7 = 6$, but not every such computation will lead to an answer which is a natural number. To be able to calculate $12-36$ or $18/4$, we need to *extend* our number system to include other types of numbers, such as negative numbers and fractions.

However, before doing so, there is one more thing worth mentioning about the natural numbers: we can *order* them. We know that 3 is smaller than 7, and that 32 is greater than 19, and we even have symbols to express these facts: $3 < 7$ and $32 > 19$. Whenever we are given two natural numbers, either they can be equal to each other, or one is greater than the other. We also use the symbols $\leq$ and $\geq$ for "*less than or equal to*" and "*greater than or equal to.*" For instance, $16 \leq 16, 9 \geq 7$ and $11 \leq 111$ are all correct statements.

The *integer numbers* are obtained by joining the number 0 and the negative numbers $-1, -2, -3, \dots$ to the natural numbers. The integers are thus the numbers $\dots, -3, -2, -1, 0, 1, 2, \dots$. They form an extension of the natural numbers, and are also ordered. For instance: $-15 < -9$ and $4 \geq -4$. Every time we add, subtract or multiply two integers, the result will also be an integer: $(-2) \cdot (-6) = 12, (-8) - 3 = -11, 5 + (-4) = 1$, etc. However, that is not the case for division. The result of a division problem with integers may or may not be an integer. The answer to $(-24)/4$ is an integer, while the answer to $32/(-5)$ is not. This leads to the following definition.

Definition 1.4.1 Let a be an integer, and b a non-zero integer. We say that a is *divisible* by b, or that b *divides* a, if there exists an integer m, for which $a = m \cdot b$.

In other words, a is divisible by b, if it is an *integer multiple* of b.

Note that the definition above does not use division, and that fractions are not mentioned at all. We prefer to introduce terminology about integers without referring to numbers outside that world. Consequently, we make use of multiplication only – an operation that can be carried out with any pair of integers. A person who has not learned about fractions yet, and has no idea what is "a half" or "a third," should be able to decide whether an integer is or is not divisible by another integer.

Example 1.4.2

- 15 is divisible by 3, since $15 = 5 \cdot 3$. Note how we used Definition 1.4.1 with $a = 15, b = 3$ and $m = 5$.
- -7 is divisible by -1 as $-7 = 7 \cdot (-1)$. Here $a = -7, b = -1$ and $m = 7$.
- 0 is divisible by 13 since $0 = 0 \cdot 13$.

- Every integer a is divisible by 1 as $a = a \cdot 1$.
- 19 is *not* divisible by 4, as it is not a multiple of 4.

We use the notion of divisibility to define a few related notions.

Definition 1.4.3

- An integer is *even* if it is divisible by 2. Otherwise, it is *odd*.
- A natural number $p > 1$ is a *prime number* if the only natural numbers that divide p are p and 1.
- A natural number $n > 1$ which is not a prime number is called a *composite number*.

As we will see later, prime numbers are, in some sense, the *building blocks* of the integers. The first few prime numbers are $2, 3, 5, 7, 11, 13, 17, 19, 23, 29, \ldots$. We will later prove that there are infinitely many of them.

Exercise 1.4.4
How many even prime numbers are there? Explain.

Next we introduce fractions or, using more accurate terminology, the *rational numbers*. A rational number is a number of the form $\frac{a}{b}$, where a and b are integers, and b is non-zero. In other words, rational numbers are quotients of integers. As we have learned in elementary and middle school, rational numbers have multiple representations. For example, here are different representations of the number three-quarters:

$$\frac{3}{4}, \qquad \frac{15}{20}, \qquad \frac{-6}{-8}, \qquad 0.75, \qquad .7500.$$

Any number that *has a representation as a quotient of two integers*, with a non-zero denominator, is considered a rational number. For instance, $5\frac{2}{3}$, $-0.333\ldots$ and $\frac{1.2}{1.7}$ are all rational numbers, as we can write them as $\frac{17}{3}, \frac{-1}{3}$ and $\frac{12}{17}$, respectively.

Exercise 1.4.5
Are the integers also rational numbers? Why?

The rational numbers is a number system extending the integers. As we have previously learned, rational numbers can be added, subtracted, multiplied and divided (except that we cannot divide by zero). Also note that the rational numbers are ordered. For instance, $\frac{2}{3}$ is larger than $\frac{7}{12}$.

Rational numbers are often thought of as points on an infinite number line. To construct a number line, we start with an infinite horizontal line, and choose

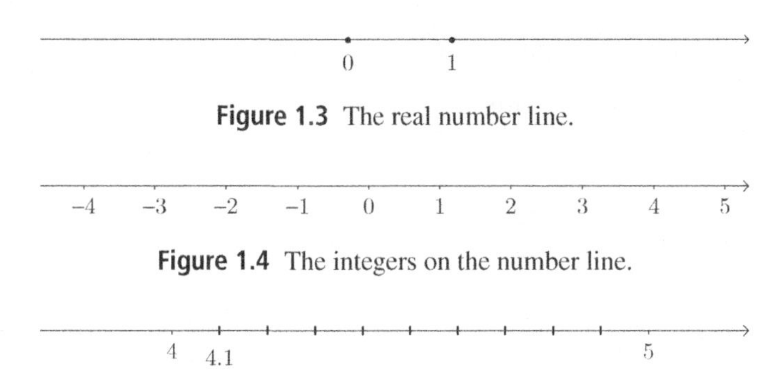

Figure 1.3 The real number line.

Figure 1.4 The integers on the number line.

Figure 1.5 Marking 4.1 on the number line.

a point to be marked as 0 (see Figure 1.3). On the right side of that point we choose another point and mark it as the number 1.

The distance between these two points is referred to as *one unit*, and can be used to mark the remaining integers (Figure 1.4). The distance between every two consecutive integers must be exactly one unit.

Rational numbers can also be identified with points on that number line. For instance, the midpoint between -2 and -1 corresponds to -1.5 or $\frac{-3}{2}$. To mark the number 4.1 on the number line, we can divide the interval from 4 to 5 into ten equal parts and choose the point closest to 4 (Figure 1.5).

A natural question comes to mind. Are the rational numbers sufficient to mark all points on our number line? In other words, does every point on the number line correspond to a quotient of two integers? The answer, which was a surprising revelation to the ancient Greek mathematicians around the fifth century BC, is No. As we will later see, most points on the number line *do not* correspond to rational numbers. We need to further extend our number system in order to "cover" the whole number line. This extension is called the *real number system*, and its elements are called real numbers. It includes the rational numbers, but also many other numbers, such as $\sqrt{7}$, π, e, $\log_2 3$ and more. These numbers cannot be represented as quotients of integers and thus are called *irrational numbers*. We will later address the issue of how to prove that a certain number is irrational.

There are ways to formally construct the real numbers from the rational numbers, but this is a topic for a more advanced course in real analysis. For our purposes, it will suffice to think of them as points on the number line. Moreover, we should remember the following.

- The natural, integer and rational numbers are all part of the real number system. In particular, they are real numbers.
- We can add, subtract, multiply and divide pairs of real numbers (expect dividing by zero). The result is always a real number.

- The real numbers are ordered. For instance, $\sqrt{2} < \pi$ and $e < 3$.

In Chapter 2 we will discuss in more detail the foundational properties of real numbers.

1.5 Problems

1.1 a. Show that if x_1 and x_2 are two solutions of a quadratic equation $ax^2 + bx + c = 0$, with $a \neq 0$, then $x_1 + x_2 = -\frac{b}{a}$ and $x_1 \cdot x_2 = \frac{c}{a}$. These are often called *Vieta's Formulas*. Also, find a formula for $x_1^2 + x_2^2$ in terms of a, b and c.

b. Use Part a to find a quadratic equation with two distinct real solutions, given that the sum of the solutions is 47 and their product is -59.

1.2 What are the dimensions of a rectangle with perimeter 12 and diagonal $\sqrt{20}$?

1.3 a. Find all the real solutions of the equation $x^2 + x + 1 = 0$, if there are any.

b. Alex, a mathematics major student, presented the following argument in his solution to Part a.

> Clearly, $x \neq 0$ as $0^2 + 0 + 1 \neq 0$, so we can divide by x to get
>
> $$x + 1 + \frac{1}{x} = 0 \quad \Rightarrow \quad x = -1 - \frac{1}{x}.$$
>
> Now, replace the term "x" in the original equation with $-1-\frac{1}{x}$, to get
>
> $$x^2 + \left(-1 - \frac{1}{x}\right) + 1 = 0 \Rightarrow x^2 - \frac{1}{x} = 0 \Rightarrow x^3 = 1 \Rightarrow x = 1$$
>
> and so the solution of the equation is $x = 1$.

Is Alex right? If not, where does the mistake occur? Explain.

1.4 For which positive numbers b does there exist a rectangle with perimeter $2b$ and area $\frac{b}{2}$?

1.5 True or False? Explain your answer briefly.

a. For any real number c, the quadratic equation $x^2 + x - c^2 = 0$ has two distinct (real) solutions.

b. If $a > 4$, then the equation $ax^2 + 4x + 1 = 0$ has no real solutions.

c. If $b^2 - 4ac \geq 0$, then the quadratic equation $ax^2 + bx + c = 0$ has at most one solution.

1.6 Provide an alternative proof of Theorem 1.1.1 by using the substitution $x = y - \frac{b}{2a}$.

1.7 a. Solve the inequality $\frac{2}{x} > 3x$.
b. Solve the equation $x^3 = x$ and the inequality $x^3 > x$.

1.8 Let a, b be two *positive* numbers. Decide whether the given statement is true or false. Give a proof or a counterexample.
a. If $a + b \leq \frac{1}{2}$, then $\frac{1-a}{a} \cdot \frac{1-b}{b} \geq 1$.
b. If $\frac{1-a}{a} \cdot \frac{1-b}{b} \geq 1$, then $a + b \leq \frac{1}{2}$.

1.9 Let $a, b > 0$.
a. Prove that $\frac{2}{\frac{1}{a}+\frac{1}{b}} \leq \sqrt{ab}$.
The quantity $\frac{2}{\frac{1}{a}+\frac{1}{b}}$ is called the *harmonic mean* of a and b.
b. **(Harder!)** In Figure 1.6 the quadrilateral $ABCD$ is a trapezoid, O is the point of intersection of its diagonals, and the line segment EF is parallel to the trapezoid's bases and passes through O. Show that the length of EF is the harmonic mean of the lengths of AB and CD. (Hint: Use similar triangles.)

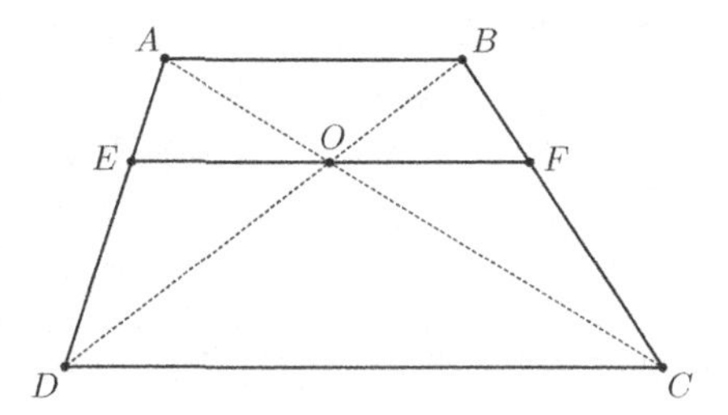

Figure 1.6 A trapezoid and the harmonic mean.

1.10 Use the Arithmetic-Geometric Mean Inequality to find the maximum value of $\left(5 + \sqrt{x^4 + 1}\right) \cdot \left(9 - \sqrt{x^4 + 1}\right)$. Do not use calculus.

1.11 Let x, y, u, v be real numbers.
a. Prove that $4xyuv \leq 2x^2y^2 + 2u^2v^2$.
b. Prove that $(xu + yv)^2 \leq (x^2 + y^2)(u^2 + v^2)$.

1.12 Prove that for every two real numbers x, y, with $x \neq 0$, we have $2y \leq \frac{y^2}{x^2} + x^2$.

1.13 Show that for every two real numbers x and y, we have $2xy \leq \frac{2}{3} \cdot x^2 + \frac{3}{2} \cdot y^2$.

1.14 Prove that if $x > y > z$, then $xy + yz > \frac{(x+y)(y+z)}{2}$.

1.15 Let x, y, z be non-negative real numbers such that $x + z \leq 2$. Prove that $(x - 2y + z)^2 \geq 4xz - 8y$. Determine when equality holds.

1.16 Show that for every two real numbers a, b and $\varepsilon > 0$, we have $ab \leq \frac{a^2}{2\varepsilon} + \frac{\varepsilon b^2}{2}$.

1.17 Show that for every two *positive* real numbers a, b, we have $\frac{a}{a+2b} + \frac{b}{b+2a} \geq \frac{1}{2}$.

1.18 a. Prove that $\sqrt[4]{xyzw} \leq \frac{x+y+z+w}{4}$ for any $x, y, z, w \geq 0$. This is the Arithmetic-Geometric Mean Inequality for four numbers.
Hint: Write $\sqrt[4]{xyzw}$ as $\sqrt{\sqrt{xy} \cdot \sqrt{zw}}$ and use Proposition 1.3.6 multiple times.

b. Prove the Arithmetic-Geometric Mean Inequality for three numbers, $\sqrt[3]{xyz} \leq \frac{x+y+z}{3}$ where $x, y, z \geq 0$, by using Part a with $w = (xyz)^{1/3}$.

1.19 When does equality hold in the Triangle Inequality? State your answer using the words "if and only if," and justify it.

1.20 Prove the following inequality for every x and y: $||x| - |y|| \leq |x - y|$. This is an alternative version of the Triangle Inequality, often called the *reverse Triangle Inequality*.

1.21 Prove that for any $x, y \geq 0$, we have $\left|\sqrt{x} - \sqrt{y}\right| \leq \sqrt{|x - y|}$.
(Hint: Treat the cases $x \geq y$ and $x \leq y$ separately.)

1.22 Find a number M, such that $|x^3 - 4x^2 + x + 1| < M$ for all $1 < x < 3$. Do not use calculus!

1.23 If $1 < x < 2$ find a bound for $\left|\frac{x^3+x^2-1}{x-6}\right|$. That is, find a number M, such that $\left|\frac{x^3+x^2-1}{x-6}\right| < M$ for all $1 < x < 2$.

1.24 Let a, b, c be three real numbers. Prove that $|a - c| \leq |a - b| + |b - c|$.

1.25 Answer the following questions. Explain your answer briefly.

a. Is -117 divisible by -13?
b. Is 3 divisible by 9?
c. Find all the integers that divide -20.
d. If k is a non-zero integer, does it divide $(k + 1)^2 - 1$?

1.26 In Definition 1.4.1, what might be the reason for requiring that b is a *non-zero* integer?

1.27 True or False? Explain your decision briefly.

a. Every non-zero integer divides zero.
b. The only integers that divide 17 are 1 and 17.

c. Let a and b be two non-zero integers. If a divides b and b divides a, then $a = b$.
d. For any integer k, the quantity $k^2 + k$ is divisible by 2.

1.28 Let a, b and c be integers, with $a \neq 0$. Show that if both b and c are divisible by a, then $b + c$ and $b - c$ are also divisible by a.

1.29 a. What are the *rational* solutions to the equation $(x^2 - 1)(x^2 - 7) = 0$?
b. Why is the number $1.111\ldots$ a rational number?
c. Show that $\frac{\sqrt{3}+\sqrt{2}}{\sqrt{3}-\sqrt{2}} - 2\sqrt{6}$ is a rational number.

1.30 The following statements are *false*. Show this by finding a counterexample for each statement.

a. The square root of every rational number is irrational.
b. The sum of two irrational numbers is irrational.
c. The product of an integer and an irrational number is irrational.
d. The quotient of two irrational numbers cannot be a natural number.
e. For every two real numbers a and b, with b non-zero, the quotient $\frac{a}{b}$ is a rational number.

1.6 Solutions to Exercises

Solution to Exercise 1.1.2

We know that $a > 0$ and $c < 0$. Moreover, $b^2 \geq 0$ and hence the discriminant

$$\Delta = b^2 - 4ac = b^2 + 4a \cdot (-c)$$

is positive. Consequently, the quadratic has two distinct solutions.

Solution to Exercise 1.2.3

We use Basic Fact 4 twice. First, $z > 0$ and $x < y$ imply $xz < yz$. Second, $y > 0$ and $z < w$, and so we get $yz < yw$. Applying Basic Fact 2 to $xz < yz$ and $yz < yw$ yields $xz < yw$, as needed.

Solution to Exercise 1.2.4

- As $0 \neq 1$, it follows from Basic Fact 1 that either $0 < 1$ or $1 < 0$. If $1 < 0$, then, using Part 1 of Proposition 1.2.2, we would need to flip the inequality sign if we multiplied both sides by 1. This will give us

$$1 \cdot 1 > 1 \cdot 0 \quad \text{or} \quad 1 > 0,$$

which violates the assumption that $1 < 0$. Therefore, we must have $0 < 1$.
- If $x > 0$, then multiplying both sides by x yields $x^2 > 0$. If $x < 0$, then $-x > 0$ (by adding $-x$ to both sides). Now, multiply both sides by $-x$ to get $(-x) \cdot (-x) > 0 \cdot (-x)$ or $x^2 > 0$.

Solution to Exercise 1.2.6
If $a = -5$ and $b = 3$, then $a < b$ but a^2 is *not* smaller than b^2, since $a^2 = 25$ and $b^2 = 9$.

Solution to Exercise 1.3.5
The triangles ADC and CDB are similar, from which it follows that

$$\frac{CD}{AD} = \frac{DB}{CD} \quad \Rightarrow \quad \frac{h}{x} = \frac{y}{h} \quad \Rightarrow \quad h^2 = xy.$$

We conclude that $h = \sqrt{xy}$. That is, h is the geometric mean of x and y.

Solution to Exercise 1.3.9
The absolute value of a non-negative number is equal to the number itself. Therefore, we get $|0| = 0$ and $|\frac{4}{7}| = \frac{4}{7}$. For negative numbers, we need to negate them, and so $|-\sqrt{2}| = -(-\sqrt{2}) = \sqrt{2}$, and similarly, $|-\pi| = \pi$.

Solution to Exercise 1.3.11
For every real number x, we have either $x = |x|$ or $x = -|x|$. As $|x| \geq 0$, it follows that $-|x| \leq x \leq |x|$.

The identity $|x \cdot y| = |x| \cdot |y|$ can be proved by considering four cases.

- If $x, y \geq 0$, then $x \cdot y \geq 0$, and so $|x \cdot y| = x \cdot y = |x| \cdot |y|$.
- If $x, y < 0$, then $-x, -y > 0$, and $x \cdot y > 0$. We thus get that $|x \cdot y| = x \cdot y = (-x) \cdot (-y) = |x| \cdot |y|$.
- If $x \geq 0$ and $y < 0$, then $x \cdot y \leq 0$ and $-y > 0$. We get $|x \cdot y| = -(x \cdot y) = x \cdot (-y) = |x| \cdot |y|$. The case where $x < 0$ and $y \geq 0$ is proved in a similar way.

Solution to Exercise 1.3.13
There are many possible examples. If $x = -4$ and $y = 7$, then $|x + y| = 3$ and $|x| + |y| = 11$. That is, $|x + y| \neq |x| + |y|$.

Solution to Exercise 1.4.4
The number 2 is a prime number, as it is divisible only by 1 and 2. It is also an even number. Any other positive even integer is not a prime number, as it will be divisible by 1, 2 and itself. Therefore, there is exactly one even prime number, which is 2.

Solution to Exercise 1.4.5
Yes. Every integer is also a rational number, as it can be written as a fraction with denominator 1. For instance: $17 = \frac{17}{1}$, $-3 = \frac{-3}{1}$, etc.

2 Sets, Functions and the Field Axioms

In this chapter we introduce and discuss fundamental notions in mathematics: sets, functions and axioms. Sets and functions show up everywhere in mathematics and science, and are common tools used in mathematical arguments. Moreover, proving statements about sets and functions can further develop our proof-writing and communication skills.

We also demonstrate, in Section 2.3, how axioms are used in mathematics as initial assumptions, from which other statements can be derived.

2.1 Sets

Sets are often considered as one of the most fundamental objects in mathematics. With some effort, nearly all of mathematics can be formulated in terms of sets. Precise definitions for functions, natural numbers, pairs of numbers, and other mathematical objects can be entirely formulated in terms of sets, and so it is essential that we become familiar with sets, and relevant terminology and notation. A formal definition of a set in mathematics is well beyond the scope of this book. Instead, we employ a naive approach, and define a set, informally, as *an unordered collection of objects*. The objects may be referred to as the *elements* or *members* of the set. We also assume that elements in a set are different from each other, and so multiple occurrences of the same element would still count as one element.

Notation and Terminology

A set will normally be labelled by an *uppercase Roman letter* such as A, B, C, S, T, etc. We use *braces* to explicitly list the elements of a set. For instance, if A is a set whose elements are the numbers $1, 7, -1$ and 5, and B is a set whose elements are the word "apple," the letter n, and the number $\frac{1}{2}$, we can write

$$A = \{1, 7, -1, 5\} \qquad \text{and} \qquad B = \left\{\text{apple}, n, \frac{1}{2}\right\}.$$

The symbol $\in$ is used to indicate that a particular object belongs to a set. The notation $y \in C$ reads "y is an element of the set C." For example, we can write

$-1 \in A$ and $n \in B$ to indicate that the number -1 belongs to the set A, and that n is a member of the set B, defined above. To indicate that a certain object is *not* an element of a set, we use the symbol $\notin$. For instance, $3 \notin A$ means that 3 is not an element of the set A.

When all the elements of a set C are also members of another set D, we say that C is a *subset* of D, and denote this fact as $C \subseteq D$ or $D \supseteq C$. Similarly, $C \nsubseteq D$ means that C is *not* a subset of D.

Naturally, two sets A and B are said to be *equal* if they have the same elements. That is, if every element of A is also in B, and vice versa. In that case, we simply write $A = B$.

There is one set that has no elements whatsoever, the *empty* or *null set*. It is commonly denoted by $\varnothing$, or by a pair of braces with nothing between them: $\{\,\}$. The symbol $\varnothing$ comes from the Norwegian alphabet, and was introduced by the mathematician André Weil (1906–1998).

We also have special symbols to denote various sets of numbers, introduced earlier in Section 1.4.

The natural numbers: $\mathbb{N} = \{1, 2, 3, \dots\}$.

The integers: $\mathbb{Z} = \{\dots, -3, -2, -1, 0, 1, 2, 3, 4, \dots\}$.

We use the symbol $\mathbb{R}$ to denote the set of all real numbers, and the symbol $\mathbb{Q}$ to denote the set of rational numbers. In symbols, we may write:

$$\mathbb{Q} = \left\{x \in \mathbb{R} : x = \frac{a}{b} \text{ for some } a, b \in \mathbb{Z} \text{ with } b \neq 0\right\}.$$

Note how the description of $\mathbb{Q}$ is different from the one used for $\mathbb{N}$ and $\mathbb{Z}$. Instead of listing the elements as an infinite sequence, we describe $\mathbb{Q}$ as a subset of real numbers satisfying a certain condition. In words, $\mathbb{Q}$ is the set of all real numbers x that have the form $\frac{a}{b}$, where a is an integer and b is a non-zero integer. We refer to such descriptions as the *set-builder notation*.

For instance, the set of *even* natural numbers $\{2, 4, 6, 8, \dots\}$ can also be written using the set-builder notation as follows:

$$\{n \in \mathbb{N} : n = 2k \text{ for some } k \in \mathbb{N}\}.$$

The sets $\mathbb{N}$, $\mathbb{Z}$, $\mathbb{Q}$ and $\mathbb{R}$ form a chain of subset inclusions: $\mathbb{N} \subseteq \mathbb{Z} \subseteq \mathbb{Q} \subseteq \mathbb{R}$. That is, $\mathbb{N}$ is a subset of $\mathbb{Z}$, $\mathbb{Z}$ is a subset of $\mathbb{Q}$, and $\mathbb{Q}$ is a subset of $\mathbb{R}$.

Example 2.1.1 A set is an *unordered* collection of elements. That is, the order in which we list the members of a set does not matter. Repeated occurrences of an element would still count as one element. For instance, if $A = \{1, 2, 3\}$, then we may also write

$$A = \{3, 2, 1\} \qquad \text{and} \qquad A = \{1, 2, 1, 2, 3\}.$$

Example 2.1.2 Here are a few examples in which we use the above symbols. Convince yourself that these statements are all correct.

$$1 \in \mathbb{N}, \quad 0 \notin \mathbb{N}, \quad \frac{2}{3} \in \mathbb{Q}, \quad \sqrt{2} \in \mathbb{R}, \quad \sqrt{2} \notin \mathbb{Q},$$
$$\{2,4,6\} \subseteq \mathbb{N}, \quad \mathbb{Z} \subseteq \mathbb{R}, \quad \mathbb{R} \nsubseteq \mathbb{Z}.$$

Example 2.1.3 Set notation and the symbols introduced must be used carefully. For instance, what is the difference between $\mathbb{N}$ and $\{\mathbb{N}\}$? The set $\mathbb{N}$ is simply the set of natural numbers, while $\{\mathbb{N}\}$ represents a set with a single element. That element happens to be itself a set – the set of all natural numbers.

Similarly, $\varnothing$ denotes the empty set, while the set $\{\varnothing\}$ is not empty, as it has one element. That element happens to be the empty set.

The set $\mathbb{Z}$ is the set of all integers, while $\{\mathbb{Z}\}$ is a set containing one element, the set of all integers. Note that $\mathbb{Z} \subseteq \mathbb{R}$, while $\{\mathbb{Z}\} \nsubseteq \mathbb{R}$.

Remarks.

- For every set B, we have $\varnothing \subseteq B$ and $B \subseteq B$. In other words, the empty set is a subset of any set, and every set is a subset of itself. Agreeing with the second statement, $B \subseteq B$, is probably easier. Every element of B is obviously also an element of B, which implies that $B \subseteq B$, but why is it true that $\varnothing \subseteq B$ for any set B? Well, let us think carefully about the meaning of being a subset. $A \subseteq B$ means that every element of A is also an element of B. Consequently, $A \nsubseteq B$ if there is at least one element in A which does not belong to B. As there are no elements in the empty set, we cannot argue that it is *not* a subset of B, and hence $\varnothing \subseteq B$. In other words, it is correct to say that every element in $\varnothing$ is also in B, as there are no elements at all in $\varnothing$.

  This might be confusing, but it is a crucial point in understanding mathematical reasoning. A statement of the form "For every $x \in A$, we have..." is considered true when the set A is empty. In Chapter 3, we will take a closer look at the language of mathematics and discuss similar statements in detail.
- As we have seen, a set can be defined or described using the *set-builder notation*. That is, using a rule or a condition its elements must satisfy. This is particularly useful when a set has many elements, or is infinite.

  For instance, the set $C = \{t \in \mathbb{R} : t^2-9 = 0\}$ is the set of all real solutions to the equation $t^2-9 = 0$. As there are only two solutions, we can simply write $C = \{-3, 3\}$.

  The set $D = \{y \in \mathbb{R} : |y| > 5\}$, on the other hand, is an infinite set, and is defined again using the set-builder notation. Here, trying to list the elements separated by commas does not seem to be possible. In Chapter 5 we will develop the tools to prove that indeed it is impossible to list the elements of D as an infinite sequence of numbers.

- When two sets A and B are equal, each element in one set is also a member of the other set. In other words,

$$A = B \quad \text{if and only if} \quad A \subseteq B \text{ and } B \subseteq A.$$

  Although straightforward, this observation is quite important, as it provides a *strategy* for proving equality of sets. Namely, if we are able to prove that $A \subseteq B$ and $B \subseteq A$, we may conclude that A and B are equal to each other.

Exercise 2.1.4

Write the set $A = \{1, 4, 9, 16, 25, \ldots\}$ using the set-builder notation.

Exercise 2.1.5

Here are six sets of numbers:

$$\begin{aligned}
A &= \{3, 5, 7, 9, 11, 13, 15, \ldots\} \\
B &= \{k \in \mathbb{Z} \colon |k| < 5\} \\
C &= \{x \in \mathbb{R} \colon x = 2k + 1 \text{ for some } k \in \mathbb{N}\} \\
D &= \{t \in \mathbb{Q} \colon t = (-1)^n \text{ for some } n \in \mathbb{N}\} \\
E &= \{y \in \mathbb{R} \colon y^4 - 1 = 0\} \\
F &= \{m \in \mathbb{Z} \colon -4 \leq m \leq 4\}.
\end{aligned}$$

Which sets are equal to each other? Explain.

Example 2.1.6 Consider the following two sets:

$$A = \left\{ x \in \mathbb{R} \colon x \neq -1 \text{ and } \left(\frac{x}{x+1}\right)^2 \leq x \right\} \qquad \text{and} \qquad B = \{x \in \mathbb{R} \colon x \geq 0\}.$$

Prove that, even though the definitions of A and B are quite different, the two sets are, in fact, equal to each other.

Solution

Following the remark above, we show that $A = B$ by proving the two inclusions $A \subseteq B$ and $B \subseteq A$.

- Suppose $x \in A$. According to the definition of A, we have

$$\left(\frac{x}{x+1}\right)^2 \leq x.$$

  However, the square of every real number is greater than or equal to zero, and so

$$x \geq \left(\frac{x}{x+1}\right)^2 \geq 0.$$

As $x \geq 0$, we conclude that $x \in B$, and so $A \subseteq B$. Pay close attention to what we have done here. To prove that $A \subseteq B$, we picked an arbitrary element x in A, and argued that it also belongs to B.

- To prove the other inclusion, we now begin by taking an arbitrary element $x \in B$. This means that $x \geq 0$. Our task is to show that $x \in A$. There are two requirements for being in the set A: $x \neq -1$ and $\left(\frac{x}{x+1}\right)^2 \leq x$. The first requirement, $x \neq -1$, is clearly satisfied, as we know that $x \geq 0$. To verify the second requirement, we rearrange the inequality is follows:

$$\left(\frac{x}{x+1}\right)^2 \leq x \quad \Leftrightarrow \quad x^2 \leq x \cdot (x+1)^2 \quad \Leftrightarrow \quad x^3 + x^2 + x \geq 0.$$

The last inequality, $x^3 + x^2 + x \geq 0$, is valid, as $x \geq 0$, and hence the inequality $\left(\frac{x}{x+1}\right)^2 \leq x$ is valid as well. We may now conclude that $x \in A$ and so we have proved the inclusion $B \subseteq A$.

As we have managed to show that $A \subseteq B$ and $B \subseteq A$, it follows that $A = B$, as needed.

The Interval Notation and Set Operations

Open and closed intervals are commonly used sets of numbers. Using the set-builder notation, we can define them as follows.

Definition 2.1.7 Let $a, b \in \mathbb{R}$ such that $a \leq b$.

- A *closed interval* with endpoints a and b, denoted as $[a, b]$, is the set $\{x \in \mathbb{R}: a \leq x \leq b\}$ (Figure 2.1).
- An *open interval* with endpoints a and b, denoted as (a, b), is the set $\{x \in \mathbb{R}: a < x < b\}$ (Figure 2.2).

Note how we used solid and open dots in the diagrams, to indicate whether an endpoint is or is not a member of the set.

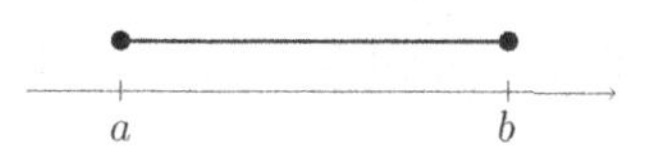

Figure 2.1 A closed interval.

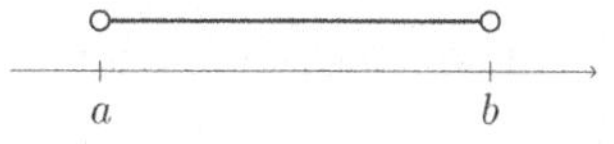

Figure 2.2 An open interval.

The definitions of half-open or half-closed and infinite intervals are given in a natural way. For instance, $[a, b)$ is the set of all real numbers between a and b, including a but not including b. The infinite interval $(-\infty, b)$ consists of all the real numbers which are smaller than b.

As with numbers, we can define operations on sets that can be used to generate new sets from existing ones. Each definition below is accompanied by a *Venn diagram*, a diagram in which the sets are represented by regions in the plane, and the shaded region represents the result of the corresponding set operation.

Definition 2.1.8 Let A and B be two sets.

- The *intersection* of A and B (Figure 2.3), denoted as $A \cap B$, is the set of all elements that belong to both sets:

$$A \cap B = \{x \colon x \in A \text{ and } x \in B\}.$$

If $A \cap B = \varnothing$, we say that the two sets are *disjoint*.

- The *union* of A and B (Figure 2.4), denoted as $A \cup B$, is the set of all elements that belong to A, to B or to both:

$$A \cup B = \{x \colon x \in A \text{ or } x \in B\}.$$

- The *difference* between A and B (Figure 2.5), denoted as $A \setminus B$, is the set of all elements of A which are *not* in B:

$$A \setminus B = \{x \colon x \in A \text{ and } x \notin B\}.$$

Another common notation for the difference between sets is $A - B$, which you may find in some other resources. In this book, we will stick to the notation $A \setminus B$.

- If A is a subset of some universal set U, we define the *complement* of A with respect to U (Figure 2.6), and denote it by A^c, as the set of all elements in U which are not in A:

$$A^c = \{x \in U \colon x \notin A\}.$$

Note that A^c is the same as $U \setminus A$. Quite often, the set U is understood to be the underlying set in our discussion, and thus thought of as our "universe." The notation A^c is used when it is clear from the context what is our underlying universe U.

- The *Cartesian product* of A and B (Figure 2.7), denoted as $A \times B$ is the set of all *ordered pairs* (x, y), in which x is an element of A and y is an element of B:

$$A \times B = \{(x, y) \colon x \in A \text{ and } y \in B\}.$$

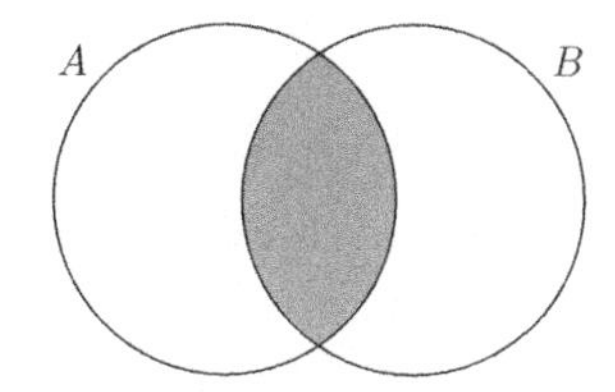

Figure 2.3 The intersection of A and B.

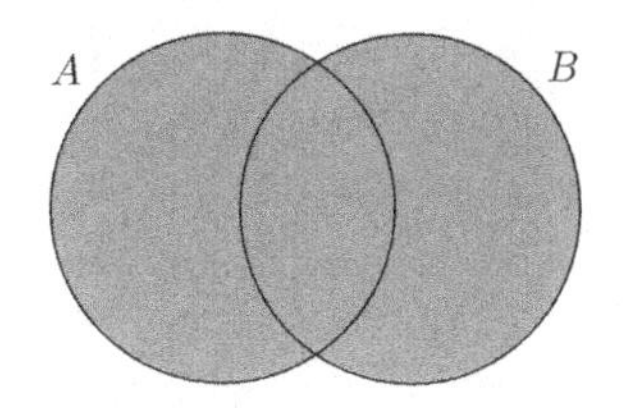

Figure 2.4 The union of A and B.

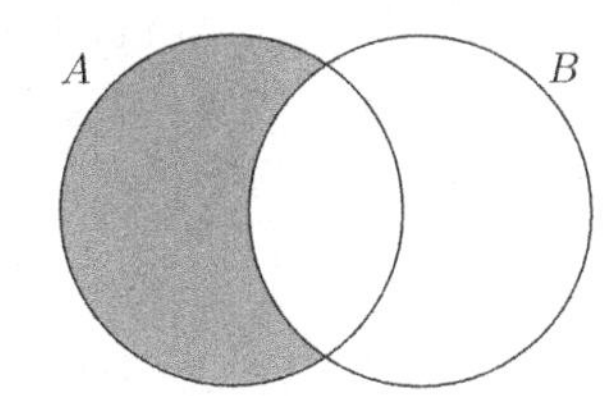

Figure 2.5 The difference between A and B.

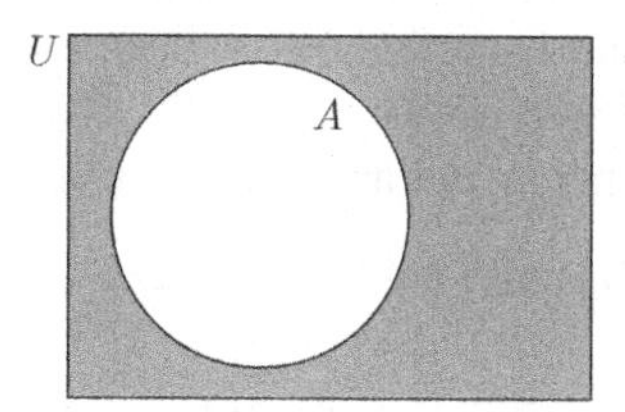

Figure 2.6 The complement of A.

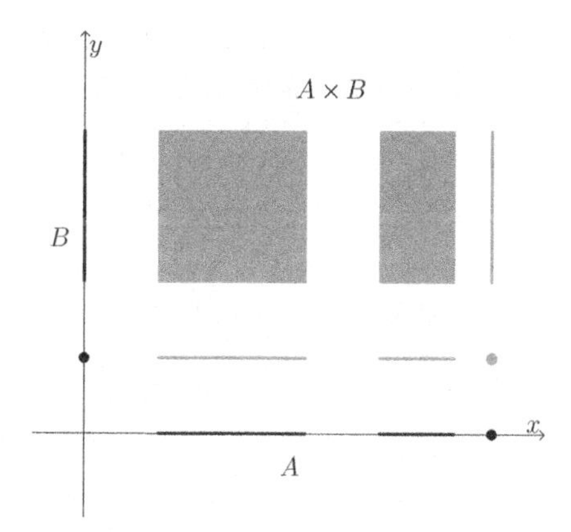

Figure 2.7 The Cartesian product of A and B.

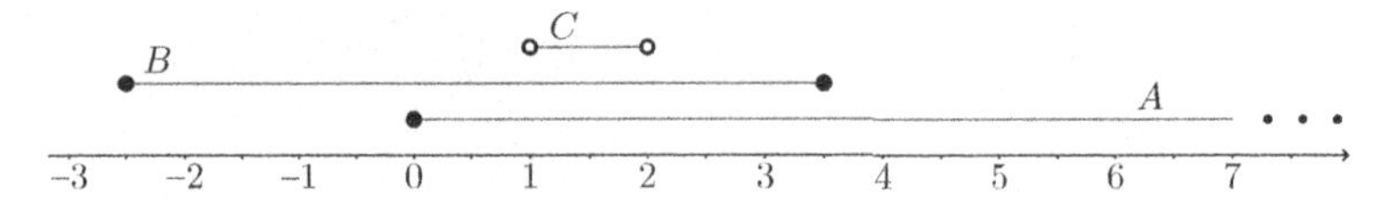

Figure 2.8 The intervals A, B and C on the number line.

Note that members of $A \times B$ are *not* elements of A or of B, and therefore we cannot draw a Venn diagram for this operation. However, we use Figure 2.7 to demonstrate the Cartesian product operation in the case where A and B are sets of numbers.

In this diagram, A and B are sets of real numbers. A is drawn on the x-axis, and B on the y-axis. Each element (x, y) of $A \times B$ is a pair of real numbers, which can be thought of as a point in the plane with coordinates x and y. The grey region in the diagram represents the Cartesian product $A \times B$.

Example 2.1.9 Consider the following sets of real numbers:

$$A = [0, \infty), \qquad B = [-2.5, 3.5] \qquad \text{and} \qquad C = (1, 2).$$

Note how we used the interval notation. The set A contains all real numbers that are greater than or equal to zero. We could also write $A = \{x \in \mathbb{R} : x \geq 0\}$. In Figure 2.8 we represent the sets on the real number line.

Using set operations, we can create the following new sets.

- $A \cup B = [0, \infty) \cup [-2.5, 3.5] = [-2.5, \infty)$.
- $B \setminus C = [-2.5, 3.5] \setminus (1, 2) = [-2.5, 1] \cup [2, 3.5]$.
- $A \cap B = [0, \infty) \cap [-2.5, 3.5] = [0, 3.5]$.
- $B \cap \mathbb{Z} = \{-2, -1, 0, 1, 2, 3\}$.
- $C \cap \mathbb{N} = \varnothing$, as there are no natural numbers between 1 and 2.
- $(B \cap \mathbb{Z}) \times \{\sqrt{2}, \sqrt{3}\} = \{-2, -1, 0, 1, 2, 3\} \times \{\sqrt{2}, \sqrt{3}\} = \{(-2, \sqrt{2}), (-1, \sqrt{2}), (0, \sqrt{2}), (1, \sqrt{2}), (2, \sqrt{2}), (3, \sqrt{2}), (-2, \sqrt{3}), (-1, \sqrt{3}), (0, \sqrt{3}), (1, \sqrt{3}), (2, \sqrt{3}), (3, \sqrt{3})\}$. Note how there are *twelve* elements in $(B \cap \mathbb{Z}) \times \{\sqrt{2}, \sqrt{3}\}$, and that each element is a *pair* of numbers.

Remark. The set $\mathbb{R} \times \mathbb{R}$, also denoted as $\mathbb{R}^2$, is the set of all pairs of real numbers, and can be thought of as representing the set of all points in an infinite two-dimensional plane with a coordinate system. Every subset of $\mathbb{R}^2$ and, in particular, every Cartesian product of two sets of numbers can be thought of as a region in that plane. For instance, the sets

$$S = \{(x, y) \in \mathbb{R}^2 : (x-5)^2 + (y-5)^2 \leq 16\} \quad \text{and} \quad T = (1, 9) \times (1, 9)$$

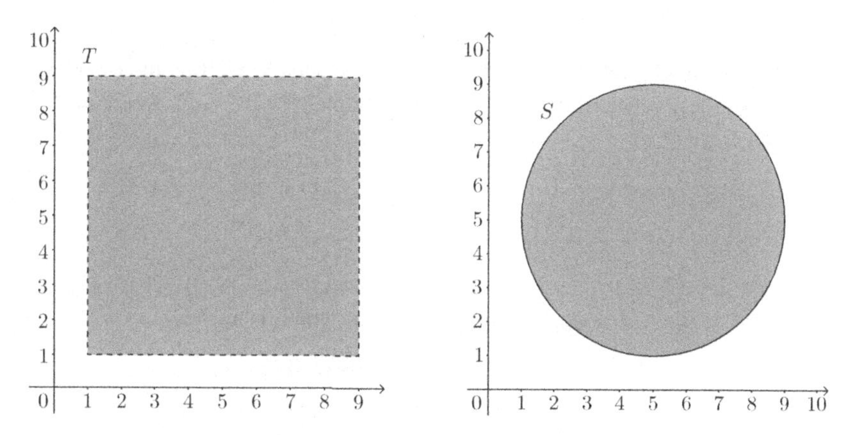

Figure 2.9 The sets T and S.

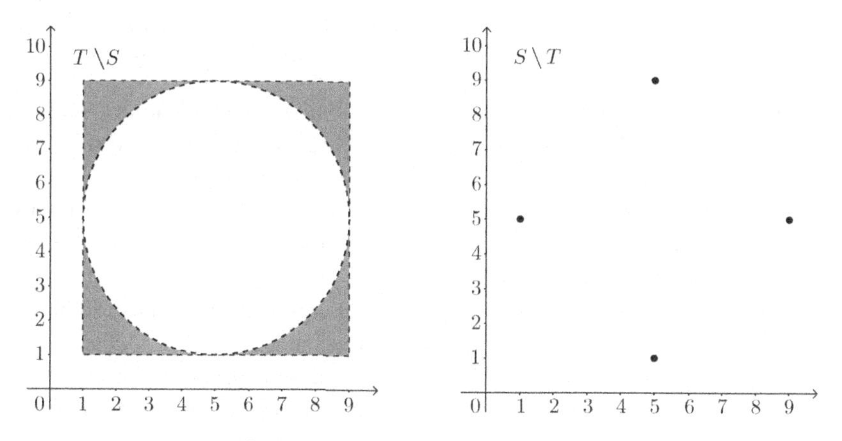

Figure 2.10 The sets $T \setminus S$ and $S \setminus T$.

are both subsets of $\mathbb{R}^2$ (see Figure 2.9). S is a full circle (or a disk) of radius 4, and T is the interior of a square. Note how we used solid and dotted lines to indicate whether the boundary is or is not part of the set.

We can perform various operations on S and T, and obtain more subsets of the plane. For instance, Figure 2.10 shows diagrams representing the sets $T \setminus S$ and $S \setminus T$.

Exercise 2.1.10

Referring to sets S and T defined above, draw diagrams for the sets $S \cap T$ and $S \cup T$.

The Cartesian product operation can be naturally extended to more than two sets.

Definition 2.1.11 Let $n \in \mathbb{N}$, and $A_1, \dots, A_n$ be sets. We define the *Cartesian product* of these sets as the collection of all n-tuples $(a_1, \dots, a_n)$, where $a_i \in A_i$ for each $1 \leq i \leq n$:

$$A_1 \times \cdots \times A_n = \{(a_1, \dots, a_n) \colon a_i \in A_i \text{ for every } 1 \leq i \leq n\}.$$

Example 2.1.12 Let $A = \{a, b\}, B = \{0, 1\}$ and $C = \{x, y\}$. The Cartesian product $A \times B \times C$ has *eight* elements, each of which is a triple:

$$\begin{aligned} A \times B \times C &= \{(a,0,x),(a,0,y),(a,1,x),(a,1,y),\\ &\quad (b,0,x),(b,0,y),(b,1,x),(b,1,y)\}. \end{aligned}$$

We end this section with the following important theorem.

Theorem 2.1.13 (De Morgan's Laws for Sets) *Suppose A and B are subsets of some universal set U. Then $(A \cup B)^c = A^c \cap B^c$ and $(A \cap B)^c = A^c \cup B^c$.*

These are examples of *set identities*. Much like algebraic identities, such as $(a+b)(a-b) = a^2 - b^2$, set identities highlight relations between various set operations, and can be used to simplify expressions involving sets.

In the Venn diagram of Figure 2.11, the region $(A \cup B)^c$ is shaded, and with some imagination one can see that the same region represents $A^c \cap B^c$, which explains the first identity $(A \cup B)^c = A^c \cap B^c$.

However, we must be cautious. An informal argument that relies too heavily on diagrams may be incomplete or flawed, and the risk is higher when there are many sets involved. Although such arguments are often accepted by mathematicians, we prefer to be more careful, and validate the claim with a proof that relies solely on the definitions of set operations, rather than on diagrams.

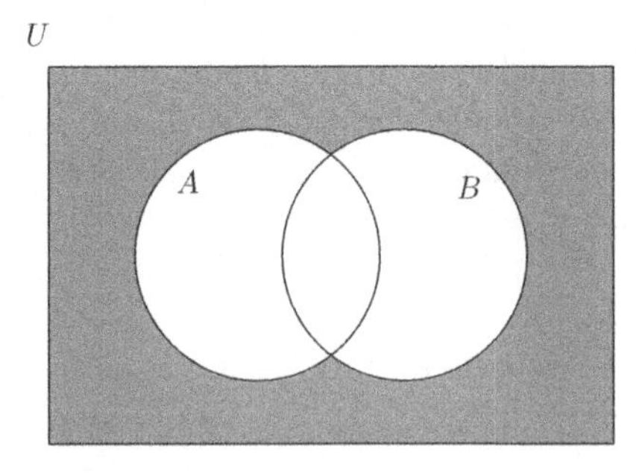

Figure 2.11 The complement of $A \cup B$.

Proof We prove the first identity by proving two inclusions: $(A \cup B)^c \subseteq A^c \cap B^c$ and $A^c \cap B^c \subseteq (A \cup B)^c$.

- Let x be an arbitrary element in $(A \cup B)^c$. According to the definitions of complements and unions, we conclude that $x \notin A \cup B$, and so $x \notin A$ and $x \notin B$. This means that x is in both A^c and B^c, and therefore $x \in A^c \cap B^c$. We have thus proved the inclusion $(A \cup B)^c \subseteq A^c \cap B^c$.
- To prove the other inclusion, we start with an arbitrary element x in $A^c \cap B^c$. The definitions of complements and intersections imply that

$$x \in A^c \text{ and } x \in B^c,$$

or, equivalently, that

$$x \notin A \text{ and } x \notin B.$$

Finally, as x is not in A nor in B, it cannot be in their union:

$$x \notin A \cup B \qquad \Rightarrow \qquad x \in (A \cup B)^c.$$

We have thus proved that $A^c \cap B^c \subseteq (A \cup B)^c$.

As we have established both inclusions, we conclude that $(A \cup B)^c = A^c \cap B^c$, as needed. The other identity is proved in a similar way, and is left to the reader. □

2.2 Functions

A function is a name we use for a "mathematical machine," creating one output for every given input. For instance, the formula $y = x^2$ can be used to describe a process that assigns to each real number x its square x^2. If the input is $x = 3$, then the output is given by

$$y = x^2 = 3^2 = 9.$$

Similarly, if $x = -1.5$, then $y = (-1.5)^2 = 2.25$, and so on. We often label a function with a lowercase letter such as f, and write $f(x) = x^2$ to indicate that f assigns to a given number x its square. To indicate the fact that f sends 3 to 9, and -1.5 to 2.25, we write

$$f(3) = 9 \qquad \text{and} \qquad f(-1.5) = 2.25.$$

Most likely, you have already encountered the notion of a function in your high school mathematics classes, and have seen many examples of functions, such as linear, trigonometric and exponential. Functions appear everywhere in mathematics and science, and are used to describe relations between two quantities, procedures and processes, geometric transformations, and more.

Nevertheless, as surprising as it may seem, the inputs and outputs of a function need not be numbers. Moreover, functions do not have to be necessarily defined or described by a formula.

For example, imagine a function g that assigns to every word its first letter:

$$g(\text{hello}) = \text{h}, \qquad g(\text{chair}) = \text{c}, \qquad g(\text{love}) = \text{l}, \ \ldots .$$

This is a perfectly legitimate function. In your future studies, you may see functions where the inputs and outputs are sets, vectors, matrices, or even other functions. This broader approach allows more flexibility in using and applying functions, while keeping common language, terminology and symbols.

With these remarks in mind, we present the definition of a function.

Definition 2.2.1 A *function* f from a set A to a set B is a rule that assigns to *each* element a in A, *exactly one* element $f(a)$ in B called the *image* of a under f.

To be honest, this definition is still somewhat informal, as we do not explicitly state what "a rule" is. Nevertheless, it is good enough for our needs in this book. A more formal definition of a function will be discussed in Chapter 5 (see page 186).

Notation and Terminology

- The sets A and B in Definition 2.2.1 are called the *domain* and the *codomain* of f, respectively.
- We use the notation $f\colon A \to B$ to indicate that f is a function with domain A and codomain B.
- The set

$$\{y \in B\colon y = f(x) \text{ for some } x \in A\}$$

is called the *range* or the *image* of f, and we denote it as $f(A)$.

Example 2.2.2 Define a function f as follows:

$$f\colon [-3,4] \to (-2,2), \qquad f(x) = \sin x \quad (\text{where } x \text{ is in radians}).$$

The image of $\frac{\pi}{2}$ is $f\left(\frac{\pi}{2}\right) = \sin\left(\frac{\pi}{2}\right) = 1$, and the image of $-\frac{\pi}{6}$ is $f\left(-\frac{\pi}{6}\right) = \sin\left(-\frac{\pi}{6}\right) = -\frac{1}{2}$.

Note that the domain of f is the closed interval $[-3,4]$, and so quantities such as $f(2\pi)$ and $f(-\pi)$ are undefined.

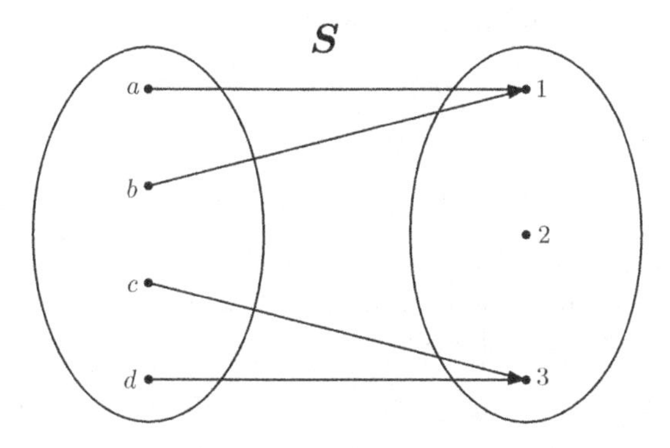

Figure 2.12 An arrow diagram for a function S.

The range of f is the set of all possible images of f. The codomain is the open interval $(-2, 2)$, but not all numbers in that interval are images. For instance, numbers such as -1.5 or 1.9 could never be reached by the sine function. Based on our knowledge of the sine function, we conclude that the image of f is the closed interval $[-1, 1]$:

$$f([-3, 4]) = [-1, 1].$$

Example 2.2.3 A function can be defined using an *arrow* or *bubble diagram*. For instance, Figure 2.12 defines a function $S:\{a, b, c, d\} \to \{1, 2, 3\}$. Note that this diagram indeed defines a function. Every element in the set $\{a, b, c, d\}$ is sent to a unique element in $\{1, 2, 3\}$. The diagram implies that $S(a) = S(b) = 1$ and $S(c) = S(d) = 3$. The image of the function S is the set $\{1, 3\}$.

Example 2.2.4 We now define a function $g\colon \mathbb{Z} \times \mathbb{Z} \to \mathbb{Z}$, which assigns to every pair of integers its sum of squares. In a formula, we write $g(m, n) = m^2 + n^2$. Here, elements in the domain are pairs of integers. To be consistent with Definition 2.2.1 we should have written $g((m, n))$, but it is standard practice to shorten this notation and simply write $g(m, n)$.

Let us compute the images of the pairs $(2, 5)$ and $(-1, 3)$:

$$g(2, 5) = 2^2 + 5^2 = 4 + 25 = 29, \qquad g(-1, 3) = (-1)^2 + 3^2 = 10.$$

Finding an explicit description of the range of g requires advanced tools which are beyond the scope of this book.

Example 2.2.5 We can define a function T on the set of all calendar months, that returns the number of days in a given month:

$$T:\{\text{Jan., Feb., ..., Dec.}\} \to \{25, 26, 27, ..., 35\}$$

$$T(y) = \text{(number of days in month } y\text{, in a non-leap year)}.$$

For example,

$$T(\text{Jul.}) = 31, \qquad T(\text{Apr.}) = 30 \quad \text{and} \quad T(\text{Feb.}) = 28.$$

The image of T is the set of possible outcomes, which is $\{28, 30, 31\}$.

Remark. If A is a set, and $f, g\colon A \to \mathbb{R}$ are functions with codomain $\mathbb{R}$, we can naturally use basic arithmetic to create new functions. For instance, the *sum* of f and g will be defined as follows:

$$f + g\colon A \to \mathbb{R}, \qquad (f + g)(a) = f(a) + g(a) \text{ for every } a \in A.$$

Other operations, such as differences, products, quotients and powers of functions are defined similarly.

Exercise 2.2.6
Suppose that two functions $f, g\colon [-1, 1] \to \mathbb{R}$ are given by

$$f(x) = \sqrt{1 - x^2} \qquad \text{and} \qquad g(x) = \sin(3x).$$

Find expressions for the functions $f + g, f \cdot g^2$ and $-5f^3 - 4g$.

Next, we present the definition of the graph of a function.

Definition 2.2.7 Let A and B be two sets, and $f\colon A \to B$ a function. The *graph* of f is the set

$$\{(x, y) \in A \times B\colon y = f(x)\}.$$

In other words, the graph of f is the set of all pairs of the form $(x, f(x))$ where $x \in A$. This is a *subset* of the Cartesian product $A \times B$.

Interestingly, this definition applies to all functions, even those involving non-numerical elements. For instance, the graph of the function T from Example 2.2.5 is the following set:

$$\{(\text{Jan.}, 31), (\text{Feb.}, 28), (\text{Mar.}, 31), (\text{Apr.}, 30), (\text{May.}, 31), (\text{Jun.}, 30),$$
$$(\text{Jul.}, 31), (\text{Aug.}, 31), (\text{Sep.}, 30), (\text{Oct.}, 31), (\text{Nov.}, 30), (\text{Dec.}, 31)\}.$$

Note how we use the term "graph," even when it is impossible to draw the graph of the function in the usual sense.

If the domain and the codomain of a function are sets of numbers, then every element in the graph is a pair of numbers, which can be drawn as a point in the two-dimensional plane $\mathbb{R}^2$.

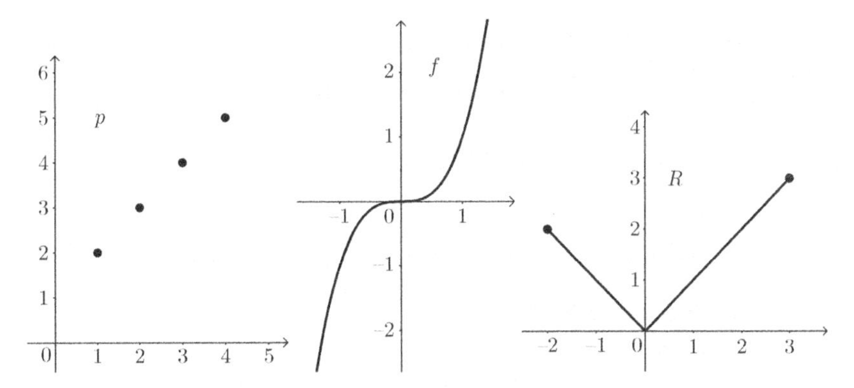

Figure 2.13 The graphs of the functions p, f and R.

Example 2.2.8 Consider the following three functions:

$$p\colon \mathbb{N} \to \mathbb{N}, \qquad p(n) = n + 1$$

$$f\colon \mathbb{R} \to \mathbb{R}, \qquad f(x) = x^3$$

$$R\colon [-2, 3] \to \mathbb{R}, \qquad R(x) = |x|.$$

Figure 2.13 shows the geometric representation of their graphs in a two-dimensional coordinate system. Each point on the graph of a function corresponds to an element in its domain. For instance, the domain of p is the set of all natural numbers, and so every point on its graph must correspond to a pair (x, y) where $x \in \mathbb{N}$ and $y = x + 1$. Connecting the points with a straight line would be incorrect!

Note that for the functions p and f, the diagram actually represents only *part* of the graph, as the domains of p and f contain arbitrarily large numbers.

Our next example requires a careful proof, and involves several ideas from previous sections.

Example 2.2.9 Consider the function $f\colon \mathbb{R} \to \mathbb{R}$, given by $f(x) = \frac{x}{1+x^2}$. Prove that $f(\mathbb{R})$, the image of f, is the closed interval $\left[-\frac{1}{2}, \frac{1}{2}\right]$.

Solution

Our task is to prove that $f(\mathbb{R}) = \left[-\frac{1}{2}, \frac{1}{2}\right]$, which is an equality between two sets: the range or image of f, and a closed interval. As we did previously, we prove this by showing mutual subset inclusions.

- Suppose that $y \in f(\mathbb{R})$. Then, by the definition of the image of a function,

$$y = f(x) = \frac{x}{1 + x^2}$$

for some $x \in \mathbb{R}$. We need to show that $y \in \left[-\frac{1}{2}, \frac{1}{2}\right]$. As $y = \frac{x}{1+x^2}$, the condition $y \in \left[-\frac{1}{2}, \frac{1}{2}\right]$ is equivalent to the following inequalities:

$$-\frac{1}{2} \leq y \leq \frac{1}{2} \quad \Leftrightarrow \quad -\frac{1}{2} \leq \frac{x}{1+x^2} \leq \frac{1}{2} \quad \Leftrightarrow \quad -(1+x^2) \leq 2x \leq 1+x^2$$

$$\Leftrightarrow \quad -x^2-2x-1 \leq 0 \leq x^2-2x+1 \quad \Leftrightarrow \quad -(x+1)^2 \leq 0 \leq (x-1)^2.$$

Note that the inequalities

$$-(x+1)^2 \leq 0 \leq (x-1)^2$$

are valid for all real numbers x. Hence, we conclude that $y \in \left[-\frac{1}{2}, \frac{1}{2}\right]$, which proves the inclusion $f(\mathbb{R}) \subseteq \left[-\frac{1}{2}, \frac{1}{2}\right]$.

- To prove the other inclusion, consider an arbitrary element $y \in \left[-\frac{1}{2}, \frac{1}{2}\right]$. Our goal is to show that $y \in f(\mathbb{R})$. More explicitly, we need to show that $y = f(x) = \frac{x}{1+x^2}$ for some real number x. Let us think of the equality $y = \frac{x}{1+x^2}$ as an equation in the variable x, treating y as a fixed constant. We may rearrange the equation as follows:

$$y \cdot (1+x^2) = x \quad \Leftrightarrow \quad y \cdot x^2 - x + y = 0.$$

Showing that $y \in f(\mathbb{R})$ is equivalent to proving that the equation

$$y \cdot x^2 - x + y = 0$$

has a solution. If $y \neq 0$, the above equation is a quadratic, with discriminant

$$\Delta = (-1)^2 - 4 \cdot y \cdot y = 1 - 4y^2.$$

As $-\frac{1}{2} \leq y \leq \frac{1}{2}$, we have

$$|y| \leq \frac{1}{2} \quad \Rightarrow \quad y^2 \leq \frac{1}{4} \quad \Rightarrow \quad 1-4y^2 \geq 0.$$

We conclude that $\Delta = 1-4y^2 \geq 0$, and hence, by Theorem 1.1.1, the equation has at least one real solution. This proves that $y \in f(\mathbb{R})$. If $y = 0$, then the equation $y \cdot x^2 - x + y = 0$ becomes $-x = 0$, which is not a quadratic. However, it still has a solution, namely, $x = 0$. As we have showed that if $y \in \left[-\frac{1}{2}, \frac{1}{2}\right]$ then $y \in f(\mathbb{R})$, we have proved the other inclusion $\left[-\frac{1}{2}, \frac{1}{2}\right] \subseteq f(\mathbb{R})$, as needed.

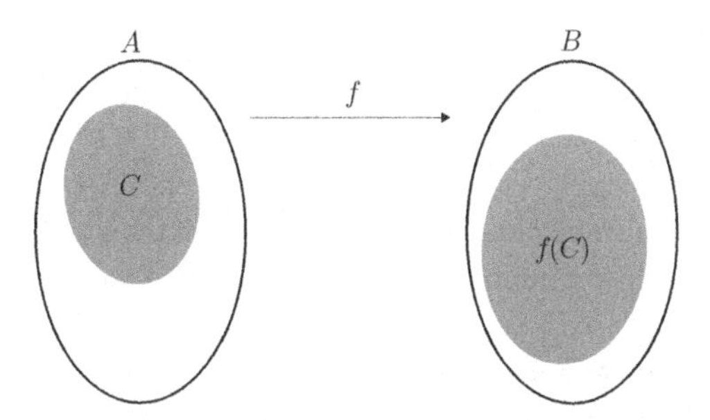

Figure 2.14 The image of a set.

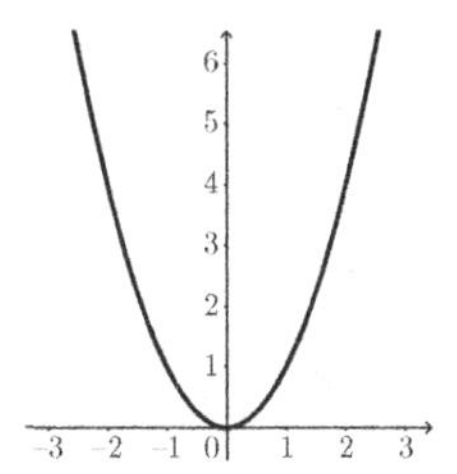

Figure 2.15 The graph of $f(x) = x^2$.

We end this section with another proof. This time, we prove a general statement about functions and sets, rather than a statement about a specific function. First, we introduce the notion of the *image of a set* under a function.

Definition 2.2.10 Let $f\colon A \to B$ be a function, and C a subset of A. The *image of C under f*, denoted as $f(C)$ is the set

$$\{y \in B\colon y = f(x) \text{ for some } x \in C\}.$$

The image of C under f is the set of all images of elements in C. Note that $f(C)$ is always a subset of B, and when C is the whole domain of f, the image of C under f is what we previously referred to as the *image of f*. Figure 2.14 visualizes the notion of the image of a set.

Example 2.2.11

Consider the function $f\colon \mathbb{R} \to \mathbb{R}$ given by $f(x) = x^2$ (Figure 2.15).

- If $C = \{-1, 1, 2, 3\}$, then the image of C under f is the set of all numbers that are obtained by squaring the elements of C. We get:

$$f(C) = \{(-1)^2, 1^2, 2^2, 3^2\} = \{1, 4, 9\}.$$

- What is $f([2,4])$? If $2 \leq x \leq 4$, then $4 \leq x^2 \leq 16$. Conversely, if $4 \leq y \leq 16$, then $2 \leq \sqrt{y} \leq 4$. We therefore conclude that

$$f([2,4]) = [4,16].$$

  This result can also be obtained informally from the graph of f, by restricting the domain to the interval $[2,4]$ and looking at the y-values obtained in this case.
- Here are a few more images of sets under f. Make sure you can justify each of them algebraically and also by referring to the graph:

$$f([-5,-4)) = (16,25]$$
$$f((-1,1)) = [0,1)$$
$$f(\mathbb{Z}) = \{0,1,4,9,16,25,\dots\}$$
$$f(\mathbb{R}) = [0,\infty).$$

We are now ready to state and prove the following proposition.

Proposition 2.2.12 *If $f: A \to B$ is a function, and C, D are subsets of A, then $f(C \cup D) = f(C) \cup f(D)$.*

In words, the proposition says that the image of a union of two sets is equal to the union of their images under the function f. Note that both $f(C \cup D)$ and $f(C) \cup f(D)$ are subsets of B. To prove that they are equal, we show that each is a subset of the other.

Proof

- Let $y \in f(C \cup D)$. According to the definition of images of sets, we conclude that $y = f(x)$ for some $x \in C \cup D$. As x is in the union of C and D, it must be an element of at least one of the sets. That is,

$$x \in C \qquad \text{or} \qquad x \in D.$$

  If $x \in C$, then $y = f(x) \in f(C)$, and if $x \in D$, then $y = f(x) \in f(D)$. Either way, we have $y \in f(C) \cup f(D)$, and so the inclusion $f(C \cup D) \subseteq f(C) \cup f(D)$ is proved.
- To prove the other inclusion, suppose that $y \in f(C) \cup f(D)$. Then, by the definition of unions, $y \in f(C)$ or $y \in f(D)$. If $y \in f(C)$, then $y = f(x_1)$ for some $x_1 \in C$, and if $y \in f(D)$, then $y = f(x_2)$ for some $x_2 \in D$. In either case, we see that $y = f(x)$ for some $x \in C \cup D$, and thus $y \in f(C \cup D)$. This establishes the inclusion $f(C) \cup f(D) \subseteq f(C \cup D)$.

The two inclusions proved above imply that $f(C \cup D) = f(C) \cup f(D)$, as needed. □

2.3 The Field Axioms

In mathematics, we use proofs to validate various statements. In most cases, these proofs rely on earlier established facts, things we have already proved or that we are convinced are true. For example, to prove the quadratic formula (Theorem 1.1.1), we relied on the known formula $(x + y)^2 = x^2 + 2xy + y^2$ and on the fact that squares of real numbers cannot be negative. Looking more carefully at our work so far, we notice that many other "known" algebraic and arithmetic rules, such as $(a \cdot b)^2 = a^2 \cdot b^2$ and $2 \cdot \frac{1}{2} = 1$, were used. In Chapter 1, we used, without proof, several Basic Facts (see page 7) to prove the Arithmetic-Geometric Mean Inequality. This raises an important fundamental question: how can we rely on "facts" or "rules" that were not properly verified? Basing our arguments on rules that were not fully justified can be risky, and jeopardize the credibility of our conclusions. After all, if we are relying on invalid rules to prove a theorem, the proof may be flawed.

How can we resolve this issue? One possible approach would be to go back, and attempt to construct proofs for all these rules and facts we relied on. For instance, we may want to try and supply proofs for the Basic Facts from page 7.

This, however, raises another question: do we expect to be able to prove *every* single algebraic identity and arithmetic rule from scratch? For example, can we prove that $x + y = y + x$ for every two real numbers x, y? Or that $(-1) \cdot (-1) = 1$? Or that $0 \cdot a = 0$ for every real number a?

Mathematical proofs depend on results that have been previously established, and so proving a fundamental result without having previous statements to rely on seems impossible. Indeed, there must be a starting point. We have no choice but to accept some mathematical facts as true (or assume that they are true), and then build our theory on this list of assumptions, from which other conclusions follow.

We may choose to assume rules such as $x + y = y + x$ or $x + 0 = x$ for all $x, y \in \mathbb{R}$. These rules are often called *axioms*, and we use them to prove more advanced results.

This approach is typical in modern mathematics. We start with a set, equipped with one or more operations, and require that certain rules, or axioms, are satisfied. The real numbers, for instance, are often defined in this way, and the axioms, which are not proved, serve as a starting point.

You may ask: how can we assume something without proving it? How do we know our assumptions are even correct? The answer to this question will probably surprise you:

It *does not matter* whether our assumptions (or axioms) are correct or not.

In fact, the question of whether an axiom is correct or not is problematic. What do we mean by "correct"? An axiom may be a reasonable assumption in some contexts, and inappropriate in others. The main point here is that mathematics is not concerned with absolute truth, whether there is such a thing or not. Mathematics is all about conclusions one can make from a given set of assumptions or axioms.

To dive in, and make the discussion more explicit, we present the definition of an important algebraic structure called *a field*.

Definition 2.3.1 A set $\mathbb{F}$, with two operations, $+$ (addition) and $\cdot$ (multiplication), and distinguished elements 0 and 1, with $0 \neq 1$, is called a *field*, if the following list of axioms hold.

0. $x + y \in \mathbb{F}$ and $x \cdot y \in \mathbb{F}$ for every $x, y \in \mathbb{F}$ (*closure* under addition and multiplication).
1. $x + (y + z) = (x + y) + z$ and $x \cdot (y \cdot z) = (x \cdot y) \cdot z$ for every $x, y, z \in \mathbb{F}$ (*associativity* of addition and multiplication).
2. $x + y = y + x$ and $x \cdot y = y \cdot x$ for every $x, y \in \mathbb{F}$ (*commutativity* of addition and multiplication).
3. $x + 0 = x$ and $x \cdot 1 = x$ for all $x \in \mathbb{F}$. The elements 0 and 1 are called the *additive identity* and the *multiplicative identity*, respectively.
4. For every $x \in \mathbb{F}$, there is a $w \in \mathbb{F}$ such that $x + w = 0$ (existence of *negatives*).
   Moreover, if $x \neq 0$, there is also an $r \in \mathbb{F}$ such that $x \cdot r = 1$ (existence of *reciprocals*).
5. $x \cdot (y + z) = x \cdot y + x \cdot z$ for every $x, y, z \in \mathbb{F}$ (*distributivity* of addition over multiplication).

From previous experience, we know that the real numbers, together with the usual addition and multiplication of numbers, satisfy all the field axioms. Consequently, $\mathbb{R}$ is a field. As we will soon see, many other familiar properties of numbers can be derived from the field axioms. These axioms serve as part of our *initial assumptions* or definition of real numbers.

Remarks.

- What exactly do we mean when we say that $+$ and $\cdot$ are *operations* in $\mathbb{F}$? The word "operation" is often used in mathematics as a synonym for a function. That is, $+$ is a function from $\mathbb{F} \times \mathbb{F}$ to $\mathbb{F}$, sending a pair of elements $(x, y) \in \mathbb{F} \times \mathbb{F}$, to exactly one element in $\mathbb{F}$ that we denote as $x + y$. The multiplication operations can be described similarly as a function. When we think of addition and multiplication as functions from $\mathbb{F} \times \mathbb{F}$ to $\mathbb{F}$, the first axiom (closure under addition and multiplication) becomes redundant. It is there to merely emphasize that sums and products of elements from $\mathbb{F}$ must also be elements of $\mathbb{F}$.

- As we will shortly see, the elements 0 and 1 in the definition above need not be the familiar real numbers 0 and 1. Some prefer to emphasize that fact and write $0_{\mathbb{F}}$ and $1_{\mathbb{F}}$ to indicate that these are the zero and the one *of the field* $\mathbb{F}$.
- We know well that every real number x has *exactly one* negative. For instance, there is only one number w such that $2 + w = 0$. Axiom 4, which requires the existence of negatives and reciprocals, does not mention uniqueness of these elements. This means that we can *potentially* have two different negatives (or reciprocals) to an element x in $\mathbb{F}$.

  Luckily, the uniqueness of negatives and reciprocals can be derived from the field axioms. Suppose that $x \in \mathbb{F}$ and that w_1, w_2 are two, potentially different negatives of x. That is,

  $$x + w_1 = 0 \qquad \text{and} \qquad x + w_2 = 0.$$

  We now show that $w_1 = w_2$ as follows:

  $$w_1 \overset{(3)}{=} w_1 + 0 = w_1 + (x + w_2) \overset{(1)}{=} (w_1 + x) + w_2 \overset{(2)}{=} w_2 + (x + w_1) = w_2 + 0 \overset{(3)}{=} w_2.$$

  The numbers above the equal signs indicate the field axiom that is used in each step. Uniqueness of reciprocals is proved in a similar way (see Exercise 2.3.2 below).

  The uniqueness of negatives and reciprocals means that we can refer to w and r in Axiom 4 as *the* negative and *the* inverse of x, and use the common notation $-x$ and x^{-1}. The elements $-x$ and x^{-1} are often referred to as the *additive inverse* and *multiplicative inverse* of x, respectively.
- In the definition of a field, the operations of *subtraction* and *division* are not mentioned. However, Axiom 4 requires the existence of negatives and reciprocals, which allows us to define subtraction and division in terms of addition and multiplication.

  If $x, y \in \mathbb{F}$, we define

  $$x - y = x + (-y).$$

  Moreover, if $y \neq 0$, we define

  $$\frac{x}{y} = x \cdot y^{-1}.$$

Exercise 2.3.2

Prove that in every field $\mathbb{F}$ reciprocals are unique. That is, show that if x is a non-zero element in a field $\mathbb{F}$, and $r_1, r_2 \in \mathbb{F}$ satisfy $x \cdot r_1 = x \cdot r_2 = 1$, then $r_1 = r_2$.

There are many other basic properties of numbers that were not included in the definition of a field. We know, for instance, that if a is a real number, then $a \cdot 0 = 0$ and $-(-a) = a$. Why are these not on the above list of axioms? Mathematicians often try to keep the list of axioms as short as possible, and not include facts that can be derived from other axioms. Here is how the property $a \cdot 0 = 0$ can be proved.

Claim 2.3.3 *If $\mathbb{F}$ is a field, and $a \in \mathbb{F}$, then $a \cdot 0 = 0$.*

Proof

$$\begin{aligned} a \cdot 0 &\overset{(3)}{=} (a \cdot 0) + 0 \overset{(4)}{=} (a \cdot 0) + [a \cdot 0 + (-a \cdot 0)] \\ &\overset{(1)}{=} [(a \cdot 0) + (a \cdot 0)] + (-a \cdot 0) \overset{(5)}{=} a \cdot (0 + 0) + (-a \cdot 0) \\ &\overset{(3)}{=} (a \cdot 0) + (-a \cdot 0) \overset{(4)}{=} 0. \end{aligned}$$

Again, the numbers above the equal signs indicate the field axiom used in each step. □

Example 2.3.4 As we mentioned, the set of real numbers with the familiar addition and multiplication is a field.

Example 2.3.5 The set of rational numbers $\mathbb{Q}$, with the usual addition and multiplication, is also a field. Axioms 1, 2, 3 and 5 are automatically satisfied as rational numbers are also real numbers. Axiom 0 is satisfied as sums and products of rational numbers are also rational. Axiom 4 is satisfied as negatives and reciprocals of rational numbers are also rational.

Example 2.3.6 The set of integers $\mathbb{Z}$, with the usual addition and multiplication is *not* a field, as Axiom 4 fails. Numbers such as 2 and -5 have no integer reciprocals.

Example 2.3.7 The interval $[-1, 1]$, with the usual addition and multiplication, is *not* a field, as it is not closed under addition (i.e., the closure axiom fails). For instance, $\frac{1}{2}$ and $\frac{2}{3}$ are in $[-1, 1]$, but their sum is not.

Let us take another look at Claim 2.3.3. What do we learn from this claim and its proof? The fact that $a \cdot 0 = 0$ is well known to us, and we have been using it for years, so why bother proving it? Here are two reasons.

- The content of the claim is not new to us, but we now see how it can be derived from the field axioms only. Therefore, there is no need to present it as an additional axiom or assumption about real numbers.
- Furthermore, note that this claim is, in fact, quite general, and can be used in fields whose elements are not necessarily numbers!

In the proof, we relied *solely* on the field axioms, and so whenever we happen to notice a "world" in which the field axioms hold, the claim will be valid.

But what other "worlds" are fields? Here are a few examples.

Example 2.3.8 Rational Functions. Recall that a polynomial with real coefficients is a sum of expressions of the form $a \cdot x^k$, where k is a non-negative integer and a is any real number. For instance, $3x^2+2x+5$, x^7-9 and x^3-6x^{10} are polynomials.

Note how we often shorten expressions in a natural way. We write:

$2x$	instead of	$2x^1$,
5	instead of	$5 \cdot x^0$,
x^3	instead of	$1 \cdot x^3$,
x^7-9	instead of	$1 \cdot x^7 + (-9) \cdot x^0$, and so on.

Polynomials can be added and multiplied. For example:

$$(x^5 + 2x^2 + x-2) + (3x^4-7x^2 + 9) = x^5 + 3x^4-5x^2 + x + 7$$

and

$$(5x^3 + 3) \cdot (4x^8-2x) = 20x^{11}-10x^4 + 12x^8-6x.$$

We also have "zero" and "one" polynomials:

$$0 = 0 \cdot x^0 \qquad \text{and} \qquad 1 = 1 \cdot x^0.$$

Is the set of polynomials a field? No, it is not, as most polynomials have no reciprocals (which are also polynomials). Consequently, Axiom 4 fails. In fact, Axiom 4 is the only one that fails for polynomials with real coefficients. All the other axioms are satisfied. We omit the proof of this fact, which is quite long and technical.

In order to fix this issue, we consider, instead, the set of all rational functions,

$$\left\{\frac{f}{g} : f \text{ and } g \text{ are polynomials with real coefficients, and } g \neq 0\right\}.$$

Rational functions can be added and multiplied, and with these operations they do form a field. The proof of that is quite long and technical, and so we omit it.

Example 2.3.9 A Field with Two Elements. What is the smallest possible field? Every field, according to the definition, must include at least two elements: 0 and 1. Is it possible to define addition and multiplication on the set $\mathbb{F} = \{0, 1\}$ in such a way that all the field axioms hold true? We must set $0 + 0 = 0$, $0 + 1 = 1 + 0 = 1$, $0 \cdot 0 = 0 \cdot 1 = 1 \cdot 0 = 0$ and $1 \cdot 1 = 1$. Otherwise, we violate Axiom 3, or the Claim 2.3.3. But what should be $1 + 1$? If 0 and 1 are the only elements in our universe, then 2 is not available, and hence setting $1+1=2$ is not possible. We must therefore set $1+1$ to be either 0 or 1. However, if we set $1+1=1$, then Axiom 4 fails, as 1 will not have a negative (i.e., there is no $w \in \mathbb{F}$ for which $1+w$ is zero). Therefore, we must set $1+1=0$. We summarize addition and multiplication in $\mathbb{F}$ in the following two tables.

+	**0**	**1**
0	0	1
1	1	0

·	**0**	**1**
0	0	0
1	0	1

As strange as it may seem, the set $\mathbb{F} = \{0, 1\}$, with the addition and multiplication defined above, does satisfy the field axioms, and hence is a field. This is *a field with two elements*, and it is widely used in mathematics and applications. There are many other fields that have only finitely many elements. We call them *finite fields*.

Building addition and multiplication tables for a finite field requires some effort and more advanced tools. In small finite fields, we can often prove certain results by using an elimination strategy. Here is an example.

Claim 2.3.10 *Let $\mathbb{F} = \{0, 1, a, b\}$ be a field with four elements. Then $a \cdot b = 1$.*

Proof $\mathbb{F}$ is a field, and so every two elements can be multiplied, and the product must also be one of the field elements. In particular, the product $a \cdot b$ is equal to either 0, 1, a or b. We eliminate the options 0, a and b, from which it follows that $a \cdot b = 1$.

- If $a \cdot b = a$, we can "divide" both sides by a, to obtain $b = 1$, which is impossible, as 0, 1, a and b are *distinct*. In fact, we should be more careful, and make sure that "dividing by a" can be fully justified by the field axioms. If $a \cdot b = a$, then

  $$a^{-1} \cdot (a \cdot b) = a^{-1} \cdot a \quad \Rightarrow \quad (a \cdot a^{-1}) \cdot b = a \cdot a^{-1} \quad \Rightarrow \quad 1 \cdot b = 1,$$

  from which it follows that $b = 1$. Note that each step can be justified by the field axioms (e.g., associativity and commutativity of multiplication).
- A similar argument shows that $a \cdot b = b$ implies $a = 1$, which is impossible.
- Finally, if $a \cdot b = 0$, we can multiply both sides by b^{-1}, and use the field axioms to get

  $$(a \cdot b) \cdot b^{-1} = 0 \cdot b^{-1} \quad \Rightarrow \quad a \cdot (b \cdot b^{-1}) = b^{-1} \cdot 0 \quad \Rightarrow \quad a \cdot 1 = 0.$$

  We conclude that $a = 0$, which is impossible. Note that we also used Claim 2.3.3 to assert that $b^{-1} \cdot 0 = 0$.

We have eliminated the options 0, a and b, and hence $a \cdot b = 1$, as needed. □

The Real Numbers System

In Section 1.4 we mentioned that the set of all rational and irrational numbers on the number line forms the set of real numbers. This description relies on the geometric view of real numbers as points on an infinite number line.

However, it is quite common nowadays to regard set theory as a foundational theory (rather than geometry). That is, we prefer to define real numbers as a set, equipped with additional structure, such as addition and multiplication, that satisfies a list of axioms.

We already noted that the real numbers satisfy the field axioms. Can we perhaps define the real numbers as a set, equipped with two operations $+$ and $\cdot$, that satisfies the field axioms? Well, not quite.

The examples above show that there are other fields out there, some of which are very different from the real numbers we are so used to. For instance, finite fields, the rational numbers and rational functions are fields which are quite different than the reals. To fully characterize the real numbers, we will need to add a few assumptions to the field axioms, that distinguish them from other possible fields. One way to do that is to add the *order* and *completeness axioms*.

The order axioms outline a few basic properties regarding ordering of numbers, allowing us to work with inequalities, and make sense of statements such as $a < b$ or $c \geq d$. We do not outline the axioms and their implications, but some details are provided in Problem 2.61.

The completeness axiom is probably the most complicated one. Roughly speaking, it says that there are no holes or gaps on the number line. We omit the precise statement, which can be found in most textbooks on real analysis.

The main point we want to make here, is that once we combine the field axioms with the order and completeness axioms, we do obtain a full characterization of the real number system, and can formally define the reals by means of the axioms. Once done, any other property of the real numbers can be derived from the axioms (such as existence of square roots and the fact that $1 > 0$).

2.4 Appendix: Infinite Unions and Intersections

In Section 2.1 we introduced operations between sets. In particular, we defined the union and intersection of two sets. In this section, we extend these definitions and discuss unions and intersections of more than two, and even of an infinite collection of sets.

Suppose that $A_1, A_2, A_3, \ldots, A_N$, where $N \in \mathbb{N}$, is a finite collection of sets. We naturally define

$$A_1 \cup A_2 \cup \cdots \cup A_N = \{x \colon x \text{ is in at least one of the sets } A_1, \ldots, A_N\}$$

and

$$A_1 \cap A_2 \cap \cdots \cap A_N = \{x\colon x \text{ is in each of the sets } A_1, \ldots, A_N\}.$$

When dealing with long unions and intersections, we often use an index, or a counter, to shorten our notation. We write

$$\bigcap_{k=1}^{N} A_k \quad \text{or} \quad \bigcap_{1 \le k \le N} A_k$$

instead of $A_1 \cap A_2 \cap \cdots \cap A_N$, and

$$\bigcup_{k=1}^{N} A_k \quad \text{or} \quad \bigcup_{1 \le k \le N} A_k$$

instead of $A_1 \cup A_2 \cup \cdots \cup A_N$.

These definitions can be further extended to infinite collections of sets. For instance, the union of an infinite sequence of sets $B_1, B_2, B_3, \ldots$ is denoted as $\bigcup_{k=1}^{\infty} B_k$ or $\bigcup_{k \in \mathbb{N}} B_k$, and defined as

$$\{x\colon x \in B_k \text{ for some } k \in \mathbb{N}\}.$$

Similarly, suppose that for every real number α in the interval $[0, 1]$, we are given a set C_α. The intersection of the collection $\{C_\alpha : \alpha \in [0, 1]\}$ is naturally defined as follows:

$$\bigcap_{\alpha \in [0,1]} C_\alpha = \{x\colon x \in C_\alpha \text{ for every } \alpha \in [0, 1]\}.$$

We say that α is an *index*, ranging over all the elements of $[0, 1]$, our *index set*.

The most general definition of unions and intersection is given below. In this definition, the index set is allowed to be any set whatsoever.

Definition 2.4.1 Let J be a set (the *index set*). Suppose that for each $\alpha \in J$, A_α is a set.

- The *union* of the sets A_α is

$$\bigcup_{\alpha \in J} A_\alpha = \{x\colon x \in A_\alpha \text{ for some } \alpha \in J\}.$$

- The *intersection* of the sets A_α is

$$\bigcap_{\alpha \in J} A_\alpha = \{x\colon x \in A_\alpha \text{ for every } \alpha \in J\}.$$

Example 2.4.2 For every $n \in \mathbb{N}$, let A_k be the interval $\left(-\frac{1}{k}, \frac{1}{k}\right)$. We show that $\bigcap_{k=1}^{\infty} A_k = \{0\}$.

Proof For each $k \in \mathbb{N}$, we have $-\frac{1}{k} < 0 < \frac{1}{k}$, and so $0 \in A_k = \left(-\frac{1}{k}, \frac{1}{k}\right)$. Consequently, $0 \in \bigcap_{k=1}^{\infty} A_k$, or, put it differently, $\{0\} \subseteq \bigcap_{k=1}^{\infty} A_k$.

To prove the other inclusion, we use an intuitive result from real analysis:

> "For every positive number r, there exists a natural number k, such that $\frac{1}{k} < r$."

Informally, the above statement says that we can make the fraction $\frac{1}{k}$ smaller than any positive number, by making k large enough. This result and its proof can be found in texbooks on real analysis.

Now back to our proof. Suppose that $r \in \bigcap_{k=1}^{\infty} A_k$. We argue, by contradiction, that $r = 0$. Assume that $r > 0$. Then $\frac{1}{k} < r$ for some $k \in \mathbb{N}$, and hence $r \notin A_k$. This is a contradiction to the fact that $r \in \bigcap_{k=1}^{\infty} A_k$, and so r cannot be a positive number. Similarly, if $r < 0$, then $-r > 0$ and hence $\frac{1}{k} < -r$ for some $k \in \mathbb{N}$. We conclude that $r < -\frac{1}{k}$ and so $r \notin A_k$, which is a contradiction.

By ruling out the options $r > 0$ and $r < 0$ we are left with $r = 0$, which completes our proof. □

Exercise 2.4.3

For the sets A_k defined in Example 2.4.2, what is $\bigcup_{k=1}^{\infty} A_k$?

Here is another example, involving a collection of sets indexed by all real numbers in the interval $(0, 1)$.

Example 2.4.4 For each $\alpha \in (0, 1)$, consider the half-open and half-closed interval $I_\alpha = (2-\alpha, 4-\alpha]$. We show that $\bigcup_{\alpha \in (0,1)} I_\alpha = (1, 4)$.

Proof For every $\alpha \in (0, 1)$, we have

$$1 < 2 - \alpha < 4 - \alpha < 4 \qquad \Rightarrow \qquad I_\alpha = (2 - \alpha, 4 - \alpha) \subseteq (1, 4),$$

and so $\bigcup_{\alpha \in (0,1)} I_\alpha \subseteq (1, 4)$.

Conversely, let $x \in (1,4)$. To show that $x \in \bigcup_{\alpha\in(0,1)} I_\alpha$, we need to argue that $x \in I_\alpha$ for some $\alpha \in (0,1)$. The condition $x \in I_\alpha$ translates to

$$2-\alpha < x \leq 4-\alpha.$$

If $1 < x \leq 2$, choose an $\alpha \in (0,1)$ such that $2-\alpha < x$ (for instance, we can take $\alpha = \frac{3-x}{2}$). Then,

$$2-\alpha < x \leq 2 \leq 4-\alpha \qquad \Rightarrow \qquad x \in I_\alpha.$$

If $3 < x < 4$, choose an $\alpha \in (0,1)$ such that $x < 4-\alpha$. This will result in $x \in I_\alpha$. Finally, if $2 < x \leq 3$, then we can take $\alpha = 0.5$. Then $I_\alpha = (1.5, 3.5]$ and again $x \in I_\alpha$.

We have shown that in each case, $x \in I_\alpha$ for some $\alpha \in (0,1)$, and hence $(1,4) \subseteq \bigcup_{\alpha\in(0,1)} I_\alpha$.

We conclude that $\bigcup_{\alpha\in(0,1)} I_\alpha = (1,4)$, as needed. □

Exercise 2.4.5
For the intervals I_α in Example 2.4.4, show that $\bigcap_{\alpha\in(0,1)} I_\alpha = [2,3]$.

2.5 Appendix: Defining Functions

According to Definition 2.2.1, three pieces of information need to be specified in order to describe a function:

- a set A (the *domain*),
- another set B (the *codomain*),
- a rule, assigning to *each* element of A *exactly one* element from B.

If any of the three ingredients above is not clearly specified or understood from the context, or if the rule is ambiguous or problematic and cannot be applied to some elements of A, then our function is *not properly defined*. Here are a few examples.

Example 2.5.1 Consider the following flawed definition of a function f:

$$f\colon \mathbb{N} \to \mathbb{N}, \quad f(n) = n-10.$$

Here, some of the images produced by the formula $f(n) = n-10$ do not lie in the stated codomain. For instance, $f(6) = 6-10 = -4$ is not a natural number.

One way to fix this issue is to enlarge the codomain to include all integers. The description

$$f\colon \mathbb{N} \to \mathbb{Z}, \quad f(n) = n-10$$

is a valid definition of a function.

Alternatively, we may remove a few elements from the domain to obtain a properly defined function:

$$f\colon \{11, 12, 13, \dots\} \to \mathbb{N}, \quad f(n) = n-10.$$

Example 2.5.2 The following definition is also flawed:

$$g\colon \mathbb{R}^2 \to \mathbb{R}, \quad g(x,y) = \frac{1}{x^2+y^2}.$$

The pair $(0,0)$ is in the domain, but the rule for computing images cannot be applied to this pair, as we obtain a denominator of zero. Whenever we define a function, we must make sure that the rule or formula can be applied to all the elements in the domain.

We can fix this issue by using $\mathbb{R}^2 \setminus \{(0,0)\}$ as our domain. Another option is to explicitly assign a value to $g(0,0)$. For instance,

$$g\colon \mathbb{R}^2 \to \mathbb{R}, \quad g(x,y) = \begin{cases} \frac{1}{x^2+y^2} & \text{if } (x,y) \neq (0,0) \\ -2 & \text{if } (x,y) = (0,0) \end{cases}$$

is a valid definition.

Example 2.5.3 Sometimes, we define a function through *representatives of elements in its domain*. Consider, for instance, the following definition:

$$h\colon \mathbb{Q} \to \mathbb{R}, \quad h\left(\frac{a}{b}\right) = a+b.$$

The function h, whose domain is $\mathbb{Q}$, is supposed to assign to every rational number a real number. However, there are multiple ways to represent a rational number. For example, what would be $h(0.25)$?

Well, we can write 0.25 as $\frac{1}{4}$, and conclude that

$$h(0.25) = h\left(\frac{1}{4}\right) = 1+4 = 5.$$

On the other hand, we can represent 0.25 as $\frac{3}{12}$ and conclude that

$$h(0.25) = h\left(\frac{3}{12}\right) = 3+12 = 15.$$

We see that the definition of h does not clearly specify a single image for 0.25, and that the "rule" depends on the way 0.25 is represented as a fraction. For that reason, this is not a valid definition of a function.

In such cases, where the "rule" depends on the way an element is represented, we often say that the function is *not well defined.*

Example 2.5.4 Here is another example in which we use representatives:

$$p\colon \mathbb{Q} \to \mathbb{R}, \quad p\left(\frac{a}{b}\right) = \frac{a^2 - b^2}{a^2 + b^2}.$$

Let us compute $p(0.25)$ using the two representatives $\frac{1}{4}$ and $\frac{3}{12}$ for 0.25:

$$p\left(\frac{1}{4}\right) = \frac{1^2 - 4^2}{1^2 + 4^2} = \frac{1 - 16}{1 + 16} = -\frac{15}{17}$$

$$p\left(\frac{3}{12}\right) = \frac{3^2 - 12^2}{3^2 + 12^2} = \frac{9 - 144}{9 + 144} = -\frac{135}{153} = -\frac{15}{17}.$$

In both cases we got the same value. In fact, we can prove that the formula for p does not depend on the way we represent a rational number.

Claim 2.5.5 *The function p, defined above, is well defined. That is, if $\frac{a}{b}$ and $\frac{c}{d}$ represent the same rational number, then $p\left(\frac{a}{b}\right) = p\left(\frac{c}{d}\right)$.*

Proof If $\frac{a}{b}$ and $\frac{c}{d}$ represent the same rational number, then

$$p\left(\frac{a}{b}\right) = \frac{a^2 - b^2}{a^2 + b^2} = \frac{\left(\frac{a}{b}\right)^2 - 1}{\left(\frac{a}{b}\right)^2 + 1} = \frac{\left(\frac{c}{d}\right)^2 - 1}{\left(\frac{c}{d}\right)^2 + 1} = \frac{c^2 - d^2}{c^2 + d^2} = p\left(\frac{c}{d}\right).$$

We conclude that p is well defined. □

Exercise 2.5.6
What is $p(0)$, where p is the function defined in Example 2.5.4?

Finally, one has to be careful when constructing piecewise definitions of functions. Again, we need to make sure that our definition clearly specifies a *single image* for *every* element in the domain.

Example 2.5.7 Consider the following two definitions of functions F and G from $\mathbb{R}$ to $\mathbb{R}$:

$$F(x) = \begin{cases} x^2 + 1 & \text{if } x \geq 3 \\ 3x - 5 & \text{if } x \leq 3 \end{cases} \qquad G(x) = \begin{cases} \sqrt{x} + 3 & \text{if } x \geq 4 \\ 2x - 3 & \text{if } x \leq 4. \end{cases}$$

For both functions, the two cases in their definition overlap. For F, if $x = 3$, we should be able to use any of the two formulas to compute F(3). However,

$$x^2 + 1 = 3^2 + 1 = 10 \qquad \text{and} \qquad 3x - 5 = 3 \cdot 3 - 5 = 4,$$

and hence F is *not* properly defined at $x = 3$.

On the other hand, the two ways to compute $G(4)$ yield the same number. Indeed, if $x = 4$, we have

$$\sqrt{x}+3 = \sqrt{4}+3 = 5 \qquad \text{and} \qquad 2x-3 = 2\cdot 4-3 = 5,$$

and so the description of G gives a valid definition of a function.

2.6 Problems

2.1 Let $S = \{(x, y) \in \mathbb{N}^2 : (2-x)(2-y) < 2(4-x-y)\}$. Prove that $S = T$, where $T = \{(1, 1), (1, 2), (2, 1), (1, 3), (3, 1)\}$.

2.2 Let $S = \{2, 3\} \times \{-2, -1, 0\}$, and let T be the set of all ordered pairs $(x, y) \in \mathbb{Z} \times \mathbb{Z}$ such that $-2 \le x + 2y \le 3$. Prove that $S \subseteq T$. Does equality hold? Explain.

2.3 Write either $\subseteq$ or $\not\subseteq$ in the space provided. Explain your answer briefly.

$$\varnothing \underline{\qquad} \mathbb{N} \qquad\qquad \{\varnothing\} \underline{\qquad} \mathbb{R}$$

$$\mathbb{Z} \underline{\qquad} \mathbb{N} \cup \mathbb{Q} \qquad\qquad \mathbb{Z} \underline{\qquad} \mathbb{N} \cap \mathbb{Q}$$

2.4 Describe the "intervals" $[a, b]$ and (a, b), in the case where $a = b$.

2.5 For each part, draw a copy of the Venn diagram in Figure 2.16, and shade the given set.

a. $(A \cup B) \setminus C$
b. $C \setminus (A \setminus B)$
c. $(B \setminus C) \cap A$

2.6 For each part, draw a copy of the Venn diagram in Figure 2.17, and shade the given set.

a. $A \cup B^c$
b. $B \setminus A^c$
c. $A^c \cap B^c$

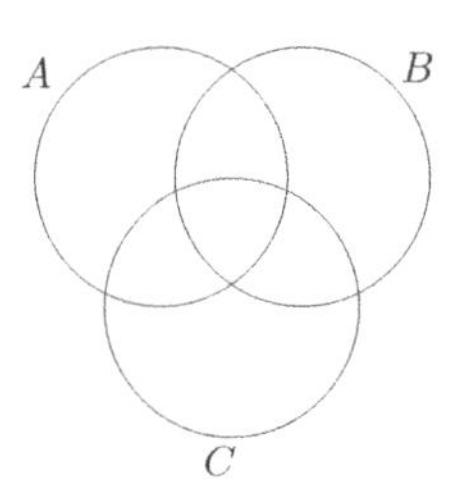

Figure 2.16 A Venn diagram for three sets.

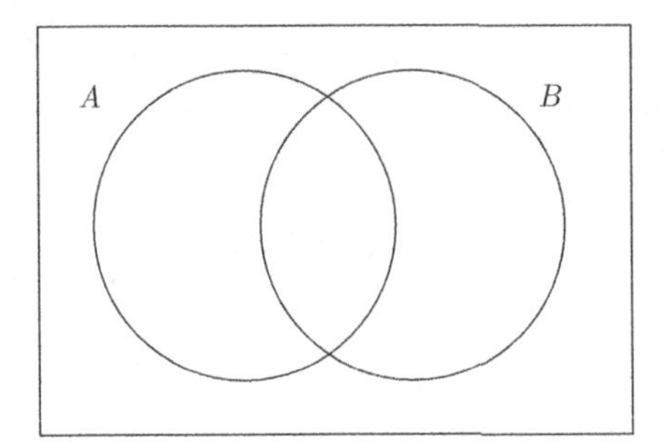

Figure 2.17 A Venn diagram for two sets.

2.7 Explain why each of the following equalities is *incorrect*.

a. $[0.5, 7.5] \cap \mathbb{N} = [1, 7]$
b. $\{4\} \times \{5\} = \{20\}$
c. $\{a, b, c\} \cup \varnothing = \{a, b, c, \varnothing\}$

2.8 Let $A = [-1, 1]$, $B = (-\pi, \pi)$, $C = [2, \infty)$ and $U = \mathbb{R}$ be the universal set.
Find the sets $A \cap C$, $A^c \cap B$, $(B \cap C) \cap \mathbb{Z}$ and $B \setminus A$. Use the interval notation, and the symbols $\{\ ,\ \}$, $\cup$, $\varnothing$, ∞ and π only.

2.9 For each statement, decide whether it is *true* or *false*. Justify your answer briefly.

a. $\{(x, y) \in \mathbb{R} \times \mathbb{R}\colon\ x-1 = 0\} \subseteq \left\{(x, y) \in \mathbb{R} \times \mathbb{R}\colon\ x^2 - x = 0\right\}$
b. $\{x \in \mathbb{R} : x^3-2x = 0\} \subseteq \mathbb{Q}$
c. $\mathbb{N} \in \mathbb{R}$
d. $\mathbb{Z} \times \mathbb{R} \subseteq \mathbb{R} \times \mathbb{Z}$
e. $\mathbb{N} \times \mathbb{Z} \subseteq \mathbb{Q} \times \mathbb{R}$

2.10 Let A, B, C be three sets. For each statement, decide whether it is necessarily *true* or could be *false*. Draw a Venn diagram to support your answer.

a. $A \setminus B \subseteq A$
b. $A \setminus B \subseteq B$
c. $(A \cup B) \cap C = A \cup (B \cap C)$
d. $(A \setminus B = A \cap B^c)$
e. $A \subseteq A \cap B$
f. $(A \setminus B)^c = A^c \setminus B^c$

2.11 a. Prove that for every three sets A, B, C, we have

$$(A \cap B) \cap C = A \cap (B \cap C) \qquad \text{and} \qquad (A \cup B) \cup C = A \cup (B \cup C).$$

The above identities allow us to drop the parentheses, and simply write $A \cap B \cap C$ and $A \cup B \cup C$, without creating any ambiguity.

b. Show that for sets A, B, C, the expressions $(A \setminus B) \setminus C$ and $A \setminus (B \setminus C)$ may yield different results.

2.12 If A and B are two sets satisfying $A \setminus B = B \setminus A$, what can we conclude about A and B? Explain.

2.13 Is it true that $(A \times A) \setminus (B \times B) = (A \setminus B) \times (A \setminus B)$ for every two sets A, B? If it is true, prove it. Otherwise, find a counterexample.

2.14 a. Given that $A = \left\{-3, -1, \frac{1}{2}, 10, \sqrt{2}\right\}$ and $B = \{2, 4\}$, list all the elements in the set $C = B \times (A \cap \mathbb{Z})$.
b. Given the intervals $I = [-1, 8]$ and $J = (3,5)$, list all the elements in the set $B = (I \setminus J) \cap \mathbb{N}$.
c. Write the set $\{x \in \mathbb{R} : 0 < x^2 \leq 25\}$ as a *union of two intervals*.
d. Express the set $A = \{x : 1 < x^2 < 4\}$ both as a union of two intervals, and as a difference of two intervals.

2.15 Give an example of a set A, for which $A \cap [1, 4] = A \cap \mathbb{N}$ and $A \setminus \mathbb{Z} \neq \varnothing$.

2.16 Consider the following two subsets of $\mathbb{R}^2$:

$$A = [0, 2] \times [0, 2] \qquad \text{and} \qquad B = [-1, 1] \times [-1, 1].$$

Draw the sets A, B, $A \cap B$ and $A \setminus B$ in the plane. Use a *solid* or a *dotted line* to indicate whether the boundary is or is not part of the set.

2.17 Consider the sets $S = \{(x, y) : y \geq x^2\}$ and $T = \{(x, y) : y \leq x + 2\}$. Draw the sets S, T, and $S \cap T$ in the two-dimensional plane.

2.18 Consider the following two subsets of $\mathbb{R}^2$:

$$T = [-1, 1] \times [-1,1] \qquad \text{and} \qquad S = \left\{ (x, y) : x^2 + y^2 \leq 4 \right\}.$$

Sketch the set $S \setminus T$.

2.19 Let $D = [1, 3] \cup \{4\}$ (note that this is a subset of $\mathbb{R}$). Sketch the set $D \times D$ in $\mathbb{R}^2$.

2.20 Let a, b, c, d be real numbers with $a < b < c < d$. Express the set $[a, b] \cup [c, d]$ as a difference of two sets.

2.21 Let $S = \{x \in \mathbb{R} : (x-2)(x + 3) < 0\}$, T the interval $(-4, 2)$ and U the interval $(-3, 5)$. Use set operations to write a simple relation between the sets S, T and U.

2.22 Prove the following set identities.

a. $(A \cap B)^c = A^c \cup B^c$
b. $A \setminus (B \cup C) = (A \setminus B) \cap (A \setminus C)$
c. $A \setminus (B \setminus C) = (A \setminus B) \cup (A \cap C)$
d. $(A \cap B) \setminus (B \cap C) = A \cap (B \setminus C)$

2.23 Let A, B, C be three sets. Prove that if $A \setminus B \subseteq C$, then $A \setminus C \subseteq B$.

2.24 Let A, B be two subsets of some universal set U. Prove that if $(A \cup B)^c = A^c \cup B^c$, then $A = B$.

2.25 Let A, B, C and D be four sets. Prove that if $A \cup B \subseteq C \cup D$, $A \cap B = \varnothing$ and $C \subseteq A$, then $B \subseteq D$.

2.26 Let A, B, C be three sets. Look carefully at Definitions 2.1.8 and 2.1.11, and explain why $A \times B \times C$ is *not* the same as $(A \times B) \times C$.

2.27 Find the images of the following functions. Explain your answer briefly.

a. $f\colon \mathbb{R} \to \mathbb{R},\ f(x) = |x|$
b. $r\colon \mathbb{R} \to \mathbb{R},\ r(x) = \frac{1}{x^2+2}$
c. $g\colon \mathbb{N} \to \mathbb{R},\ g(n) = (-1)^n$
d. $h\colon \mathbb{Z} \to \mathbb{Z},\ h(k) = 3k + 1$

2.28 Consider the function $f\colon \mathbb{N} \times \mathbb{N} \to \mathbb{R},\ f(m, n) = m - n$.

a. Find $f(3, 5)$ and $f(5, 10)$.
b. Find two pairs (m, n) for which $f(m, n) = 9$.
c. What is the image of f?

2.29 What is the image of the function $f\colon \mathbb{Z}\times\mathbb{N} \to \mathbb{R},\ f(a, b) = \frac{a}{b}$? Explain.

2.30 What is the image of the function $f\colon \mathbb{Z} \times \mathbb{Z} \to \mathbb{R},\ f(a, b) = \frac{a+b}{2}$? Explain.

2.31 A function $f\colon \mathbb{R} \to \mathbb{R}$ is said to be *bounded*, if there is a positive number M, such that $|f(x)| \leq M$ for all $x \in \mathbb{R}$. Which of the following statements are true for any two functions $f, g\colon\mathbb{R} \to \mathbb{R}$? Give a *proof* or a *counterexample*.

a. If f and g are bounded, then $f + g$ is bounded.
b. If f and g are bounded, then $f^2 - g^2$ is bounded.
c. If $f + g$ is bounded, then $f - g$ is bounded.
d. If f and g are bounded, then $f \cdot g$ is bounded.
e. If $f \cdot g$ is bounded, then both f and g are bounded.
f. If $|f| + |g|$ is bounded, then both f and g are bounded.

2.32 Is the function $f\colon \mathbb{R} \to \mathbb{R},\ f(x) = \frac{|x+5|}{|x|+5}$ bounded (see Problem 2.31)? Explain.
Do *not* use calculus.

2.33 Consider the following subsets of $\mathbb{R}$:

$$A = [1, 4], \qquad B = (-3, 3), \qquad D = \left\{1, 5, \frac{3}{7}, \frac{11}{2}\right\},$$

and the function $g\colon \mathbb{R} \to \mathbb{R},\ g(x) = 2x + 1$. Compute the following sets (you can use symbols like {, }, $\varnothing$ and the interval notation).

a. $A \cup B$
b. $A^c \cap B$
c. $A \setminus B$
d. $D \cap \mathbb{Z}$
e. $D \setminus \mathbb{N}$
f. $g(D)$
g. $g(B \cap D)$
h. $g(A) \cap D$

2.34 Consider the intervals $A = (-1, 2)$ and $B = [1, 3]$, and the function $f{:}\mathbb{R} \to \mathbb{R},\ f(x) = |x|$. Compute the following sets (you can use symbols like {, }, $\varnothing$ and the interval notation).

a. $A \cap B$
b. $A \cup B$
c. $A \setminus B$
d. $(A \cap \mathbb{Z}) \times (B \cap \mathbb{Z})$
e. $A \cap B \cap \mathbb{Z}^c$
($\mathbb{Z}^c$ is the complement of $\mathbb{Z}$ in $\mathbb{R}$)
f. $f(A)$
g. $f(B \cap \mathbb{Z})$
h. $f(A \cup B) \cap \mathbb{Z}$

2.35 Prove that the image of the function $f\colon \mathbb{R} \to \mathbb{R},\ f(x) = \frac{x^2}{1+x^2}$ is the interval $[0\,,1)$. Do *not* use calculus.

2.36 What is the image of the function $f{:}(0, \infty) \to \mathbb{R},\ f(x) = \frac{4x}{x+1}$? Prove your answer (and do *not* use calculus).

2.37 Prove (without using calculus), that the image of the function $f\colon (0, \infty) \to \mathbb{R},\ f(x) = x + \frac{1}{x}$ is the interval $[2, \infty)$.

2.38 Let $f\colon A \to B$ be a function.

a. Prove that for any two sets $C, D \subseteq A$, we have $f(C \cap D) \subseteq f(C) \cap f(D)$.
b. Give an example of a function f, and sets C, D, for which $f(C \cap D) \neq f(C) \cap f(D)$.

2.39 Is it true that for any function $f\colon A \to B$, and $C, D \subseteq A$, if $C \cap D = \varnothing$, then $f(C) \cap f(D) = \varnothing$? Give a proof or a counterexample.

2.40 Let $f\colon A \to B$ be a function.

a. Prove that for any two sets $C, D \subseteq A$, we have $f(C) \setminus f(D) \subseteq f(C \setminus D)$.
b. Give an example of a function f, and sets C, D, for which $f(C) \setminus f(D) \neq f(C \setminus D)$.

2.41 Let $f{:}X \to Y$ be a function, and $A, B \subseteq X$ two subsets.

a. Must $f(A \cap B) \subseteq f(A) \cup f(B)$? Why?
b. Must $f(A \cup B) \subseteq f(A) \cap f(B)$? Why?

2.42 Let $f\colon A \to B$ and $C\,, D \subseteq A$.

a. Is it necessarily true that if $C \subseteq D$, then $f(C) \subseteq f(D)$? Why?
b. Is it necessarily true that if $f(C) \subseteq f(D)$, then $C \subseteq D$? Why?

2.43 Let $f: A \to B$ be a function and $D \subset B$ a subset. We define the *pre-image* of D under f as the set

$$f^{-1}(D) = \{x \in A: f(x) \in D\}.$$

a. Consider the function $f: \mathbb{R} \to \mathbb{R}$ given by $f(x) = 2 - x^2$. Find, with proof, the following sets:

$$f^{-1}([-3,-2]), \quad f^{-1}(\{0,1,2,3\}), \quad f^{-1}(\mathbb{N}) \quad \text{and} \quad f^{-1}((-\infty,0]).$$

b. Let $g: \mathbb{R} \to \mathbb{R}$ be given by $g(x) = x^2$, and let $I = [-1,2]$. Find, with proof, the sets $g^{-1}(g(I))$ and $g(g^{-1}(I))$.

2.44 Let $f: A \to B$ be a function, $C \subseteq A$ and $D \subseteq B$. Prove that $C \subseteq f^{-1}(f(C))$ and $f(f^{-1}(D)) \subseteq D$.
See Problem 2.43 for the definition of the pre-image of a set under a function.

2.45 Read the definition of a pre-image in Problem 2.43 . Let $f: A \to B$ be a function, and D_1, D_2 two subsets of B.

a. Prove that $f^{-1}(D_1 \cup D_2) = f^{-1}(D_1) \cup f^{-1}(D_2)$.
b. Prove that $f^{-1}(D_1 \cap D_2) = f^{-1}(D_1) \cap f^{-1}(D_2)$.

2.46 Is the set of natural numbers $\mathbb{N}$, with the usual addition and multiplication of numbers, a field? How about the interval $[0,\infty)$?

2.47 Is the set $\mathbb{Q} \cup [-1,1]$, with the usual addition and multiplication, a field? Explain.

2.48 Prove that in a field $\mathbb{F}$, reciprocals are unique. Namely, show that if x is a non-zero element of $\mathbb{F}$, and $x \cdot r_1 = x \cdot r_2 = 1$, then $r_1 = r_2$.

2.49 Let $\mathbb{F}$ be a field. Prove the following statements. Justify each step in your proofs, and make sure to use only the field axioms, or claims that have been already proved.

a. For any $x \in \mathbb{F}$, $(-1) \cdot x = -x$.
(Hint: It is enough to prove that $x + (-1) \cdot x = 0$.)
b. If $x, y \in \mathbb{F}$, and $x \cdot y = 0$, then $x = 0$ or $y = 0$.
c. If $x, y, z \in \mathbb{F}$, and $x + z = y + z$, then $x = y$.

2.50 Let $\mathbb{F} = \{0, 1, a\}$ be a field with three elements.

a. Prove that $a + 1 = 0$, and use it to conclude that $a + a = 1$.
b. Prove that $a \cdot a = 1$.
c. Draw the addition and multiplication table for $\mathbb{F}$.

2.51 Let $\mathbb{F} = \{0, 1, a, b\}$ be a field with four elements.

a. Prove that $a^2 = b$, and that $b^2 = a$.
b. What are a^3 and b^3?
c. (**Harder**!) Prove that $1 + 1 = 0$.
(Hint: Assume that $1 + 1 = a$. What does that imply about $1 + a$ and $1 + b$? Show that the assumption leads to $a^2 = 0$, which is impossible.)

2.52 Let $\mathbb{F} = \{0, 1, a, b, c\}$ be a field with five elements. It is known that $a \cdot a = c$. Prove that $ab = 1$.

2.53 Here is the addition and multiplication table for a field with *four* elements.

+	**0**	**1**	**a**	**b**
0	0	1	a	b
1	1	0	b	a
a	a	b	0	1
b	b	a	1	0

·	**0**	**1**	**a**	**b**
0	0	0	0	0
1	0	1	a	b
a	0	a	b	1
b	0	b	1	a

Complete with either $0, 1, a$ or b.

$a + b =$ ______ $\quad -b =$ ______ $\quad b^{-1} =$ ______ $\quad a \cdot (1 + b) =$ ______

2.54 Is the set $\mathbb{R}^2$, with addition and multiplication defined below, a field? Explain.

$$(a, b) + (c, d) = (a + c, b + d) \qquad (a, b) \cdot (c, d) = (ac, bd)$$

2.55 Let $A = \{0, 1, x, y\}$ be a set with four elements. Explain why the set A, with addition and multiplication defined by the two tables below, is *not* a field.

+	**0**	**1**	$\boldsymbol{x}$	$\boldsymbol{y}$
0	0	1	x	y
1	1	x	y	0
$\boldsymbol{x}$	x	y	0	1
$\boldsymbol{y}$	y	0	1	x

·	**0**	**1**	$\boldsymbol{x}$	$\boldsymbol{y}$
0	0	0	0	0
1	0	1	x	y
$\boldsymbol{x}$	0	x	0	x
$\boldsymbol{y}$	0	y	x	1

2.56 Show that in any field $\mathbb{F}$, the equation $x^2 = 1$ can have *at most* two solutions. Can you think of a field in which the equation $x^2 = 1$ has exactly one solution?

2.57 Let $\mathbb{F}$ be a field in which $1 + 1 = 0$ (there are many such fields, some of which are infinite). Prove that for any $x \in \mathbb{F}$, we have $x = -x$. That is, every element in $\mathbb{F}$ is its own negative.

2.58 Let $\mathbb{F} \subseteq \mathbb{R}$ be a subset, which is also a field (with the usual addition and multiplication inherited from $\mathbb{R}$). We say that $\mathbb{F}$ is a *subfield* of $\mathbb{R}$.

a. Explain why the numbers $3, -4, \frac{1}{2}$ and $\frac{2}{5}$ must be elements of $\mathbb{F}$, while $\sqrt{2}$ and π need not be in $\mathbb{F}$.
b. Prove that $\mathbb{Q} \subseteq \mathbb{F}$.

2.59 (**Harder!**) Let $\mathbb{F}$ be a subfield of $\mathbb{R}$ (see Problem 2.58). Prove that if $\sqrt{2} + \sqrt{3} \in \mathbb{F}$, then both $\sqrt{2}$ and $\sqrt{3}$ are in $\mathbb{F}$.
(Hint: What is the reciprocal of $\sqrt{2} + \sqrt{3}$?)

2.60 Let $K = \{$All real numbers of the form $a + b \cdot \sqrt{2}$, where $a, b \in \mathbb{Q}\}$. In this exercise, we show that K is a field, with respect to the usual addition and multiplication of numbers. It is often denoted as $\mathbb{Q}(\sqrt{2})$.

a. Verify that the Field Axioms 1, 2, 3 and 5 hold for K. Why is it so easy to check these axioms?
b. Verify Axiom 0.
c. The hardest part is to check Axiom 4. If $x = a + b\sqrt{2}$ is an element of K, what would you expect its negative to be? How can you check your conjecture? Also, if $x \neq 0$, we would need to show that $x^{-1} = \frac{1}{a+b\sqrt{2}}$ is in K. To do that, multiply the numerator and denominator by $a - b\sqrt{2}$. How does that help?

2.61 **The Order Axioms**. The following definition introduces the notion of an *ordered field*.

Definition 2.6.1 An *ordered field* is a field $\mathbb{F}$, and a subset $P \subseteq \mathbb{F}$, called a *positive set*, such that

a. If $x, y \in P$, then $x + y \in P$.
b. If $x, y \in P$, then $x \cdot y \in P$.
c. For any $x \in \mathbb{F}$, *exactly one* of the following must hold true: $x \in P$, $-x \in P$ or $x = 0$.

Conditions 1, 2 and 3 are called the *order axioms*, from which many fundamental properties involving inequalities can be derived. For instance, the Basic Facts 1–4 from Section 1.2 (see page 7) follow from the field and order axioms.

Suppose that $\mathbb{F}$ is an ordered field. If $a, b \in \mathbb{F}$, we use the notation $a < b$ to mean $b - a \in P$ and $a > b$ to mean $a - b \in P$. Prove the following statements.

a. $1 \in P$ (that is, $1 > 0$).
b. For every non-zero $x \in \mathbb{F}$, we have $x^2 \in P$ (that is, $x^2 > 0$).

c. For every $x, y \in \mathbb{F}$, exactly one of the following occurs: $x > y$, $x < y$ or $x = y$.
d. For every $x, y, z \in \mathbb{F}$, if $x < y$ and $y < z$ then $x < z$.
e. For every $x, y, z \in \mathbb{F}$, if $x < y$, then $x + z < y + z$.
f. For every $x, y, z \in \mathbb{F}$, if $x < y$ and $z > 0$ then $xz < yz$.

2.62 For each $n \in \mathbb{N}$, define $A_n = \left[-\frac{2}{n}, \frac{2}{n}\right] \subseteq \mathbb{R}$. Find $\bigcup_{n\in\mathbb{N}} A_n$ and $\bigcap_{n\in\mathbb{N}} A_n$.

2.63 Let $P = (1, \infty) \subseteq \mathbb{R}$. Find $\bigcup_{r\in P} \left(\frac{1}{r}, 3r\right)$ and $\bigcap_{r\in P} \left(\frac{1}{r}, 3r\right)$.

2.64 Let $P = (2, \infty) \subseteq \mathbb{R}$, and define $B_\alpha = \left(-\frac{1}{\alpha}-1, \alpha\right) \subseteq \mathbb{R}$ for each $\alpha \in P$. Find $\bigcup_{\alpha\in P} B_\alpha$ and $\bigcap_{\alpha\in P} B_\alpha$.

2.65 Let X be a set, and $\{A_\alpha\}_{\alpha\in J}$ a collection of sets. Prove that

$$X \Big\backslash \left(\bigcap_{\alpha\in J} A_\alpha\right) = \bigcup_{\alpha\in J} (X \setminus A_\alpha) \quad \text{and} \quad X \Big\backslash \left(\bigcup_{\alpha\in J} A_\alpha\right) = \bigcap_{\alpha\in J} (X \setminus A_\alpha).$$

2.66 In each part, decide whether the given rule properly defines a function. Explain your answers.

a. $g\colon \mathbb{Z} \times \mathbb{Z} \to \mathbb{Q}, \quad g(a, b) = \frac{a}{b}$.
b. $h\colon \mathbb{R} \to \mathbb{R}, \quad h(x) = |\sqrt{x}-2|$.
c. $r\colon \mathbb{R} \to \mathbb{R}, \quad r(x) = \sqrt{|x|}-2$.
d. $f\colon \mathbb{Q} \setminus \{0\} \to \mathbb{Q}, \quad f\left(\frac{a}{b}\right) = \frac{a^2+b^2}{ab}$.
e. $p\colon \mathbb{R} \to \mathbb{R}, \quad p(x) = \begin{cases} \sin x & \text{if } x \le 0 \\ |x| & \text{if } x \ge 0. \end{cases}$
f. $q\colon \mathbb{R} \to \mathbb{R}, \quad q(x) = \begin{cases} \ln(1 - x^2) & \text{if } |x| < 1 \\ \ln(x^2-1) & \text{if } |x| > 1. \end{cases}$

2.67 Suppose that $f\colon [0, 2] \to \mathbb{R}$ and $g\colon [1, 3] \to \mathbb{R}$ are two functions, and define

$$h\colon [0, 3] \to \mathbb{R}, \qquad h(x) = \begin{cases} f(x) & \text{if } 0 \le x \le 2 \\ g(x) & \text{if } 1 \le x \le 3. \end{cases}$$

Under what condition(s) will h be a properly defined function? Explain.

2.7 Solutions to Exercises

Solution to Exercise 2.1.4

We can write

$$A = \{n \in \mathbb{N} \colon n = k^2 \text{ for some } k \in \mathbb{N}\}.$$

Solution to Exercise 2.1.5

C is the set of all real numbers of the form $2k + 1$ where $k \in \mathbb{N}$. By letting k go through the natural numbers $1, 2, 3, 4, \ldots$, the expression $2k + 1$ evaluates to $3, 5, 7, 9, \ldots$. Therefore, A and C are equal to each other.

B and F are equal to each other, as they both consist of the numbers $-4, -3, -2, -1, 0, 1, 2, 3, 4$.

Finally, D and E are both equal to the set $\{-1, 1\}$, and hence equal to each other.

Solution to Exercise 2.1.10

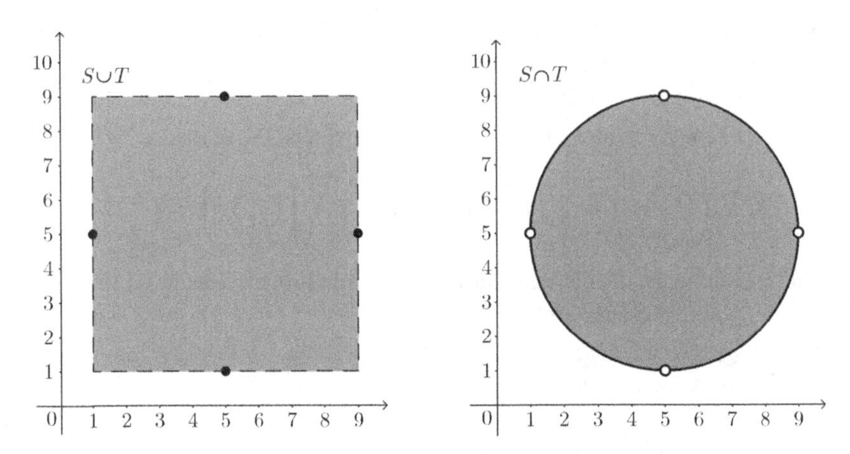

Figure 2.18 The sets $S \cup T$ and $S \cap T$.

Solution to Exercise 2.2.6

We obtain the following:

$$(f + g)(x) = f(x) + g(x) = \sqrt{1 - x^2} + \sin(3x)$$
$$(f \cdot g^2)(x) = f(x) \cdot g(x)^2 = \sqrt{1 - x^2} \cdot \sin^2(3x)$$
$$(-5f^3 - 4g)(x) = -5 \cdot f(x)^3 - 4g(x) = -5(\sqrt{1 - x^2})^3 - 4\sin(3x).$$

Solution to Exercise 2.3.2

Using the field axioms, and the fact that $x \cdot r_1 = x \cdot r_2 = 1$ we have:

$$r_1 = r_1 \cdot 1 = r_1 \cdot (x \cdot r_2) = (r_1 \cdot x) \cdot r_2 = r_2 \cdot (x \cdot r_1) = r_2 \cdot 1 = r_2.$$

Solution to Exercise 2.4.3

For every $k \in \mathbb{N}$, we have $A_k = \left(-\frac{1}{k}, \frac{1}{k}\right) \subseteq (-1, 1)$, and so $\bigcup_{k=1}^{\infty} A_k \subseteq (-1, 1)$.

Moreover, by the definition of unions, $A_1 = (-1, 1) \subseteq \bigcup_{k=1}^{\infty} A_k$. We thus conclude that $\bigcup_{k=1}^{\infty} A_k = (-1, 1)$.

Solution to Exercise 2.4.5

For every $\alpha \in (0,1)$ we have

$$2 - \alpha < 2 < 3 \leq 4 - \alpha$$

and so $[2,3] \subseteq I_\alpha$. This means that $[2,3] \subseteq \bigcap_{\alpha \in (0,1)} I_\alpha$.

Conversely, suppose that $x \in \bigcap_{\alpha \in (0,1)} I_\alpha$. Then, for each $\alpha \in (0,1)$, we have

$$x \in I_\alpha \qquad \Rightarrow \qquad 2 - \alpha < x \leq 4 - \alpha.$$

Now, assume by contradiction, that $x < 2$. If $x < 1.9$, take $\alpha = 0.05$, which implies

$$2 - 0.05 < x < 1.9,$$

contradicting the fact that $2 - 0.05 > 1.9$.

If $1.9 \leq x < 2$, take $\alpha = \frac{2-x}{2}$, which gives

$$2 - \frac{2-x}{2} < x < 2 \qquad \Rightarrow \qquad \frac{2+x}{2} < x,$$

contradicting the fact that $\frac{2+x}{2} > x$. We conclude that $x \geq 2$. A similar argument shows that $x \leq 3$, and so $x \in [2,3]$. This completes the proof of the second inclusion, $\bigcap_{\alpha \in (0,1)} I_\alpha \subseteq [2,3]$.

Solution to Exercise 2.5.6

We can represent 0 as the fraction $\frac{0}{1}$, and then

$$p(0) = p\left(\frac{0}{1}\right) = \frac{0^2 - 1^2}{0^2 + 1^2} = \frac{-1}{1} = -1.$$

3 Informal Logic and Proof Strategies

In this chapter, we take a step back to discuss, more generally, the language of mathematics, and some proof techniques and strategies. In the previous chapters, we have seen numerous mathematical notions, theorems, proofs and examples. As you have probably noticed, communicating mathematical arguments and ideas in a coherent and precise way is at the core of the subject.

In all our proofs, we used the spoken language (English, in our case), together with mathematical terms and symbols. We will continue to do so. There is, however, an important difference between using the spoken language in mathematics and in everyday life. In mathematics, we are held to higher standards of precision and accuracy, and cannot tolerate words or phrases that have multiple meanings or are too vague and imprecise.

For instance, suppose you get a phone call from FedEx. The clerk says:

"Your package has arrived. You can come and pick it up next week, on Monday *or* Thursday morning."

What would be your interpretation of the word "*or*"? Naturally, you are going to pick one of the two days offered, and collect your package on the morning of that day. Picking up the package on *both* days is not a feasible option.

On the other hand, suppose you are invited for a gathering at a friend's house. On the email invitation, the host asks that...

...each guest brings a dessert *or* an alcoholic beverage.

This time, if you show up with a chocolate cake and a bottle of wine, will the host be upset for you bringing *both* a dessert and a beverage? Probably not. Here, it is completely natural to interpret the word "*or*" as "one of the two options, or both," while in the FedEx example, it meant "exactly one of the options, and not both."

In mathematics, we cannot allow words to have multiple meanings, and so an interpretation of "or" must be chosen and used consistently throughout mathematics. As we have already mentioned before, "*or*" in mathematics means "*one of the two options, or both.*"

There are many other words, such as "and," "not," "if-then," which are commonly used in mathematics, but may have multiple interpretations in real-life.

We must clear any possible ambiguity by assigning a single interpretation to be used everywhere in mathematics. These words are called *connectives*, and play a fundamental role in communicating mathematical ideas.

However, before going into details regarding connectives and their mathematical interpretation, we first discuss a more fundamental concept – mathematical statements.

3.1 Mathematical Statements and their Building Blocks

Definition 3.1.1 A *mathematical statement*, or a *proposition*, is a declarative sentence such that, in every given context, is either true or false, but not both.

In other words, a mathematical statement, or simply a statement, is a phrase for which *it makes sense to ask whether it is true or false*. It may contain words, symbols, or both. To be honest, Definition 3.1.1 is informal, but it will be sufficient for our needs in this book. A more formal treatment can be found in advanced books on mathematical logic.

Let us look at a few examples.

Example 3.1.2

- "The square root of 9 is 3" is an example of a *true* mathematical statement.
- "The set $\{\varnothing\}$ is empty" is a *false* statement.
- "The function $f\colon \mathbb{R} \to \mathbb{R}$, $f(x) = x^2$ is a bounded function" is also a *false* statement. See Problem 2.31 for the definition of a bounded function.
- "$\mathbb{Z} \subseteq \mathbb{Q}$" is also a mathematical statement, which is *true*, as every integer is also a rational number. Note how this statement contains *only mathematical symbols* without any words.
- The phrase "$\mathbb{Z} \cup \left\{\frac{1}{2}, \frac{1}{3}\right\}$" is *not* a mathematical statement. It is simply a set of numbers. It does not make any sense to ask whether "$\mathbb{Z} \cup \left\{\frac{1}{2}, \frac{1}{3}\right\}$" is true or false.
- Similarly, the expression "$23 + 15 - 7$" is not a mathematical statement. However, both "$23 + 15 - 7 = 31$" and "$23 + 15 - 7 > 35$" are statements (the former is *true*, and the latter is *false*).
- The equality "$1 + 1 = 0$" is a mathematical statement. In the context of real numbers, this is a false statement. However, in the context of a field with two elements, it is a true statement. This example shows how important it is to make sure the context for a given statement is clearly understood.

The next example involves a variable.

Example 3.1.3 Is "$x^2 \geq x$" a mathematical statement? Does it make sense to pose the question whether $x^2 \geq x$ is true or false? Not quite. However, if x represents a real number, this will be a statement. When $x = 5$, $x^2 \geq x$ is a *true* mathematical statement, and when $x = \frac{1}{2}$ we get a *false* statement.

Such phrases are called *predicates*, and can be thought of as *templates* for creating mathematical statements. There is another way to turn predicates into statements, by using quantifiers. This method is discussed below.

Naturally, you might ask yourself why is the notion of a mathematical statement so important? And why would we ever consider false statements? The answer to the first question is simple. Every claim, proposition and theorem is an example of a mathematical statement. Mathematicians spend most of their time trying to prove or disprove statements, and therefore it is crucial that we are able to identify what is a statement, and what is not. It would be meaningless to try and prove a phrase which is not a statement.

To answer the second question, keep in mind that when mathematicians attempt to prove a new claim or theorem, they do not know, at first, whether that statement is true or false. This is why we must be prepared to work with statements that are potentially false. There are many other reasons for considering false statements. False statements are often used in proofs and in negating phrases. We will get to discuss these later.

Quantifiers

As mentioned in Example 3.1.3, the phrase "$x^2 \geq x$" is *not* a mathematical statement, unless we assign a value to x. Another way of turning this inequality into a statement is by adding words such as "for every" and "for some."

(i) For every real number x, we have $x^2 \geq x$.
(ii) $x^2 \geq x$ for some real number x.

Both of these phrases are mathematical statements. Statement (i) is false, as the inequality $x^2 \geq x$ is false for some real numbers, such as $x = \frac{1}{2}$. On the other hand, statement (ii) is true, as there do exist numbers satisfying this inequality. For instance, $5^2 \geq 5$.

Words such as "for any," "every," "for all," "there is," "there exists," "for some," etc. are called *quantifiers*, and are used to turn predicates (sentences with variables) into statements. Quantifiers play an essential role in building mathematical statements. As we have just seen, replacing one quantifier by another can change the meaning of a statement.

Connectives

A connective is a word, or a combination of words, that is applied to one or two statements to create a new statement. Here are some examples.

Example 3.1.4 "3 is an odd number *and* π is an irrational number." The word "and" is used to take the two statements "3 is an odd number" and "π is an irrational number" and create a new longer true statement. Proving that π is irrational requires some effort and is beyond the scope of this book.

Example 3.1.5 "The unit (rightmost) digit of 88^{88} is 2 *or* 8." This statement includes the connective "or" to merge two statements into one, which is, by the way, a false statement (the unit digit of 88^{88} is 6).

Example 3.1.6 Mathematical statements may contain multiple connectives and quantifiers. For instance:

"For every two real numbers x, y, if $x^2 + y^2 = 0$, then $x = 0$ and $y = 0$."
Both variables x and y are quantified by the words "For every," and the words "if-then" and "and" are connectives. This statement is true.

Table 3.1 lists commonly used connectives in mathematics, their official name, and examples.

Exercise 3.1.7
In the table above, which of the examples are true statements?

Note that the negation connective "*not*" is used with *one* statement, so it does not really connect multiple statements. Nevertheless, we still refer to it as a connective.

Exercise 3.1.8
The following is a true statement, which is a special case of a more general theorem called the Division Algorithm. It will be discussed and proved in Chapter 6. Can you identify all the quantifiers and connectives in the statement?

"For every $m \in \mathbb{Z}$, if m is not divisible by 3, then there exists a $k \in \mathbb{Z}$, such that $m = 3k + 1$ or $m = 3k + 2$."

Table 3.1 The logical connectives

Connective	Name	Example
and	conjunction	The smallest natural number is 1 *and* there is no largest natural number.
or	disjunction	For every two sets A and B, we have $A \subseteq B$ *or* $B \subseteq A$.
if-then	implication	For every function $f: \mathbb{R} \to \mathbb{R}$, *if* f is bounded, *then* f^2 is also bounded.
if and only if	equivalence	149 is a prime number *if and only if* 147 is a prime number.
not	negation	*Not* every real number is rational.

Example 3.1.9 Implicit Quantifiers. Many mathematical statements and proofs include sentences of the form "Let... be..." or "Suppose that...." For instance, Theorem 1.1.1 starts with "Let a, b, c be real numbers, with $a \neq 0$." Theorem 2.1.13 begins with "Suppose A and B are subsets of...."

What is the correct interpretation of these sentences? How are our variables quantified? Looking carefully at the above theorems, we realize that words like "Let... be..." and "Suppose that..." often refer *implicitly* to the quantifier "for all" or "for every."

For instance, consider the following (true) statements.

(i) Let A and B be two sets. If $A \setminus B = \varnothing$, then $A \subseteq B$.
(ii) Suppose that r is a positive real number. Then $3r + \frac{2}{r} \geq 2\sqrt{6}$.

Both statements assert that a certain property holds true for all sets, or for all positive real numbers. Neither statement specifies particular objects, nor talks about the existence of a certain object. We may thus rephrase the statements as follows.

(i) For every two sets A and B, if $A \setminus B = \varnothing$, then $A \subseteq B$.
(ii) For every positive real number r, we have $3r + \frac{2}{r} \geq 2\sqrt{6}$.

The meaning of the rephrased statements is the same as the original statements. We use words such as "let" and "suppose" as we often prefer to think about a fixed, but arbitrary object, such as a number or a set, rather than considering all possible objects in a certain context.

Exercise 3.1.10
Prove statements (i) and (ii) in Example 3.1.9.

3.2 The Logic Symbols

In mathematics, we sometimes use special symbols to denote logical phrases. These symbols allow us to write *compound statements* (that is, statements involving one or more connectives and quantifiers) without using the spoken language. Statements or predicates are often denoted by an uppercase Latin letter, such as P, Q, R or $P(x), Q(x, y)$ for predicates involving variables. Table 3.2 describes logic symbols associated to commonly used connectives and quantifiers.

Now let us look at a few examples.

Example 3.2.1 Our first example is non-mathematical. Assume that the set B represents a group of parents, and the set A represents the group of their

Table 3.2 The logic symbols

Connective/ Quantifier	Name	Symbol	How to use?
for all, for any, every	the universal quantifier	$\forall$	$(\forall x \in U)P(x)$ means "For every x in the set U, $P(x)$"
there is, there exists, for some	the existential quantifier	$\exists$	$(\exists x \in U)P(x)$ means "There is an x in the set U, for which $P(x)$"
not	negation	$\neg$	$\neg P$ means "Not P"
and	conjunction	$\wedge$	$P \wedge Q$ means "P and Q"
or	disjunction	$\vee$	$P \vee Q$ means "P or Q"
if-then	implication	$\Rightarrow$	$P \Rightarrow Q$ means "If P, then Q"
iff (if and only if)	equivalence	$\Leftrightarrow$	$P \Leftrightarrow Q$ means "P if and only if Q"

children. Denote by $P(x, y)$ the phrase "x is y's child." Then we can form the following two statements:

(i) $(\forall x \in A)(\exists y \in B)P(x, y)$ (ii) $(\exists y \in B)(\forall x \in A)P(x, y)$.

Statement (i) reads "For every element x in A, there is an element y in B, such that $P(x, y)$." In the context of children and their parents, the statement becomes "For every child in A, there is a person in B, who is a parent of that child." This is a true statement, as every child in A has a parent in B. This follows from the way A and B were formed.

What about statement (ii)? This statement reads "There is an element y in B, such that for every x in A, $P(x, y)$." In our context, this means that "There is a person in B, who is the parent of *every* child in A," which might be true or false, depending on the sets A and B.

This example highlights an important observation. Statements (i) and (ii) look quite similar, but they have different meanings. As statement (ii) was obtained from (i) by changing the order of the two quantifiers, we conclude that *the order of quantifiers matters*, and changing the order may change the meaning of a statement.

Example 3.2.2 Consider the mathematical statement

"Every integer is either even or odd."

How can we write this statement using the logic symbols? First, let $E(x)$ denote the phrase "x is even," and $O(x)$ denote "x is odd." Now, the statement can be written as

$$(\forall x \in \mathbb{Z})(E(x) \vee O(x)).$$

Note how the disjunction connective is used. The brackets are added to help with the reading of the statement.

Our last example is a little more complicated. We wish to represent the following statement, that we call P, with the logic symbols.

Statement P: "There is no smallest real number."

We start by representing the statement "There is a smallest real number." which we call Q. To do so, we rewrite the statement, without using the word "smallest." This can be done as follows.

Statement Q: "There is a real number x that is smaller than or equal to every real number."

Note that the words "there is" and "every" suggest that two quantifiers should be used here. Using only the logic symbols, Q can be represented as

$$(\exists x \in \mathbb{R})(\forall y \in \mathbb{R})(x \leq y).$$

As P is simply the negation of Q, the statement P becomes

$$\neg Q$$

or

$$\neg[(\exists x \in \mathbb{R})(\forall y \in \mathbb{R})(x \leq y)].$$

We will see shortly, how such complex statements can often be shortened and simplified.

Remarks.

- Notice how quantifiers are placed in a statement written with the logic symbols. A quantifier will always appear *before* the variable is mentioned. For instance, in the following statement, written in words, the quantifier "for every" appears at the end of the sentence:

  "$|a+1| \leq |a|+1$ for every real number a."

  However, when converting to the logic symbols, we must write

  $$(\forall a \in \mathbb{R})(|a+1| \leq |a|+1).$$

- When the context is clear, we sometimes omit the universal set U. For instance, if it is clear that our variables must represent real numbers, we may write

  $(\forall a)(|a+1| \leq |a|+1)$ instead of $(\forall a \in \mathbb{R})(|a+1| \leq |a|+1)$.

3.3 Truth and Falsity of Compound Statements

As we already mentioned, a mathematical statement can be either *true* or *false*, in a given context. We denote the *truth value* of a statement by T for true and by F for false. When a statement is built from other statements, often called *elementary statements*, through the use of connectives, its truth value will depend on the truth values of the elementary statements.

To clarify, let us look at an explicit case. Suppose that P and Q are two statements (our *elementary statements*). Now let R be the statement $P \vee Q$. What is the truth value of R? Is R a true or a false statement? Obviously, we cannot answer this question without knowing the truth values of P and Q. In fact, knowing the truth values of P and Q is enough to determine the truth value of R. In other words, the truth value of R will depend *only* on the truth values of P and Q, regardless of their actual content.

Since the symbol $\vee$ represents the connective "or," and its meaning in mathematics resembles the meaning of "or" in the spoken language, the statement $P \vee Q$ is true when either P, Q or both are true. If P and Q are both false, then $P \vee Q$ is false. This information can be summarized in a *truth table*, as follows:

P	Q	$P \vee Q$
T	T	T
T	F	T
F	T	T
F	F	F

In the first two columns, we list all possible combinations of truth values for the elementary statements P and Q. In the third column, we write the corresponding truth value of $P \vee Q$. For instance, the second last row says that when P is false and Q is true, then $P \vee Q$ is a true statement. The last row says that if P and Q are both false, then $P \vee Q$ is false as well.

A truth table is a convenient way to summarize all possible truth values of a statement, *as a function of the truth values of its building blocks, the elementary statements*. Of course, if more elementary statements are present, the table will be longer. Moreover, we can view the above truth table as *defining the meaning* of the disjunction connective in mathematics.

We can easily construct truth tables for most of the other connectives, as their use in mathematics is quite similar to their use in the spoken language. Here are the tables for the negation, conjunction and equivalence connectives.

P	$\neg P$
T	F
F	T

P	Q	$P \wedge Q$
T	T	T
T	F	F
F	T	F
F	F	F

P	Q	$P \Leftrightarrow Q$
T	T	T
T	F	F
F	T	F
F	F	T

Note how the truth table for negation has only two rows, as this connective is applied to a *single* statement.

There is one connective that has not been discussed yet, the implication connective if-then, as its truth table may seem a little odd at first. We discuss it separately below.

Implications

Before presenting the truth table for the implication connective, we discuss the following statement, that we label as R.

Statement R: "If it is snowing, then the temperature is less than or equal to 0°C."

This statement is an implication. Its structure is of the form $P \Rightarrow Q$, where P is the phrase "it is snowing," and Q is "the temperature is less than or equal to 0°C."

Under most circumstances, the above statement is true, and snow is seen at freezing temperature, or below. However, it turns out that under very dry conditions, this statement might be *false*.

What does it mean for R to be false? Well, if snowing *does not* imply a temperature of 0°C or less, then we could see snow at above-freezing temperatures. In other words, claiming that R is false means that it is possible to have snow (namely, P is true), while the temperature is greater than 0°C (i.e., Q is false). In a table, we indicate this case as follows:

P	Q	$P \Rightarrow Q$
T	F	F

What about the rest of the truth table? Is there any other scenario that would make R false? If P and Q are both true (i.e., it snows *and* the temperature is 0°C or less), then R holds true. Moreover, if it does not snow, then regardless of the temperature, the implication is not violated. We conclude that when P is false, the implication $P \Rightarrow Q$ holds true, no matter what the truth value of Q is, and so the full truth table for implications is the following:

P	Q	$P \Rightarrow Q$
T	T	T
T	F	F
F	T	T
F	F	T

It might be hard for you to accept, at first, the last two rows of the truth table, so let us look at a few more examples.

Example 3.3.1 The following phrase, or some other version of it, may be included in rental lease agreements.

"If the tenant severely damages the property,
the landlord has the right to terminate the lease."

Again, this is an if-then statement of the form $P \Rightarrow Q$, where P is "the tenant severely damages the property" and Q is "the landlord has the right to terminate the lease."

Can the landlord terminate the lease even when the tenant does not severely damage the property? Yes they can, for other reasons, such as not paying rent. The implication is *not* violated when P is false and Q is true. Clearly, having both P and Q true or both false (e.g., the tenant damages and the lease is terminated, or the tenant does not damage and the lease is not terminated) is consistent with the above phrase.

The only possible scenario in which the above clause is not followed, is when the tenant does damage the property (that is, P is true), but the landlord cannot terminate the agreement (Q is false). In all other cases, $P \Rightarrow Q$ is true, which is consistent with the truth table for implications.

Example 3.3.2 In everyday language, we often use the combination if-then to describe *cause and effect*. For instance:

"If you do not brush your teeth, you will have cavities."

This sentence describes cause and effect: not brushing your teeth will have a direct effect on them. But cause and effect makes little sense in mathematics. Consider, for example, the following statement:

"If money grows on trees, then cats have five legs."

This is a true statement, as both P, the hypothesis, and Q, the conclusion, are false. Clearly, there is no cause and effect involved here. It is the logical structure of the statement, and its interpretation based on the truth table above, that make it true. We often say that if-then statements in which the hypothesis is false are *vacuously true*.

Example 3.3.3 We end this discussion with one last example. Suppose you are borrowing your friend's bicycle for a one-day biking trip. When you pick up the bike, your friend says:

"If you damage or lose my bike, you will have to pay for it!"

Again, this statement has the structure $P \Rightarrow Q$, with P being "you damage or lose my bike," and Q being "you will have to pay for it!" Now imagine that when you come back from the trip, and return the undamaged bike to your friend, they say "Now I want you to pay for it!" Surprised, you reply "How come? There isn't even a scratch on your bike. What do I need to pay for?" Your friend keeps insisting that you pay, arguing that according to the truth table above, asking you to pay even when the bike is undamaged is consistent with the if-then statement: If P is false, and Q is true, the implication $P \Rightarrow Q$ is considered true! What is going on here?

The problem lies in the way we use the words "if-then" in everyday language. Occasionally, "if-then" would actually mean "if-and-only-if" in ordinary day-to-day language. When your friend said the above phrase, you probably interpreted it (as most people would) as follows:

"If you damage or lose my bike, you will have to pay for it!
But if you return it undamaged, you will not need to pay at all."

In other words, you interpreted the phrase as an if-and-only-if statement:

"You will have to pay for my bike if and only if you damage or lose it."

Consequently, you were very surprised when your friend demanded that you pay.

These kind of issues can cause great confusion in mathematics. Precision and accuracy are crucial in mathematical arguments and proofs, and so we have to make sure that we all interpret statements in the same way. Therefore, we must use "if-then" when we *actually mean* if-then, and "if-and-only-if" when our statement describes an equivalence.

A Remark on Quantifiers

In this section we have provided truth tables for the logical connectives, and used them to find out the truth value of a compound statement. How about statements that involve quantifiers? For instance, can we construct a truth table for the following statement?

"For every prime number p greater than 3, $p^2 - 1$ is divisible by 24."

This statement has the form $(\forall p \in A)S(p)$, where A is the set of prime numbers greater than 3, and $S(p)$ is the phrase "$p^2 - 1$ is divisible by 24." We might be tempted to construct a truth table for this statement, by evaluating the truth value of $S(p)$ for every prime number p greater than 3. The first few rows of such a table will look as shown in Table 3.3. However, it is impossible to create a complete table, as there are infinitely many primes greater than 3 (a fact that we will soon prove). Therefore, truth tables are rarely used to analyse statements that involve quantifiers, and other techniques must be used instead. We need to remember though, the way quantifiers are interpreted in mathematics.

- $(\forall x \in U)P(x)$ is true when $P(x)$ holds true *for every* value of x in the set U.
- $(\exists x \in U)P(x)$ is true when $P(x)$ holds true *for at least one value* of x from the set U.

Table 3.3 An *infinite* truth table for the predicate $S(p)$

p	$S(p)$	Truth value of $S(p)$
5	$5^2 - 1$ is divisible by 24	T
7	$7^2 - 1$ is divisible by 24	T
11	$11^2 - 1$ is divisible by 24	T
⋮	⋮	⋮

As you can see, the mathematical interpretation of the quantifiers is quite consistent with our everyday use of "for every" and "for some" in the spoken language. When the universe U is understood from the context, we often omit it, and simply write $\forall x P(x)$ and $\exists x P(x)$.

3.4 Truth Tables and Logical Equivalences

The truth tables for the logical connectives are often referred to as the *elementary truth tables*. These tables define the *meaning* of the connectives in mathematics. However, truth tables can be created for longer and more complex statements, involving multiple connectives. Here is an example.

Example 3.4.1 Suppose that P are Q are two unknown statements. Using P and Q, we construct a new compound statement, that we call R.

$$\text{Statement } R: \qquad (P \vee Q) \wedge [P \Rightarrow (\neg Q)].$$

Note how the brackets are used to indicate the order in which the connectives should be applied. The truth value of R will depend on the truth values of P and Q, and we would like to describe this dependency in a table. Our truth table will have four rows, one for each possible combination of truth values for P and Q.

P	Q	R
T	T	?
T	F	?
F	T	?
F	F	?

Our task is to complete the third column of the table. Namely, to find the truth values of R in each of the four possible scenarios. However, as R is a relatively complex statement, with several connectives, finding the truth values requires some work. We must carefully analyze the various parts of R and the way they

Table 3.4 A truth table for the statement R

P	Q	$P \vee Q$	$\neg Q$	$P \Rightarrow (\neg Q)$	R
T	T				
T	F				
F	T				
F	F				

Table 3.5 A completed truth table for the statement R

P	Q	$P \vee Q$	$\neg Q$	$P \Rightarrow (\neg Q)$	R
T	T	T	F	F	**F**
T	F	T	T	T	**T**
F	T	T	F	T	**T**
F	F	F	T	T	**F**

are put together, and we can do so by adding a few more "helper" columns to our truth table (see Table 3.4).

Without too much effort, we can complete the helper columns in order, from left to right, by referring to the elementary truth tables of the connectives $\neg$, $\Rightarrow$ and $\vee$. For instance, the truth values of $\neg Q$ are obtained by reversing those of Q. Once we have the truth values for $P \vee Q$ and $P \Rightarrow (\neg Q)$, we can use them, together with the truth table of the conjunction ($\wedge$), to find the desired truth values of R. Table 3.5 is the completed table. We see that R is true when P and Q have different truth values. If P and Q are both false or both true, then the statement R is false. Keep in mind that the helper columns are optional, and we add as many as we need in order to complete the rightmost column.

Example 3.4.2 Let us look at another example. Consider the following statement S, built out of two unknown statements P and Q.

$$\text{Statement } S: \qquad (P \vee Q) \Leftrightarrow [(\neg Q) \vee (\neg P)].$$

To construct the truth table of S, we set up, as before, a table with four possible cases, and some helper columns (Table 3.6). Using the elementary truth tables for negation ($\neg$) and disjunction ($\vee$), we complete the helper columns. Then, we use the truth table of equivalence ($\Leftrightarrow$) to find the truth values of S in each case (Table 3.7).

Table 3.6 A truth table for the statement S

P	Q	$P \vee Q$	$\neg Q$	$\neg P$	$(\neg Q) \vee (\neg P)$	S
T	T					
T	F					
F	T					
F	F					

Table 3.7 A completed truth table for the statement S

P	Q	$P \vee Q$	$\neg Q$	$\neg P$	$(\neg Q) \vee (\neg P)$	S
T	T	T	F	F	F	**F**
T	F	T	T	F	T	**T**
F	T	T	F	T	T	**T**
F	F	F	T	T	T	**F**

Taking a careful look at the truth tables for R and S above, we see that the rightmost column in each of the tables is the same. If we remove the helper columns from each table, we get:

P	Q	R
T	T	**F**
T	F	**T**
F	T	**T**
F	F	**F**

P	Q	S
T	T	**F**
T	F	**T**
F	T	**T**
F	F	**F**

In other words, in every possible scenario, R and S will have the same truth value. We say that R and S are *logically equivalent statements.*

Definition 3.4.3 Suppose that R and S are two compound statements, made up from elementary statements and connectives. We say that R and S are *logically equivalent*, or simply *equivalent*, if they have the same truth table, that is, if they have the same truth value in every possible scenario. We write $R \equiv S$ to indicate that R and S are equivalent.

Exercise 3.4.4

Show that the statements R and S from Examples 3.4.1 and 3.4.2 are equivalent to the statement $P \Leftrightarrow (\neg Q)$.

Informally speaking, two logically equivalent statements are statements that have the same logical meaning. That is, they say the same thing, though in a different way. This is much like two equivalent algebraic expressions, such as

$4x^2 - 6y$ and $(2x)^2 + (-2y) \cdot 3$. These expressions, although different, always produce the same value for every choice of numbers x and y, and so we treat them as being equal (or equivalent) to each other.

Logical equivalences can be used to simplify statements, in the same way that algebraic identities are used to simplify algebraic expressions. The following is a list of important and commonly used logical equivalences.

Proposition 3.4.5 *Let P and Q represent two statements. Then we have the following logical equivalences.*

1. $\neg(P \wedge Q) \equiv (\neg P) \vee (\neg Q)$
2. $\neg(P \vee Q) \equiv (\neg P) \wedge (\neg Q)$
3. $\neg(P \Rightarrow Q) \equiv P \wedge (\neg Q)$
4. $P \Leftrightarrow Q \equiv (P \Rightarrow Q) \wedge (Q \Rightarrow P)$
5. $P \Rightarrow Q \equiv (\neg Q) \Rightarrow (\neg P)$

Proof (partial) Proving the proposition does not require too much effort. We create truth tables for each pair, and verify that they are indeed the same.

For instance, here are the truth tables, including one helper column in each, for the statements in Part 3.

P	Q	$P \Rightarrow Q$	$\neg(P \Rightarrow Q)$
T	T	T	F
T	F	F	T
F	T	T	F
F	F	T	F

P	Q	$\neg Q$	$P \wedge (\neg Q)$
T	T	F	F
T	F	T	T
F	T	F	F
F	F	T	F

The rightmost columns in the truth tables are the same, which proves the equivalence of $\neg(P \Rightarrow Q)$ and $P \wedge (\neg Q)$. To prove Part 5, we construct the following two truth tables. Note that the table on the left is just the truth table for the implication connective.

P	Q	$P \Rightarrow Q$
T	T	T
T	F	F
F	T	T
F	F	T

P	Q	$\neg Q$	$\neg P$	$(\neg Q) \Rightarrow (\neg P)$
T	T	F	F	T
T	F	T	F	F
F	T	F	T	T
F	F	T	T	T

And again, as the rightmost columns are identical, the statements $P \Rightarrow Q$ and $(\neg Q) \Rightarrow (\neg P)$ are logically equivalent.

The proofs of the other parts are left as an exercise. □

Remarks.

- Equivalence of statements is often proved by constructing truth tables and comparing them, and that is what we did to prove Parts 3 and 5 of the proposition above. However, it might be useful to try and create real-life non-mathematical examples that illustrate the equivalence. This cannot serve as a mathematical proof, but it can strengthen our intuition and confidence as to why an equivalence is valid.

  For instance, denote by P the phrase "being rich" and by Q the phrase "being happy." Then the statement $\neg(P \Rightarrow Q)$ becomes "Being rich *does not* imply being happy." The statement $P \wedge (\neg Q)$, however, translates into "One can be rich and unhappy." It is quite evident, at least informally, that the two statements have the same meaning. This illustrates the equivalence of the statements in Part 3 of Proposition 3.4.5.

  For Part 5, consider the following if-then statement:

  "If a person has a driver's license, they are at least 16 years old."

  The statement has the structure $P \Rightarrow Q$, where P is "a person has a driver's license," and Q is "being at least 16 years old." Now, if we convert $(\neg Q) \Rightarrow (\neg P)$ into words, we get

  "If a person is not at least 16 years old, then they do not have a driver's license."

  Again, it is not hard to observe the equivalence of the two statements. They both say the exact same thing, which supports the equivalence in Part 5 of the proposition.
- The equivalence in Part 5 of Proposition 3.4.5 is of great importance, as it provides us with a proof technique that is commonly used in mathematics. We call $(\neg Q) \Rightarrow (\neg P)$ the *contrapositive* of $P \Rightarrow Q$, and as we have seen, it is equivalent to $P \Rightarrow Q$. This means that proving an if-then statement can be done by proving its contrapositive. Here is an example.

Example 3.4.6 Consider the following statement:

Let $n \in \mathbb{Z}$. If n^3 is odd, then n is odd.

How can we prove this statement? An odd number has the form $2k + 1$ for some integer k, so we can start by writing $n^3 = 2k + 1$. Our task is to show that n is odd, so we might try to solve for n, which gives $n = \sqrt[3]{2k+1}$. However, it seems like writing n as $\sqrt[3]{2k+1}$ is not going to be of any help, and we are stuck!

Fortunately, the contrapositive of the implication "If n^3 is odd, then n is odd." is "If n is even, then n^3 is even" (as an integer which is not odd is even), and is much easier to prove.

Proof We prove the contrapositive "If n is even, then n^3 is even." An even number is a multiple of 2, and so $n = 2k$ for some $k \in \mathbb{Z}$. Therefore,

$$n^3 = (2k)^3 = 8k^3 = 2 \cdot 4k^3.$$

We now see that n^3 is an integer multiple of 2, and hence is even, as needed. □

This example shows how sometimes implications which are hard (or even impossible) to prove directly, may have a simple proof for their contrapositive. We will see more examples of using the contrapositive method later in this chapter.

Equivalences Involving Quantifiers

The following two equivalences involve quantifiers, and so we are unable to verify them with a truth table. We state the equivalences without proof. Suppose that $P(x)$ is a predicate that depends on a variable x, and that U is our universal set. Then we have the following.

- $\neg[(\forall x \in U)P(x)]$ is logically equivalent to $(\exists x \in U)[\neg P(x)]$.
- $\neg[(\exists x \in U)P(x)]$ is logically equivalent to $(\forall x \in U)[\neg P(x)]$.

These equivalences will be used in the next section, and even though we do not provide a formal proof, they are quite intuitive. Try to say them in words. Can you see why each pair has the exact same meaning?

3.5 Negation

The negation of a statement is its opposite, usually obtained by adding the word "*not*."

Definition 3.5.1 The *negation* of a statement P is the statement "Not P" which we can also write, using the logic symbols, as $\neg P$.

For instance, the negation of

"All cats are black"

is

"*Not* all cats are black"

and the negation of

"Being tall implies having back problems"

is

"Being tall *does not* imply having back problems."

Mathematical statements can also be negated by adding words of negation, such as "not" and "no." For instance, the negation of the statement:

"There is a set A for which $A \in A$"

is

"There is *no* set A for which $A \in A$."

In mathematics, we need to work with negations and use them in various arguments and proofs regularly. However, forming a negation by simply adding the word "not," or by putting the negation symbol $\neg$ in front of it, may be insufficient, and we often simplify negations and make them as explicit as possible. For instance, the statement "Not all cats are black" can be restated as "Some cats are not black." Here is a more mathematical example.

Example 3.5.2 Let $U = \mathbb{R}$ be the set of all real numbers, and consider the following statement Q.

Statement Q: "For all $a, b \in U$, if $a \cdot b = 0$, then $a = 0$ or $b = 0$."

This is a true statement. In fact, the statement can be proved from the field axioms, and thus remains true as long as U is a field. There are, however, mathematical "universes" (which we will not discuss in this book), in which Q is false, and hence its negation $\neg Q$ is true.

In other words, there are cases in which the statement

$\neg Q \equiv$ "*Not* for all $a, b \in U$, if $a \cdot b = 0$, then $a = 0$ or $b = 0$"

is valid. To better understand the negated statement, we simplify it as much as we can. Ideally, we try to restate the negation without using the word "not" or "no." To begin with, we observe that if not all a, b satisfy the if-then phrase, then at least one pair of a, b must violate it. We can therefore restate the negation $\neg Q$ as

$\neg Q \equiv$ "There exist $a, b \in U$ for which $a \cdot b = 0$ does *not* imply $a = 0$ or $b = 0$."

Moreover, if $a \cdot b = 0$ does not imply $a = 0$ or $b = 0$, then $a \cdot b$ can be zero, while none of a, b are zero (remember, an implication is false when the hypothesis is true and the conclusion is false). Hence, we can equivalently write $\neg Q$ as

$$\neg Q \equiv \text{"There exist } a, b \in U\text{, such that } a \cdot b = 0,\ a \neq 0 \text{ and } b \neq 0.\text{"}$$

Note how we were able to restate $\neg Q$ without using any words of negation such as "not," "no," "it is false that," etc. By doing so, we have made the negation explicit, simpler, and easier to use in arguments and proofs.

In Example 3.5.2 above, we started with an unsimplified negation, and gradually replaced it by equivalent statements, which are simpler. This process can be done more mechanically, by referring to the logical equivalences mentioned in the previous section. We illustrate this procedure by re-doing Example 3.5.2.

Example 3.5.3 We negate and simplify the following statement Q:

$$(\forall a \in \mathbb{R})(\forall b \in \mathbb{R})[(a \cdot b = 0) \Rightarrow ((a = 0) \vee (b = 0))].$$

Using the equivalence of $\neg[(\forall x)P(x)]$ and $(\exists x)[\neg P(x)]$, we have

$$\neg Q \equiv (\exists a \in \mathbb{R})(\exists b \in \mathbb{R})[\, \neg[(a \cdot b = 0) \Rightarrow ((a = 0) \vee (b = 0))]\,].$$

Next, we apply the equivalence of $\neg(P \Rightarrow Q)$ and $P \wedge (\neg Q)$ (see Proposition 3.4.5), to get

$$\neg Q \equiv (\exists a \in \mathbb{R})(\exists b \in \mathbb{R})[(a \cdot b = 0) \wedge [\, \neg[(a = 0) \vee (b = 0)]\,]\,].$$

Now, we apply the equivalence of $\neg(P \vee Q)$ and $(\neg P) \wedge (\neg Q)$ on the "$a = 0$ or $b = 0$" part:

$$\neg Q \equiv (\exists a \in \mathbb{R})(\exists b \in \mathbb{R})[\, (a \cdot b = 0) \wedge [\neg(a = 0)] \wedge [\neg(b = 0)]\,].$$

Finally, we replace $\neg(a = 0)$ and $\neg(b = 0)$ by $a \neq 0$ and $b \neq 0$. After all, $\neq$ means "not equal":

$$\neg Q \equiv (\exists a \in \mathbb{R})(\exists b \in \mathbb{R})[\, (a \cdot b = 0) \wedge (a \neq 0) \wedge (b \neq 0)\,].$$

As you can see, we managed to express the negation of Q without using the negation symbol $\neg$. We obtained the exact same statement as the one we got in Example 3.5.2, except that it is written with the logic symbols, instead of with words.

An important use of negation is with definitions, as we illustrate in the next example. Showing that an object *does not* satisfy a definition is the same as showing that its negation holds true.

Example 3.5.4 Here is the definition of a bounded function.

"A function $f: \mathbb{R} \to \mathbb{R}$ is *bounded*, if there is an $M \in \mathbb{R}$, such that $|f(x)| \leq M$ for all $x \in \mathbb{R}$."

Note that $|f(x)| \leq M$ is equivalent to $-M \leq f(x) \leq M$. Geometrically, f is bounded if its graph lies between the horizontal lines $y = M$ and $y = -M$, for some real number M.

Using the logic symbols, we can rewrite the definition as follows.

"A function $f: \mathbb{R} \to \mathbb{R}$ is *bounded*, if $(\exists M \in \mathbb{R})(\forall x \in \mathbb{R})(|f(x)| \leq M)$."

Now suppose we want to prove that a function f is *unbounded* (i.e., not bounded). To do so, we need to show that the above definition does not hold. Or, equivalently, that its negation is satisfied. Using the negation symbol, we can write:

"A function $f: \mathbb{R} \to \mathbb{R}$ is *unbounded* if $\neg(\exists M \in \mathbb{R})(\forall x \in \mathbb{R})(|f(x)| \leq M)$."

We can now apply the logical equivalences involving quantifiers, and restate the definition of an unbounded function without the negation symbol:

"A function $f: \mathbb{R} \to \mathbb{R}$ is *unbounded* if $(\forall M \in \mathbb{R})(\exists x \in \mathbb{R})(|f(x)| > M)$."

This is an explicit and more practical definition. Note how the inequality $|f(x)| \leq M$ was replaced by its negation, $|f(x)| > M$.

We end this section with one final note. Negating a statement reverses its truth value, and so a statement and its negation will never be both true or both false. Keep that fact in mind, and use it to check your negations.

Example 3.5.5 The simplified negation of

$$(\exists x \in (-3, 2) \cap \mathbb{Z})\left(x^2 < \frac{1}{2}\right)$$

is

$$(\forall x \in (-3, 2) \cap \mathbb{Z})\left(x^2 \geq \frac{1}{2}\right).$$

As $0^2 < \frac{1}{2}$, the first statement is true, and its negation is thus false.

Example 3.5.6 The negation of

$$(\forall x \in \mathbb{R})[(x^2 + 1 < 0) \Rightarrow (15 < 5)]$$

is

$$(\exists x \in \mathbb{R})[(x^2 + 1 < 0) \wedge (15 \geq 5)].$$

For every real number x, $x^2 + 1 > 0$, and hence the implication

$$(x^2 + 1 < 0) \Rightarrow (15 < 5)$$

is *vacuously true*. We conclude that the given statement is true, and its negation is false.

3.6 Proof Strategies

A major focus of this book is on the notion of a mathematical proof. We discuss numerous examples of proofs, and provide guidance and tools for creating them. However, we have not discussed yet explicitly the general question of how to *construct*, or generate, mathematical proofs. It would be nice to have a clear set of guidelines, procedures or algorithms that we can use to generate proofs of mathematical statements. In Chapters 1 and 2 we have seen quite a few proofs, but with very little similarities. For instance, the proofs of the quadratic formula and inequalities were quite algebraic, while proofs involving sets were mostly done by showing two inclusions. Proving statements about fields required careful use of the field axioms, and sometimes an elimination strategy, such as in Claim 2.3.10.

How would one know the right way to tackle a claim, theorem or proposition? Is there a recipe that can be executed to generate a proof? We do have algorithms for solving quadratic equations and systems of linear equations, and for finding extreme values of a function. Is there also an algorithm for generating proofs?

Unfortunately, no. In fact, I would rather say: *Fortunately*, no!

If there was an algorithm, or an explicit set of guidelines for generating proofs, mathematics would become easier, but also a mechanical and procedural field. We could then probably have computers generate proofs for many mathematical statements, and theoretical mathematics will become as dull as adding fractions or multiplying two-digit numbers. Many are attracted to mathematics because of the *creative nature* of this field. Some compare mathematics to the arts, and see beauty in the process of discovering and creating mathematical proofs. It is the *lack* of an algorithm or procedures for creating proofs that makes mathematics so interesting (and often challenging).

For this reason, proving theorems can be difficult, at times frustrating, and can be a long process. But the journey can nevertheless be exciting, enlightening, and transformative. Discovering a proof, especially after working hard and implementing new ideas, can be extremely rewarding. The need to be creative, produce original ideas, and look at things in a unique way is at the heart of mathematics. Consequently, proving mathematical statements is far from applying a cook-book recipe, and requires effort, persistence, and risk-taking.

However, even without step-by-step instructions, there are some proof strategies that you should be aware of. These strategies are not algorithms and cannot be applied blindly to create proofs. Instead, they can be seen as commonly used approaches, or *modes of thinking*. Choosing an appropriate strategy for proving a given statement can still be challenging, and even after doing so, implementing that strategy can be tricky. With experience and practice, you will become better in deciding how to tackle mathematical statements and construct proofs.

Direct Proof

Most statements in mathematics can be regarded as implications. That is, as having the form $P \Rightarrow Q$. P can be seen as the assumptions or hypotheses, while Q is what needs to be proved. Some mathematical statements are "if-and-only-if" statements, but these can be naturally regarded as two implications, as $P \Leftrightarrow Q$ is logically equivalent to $(P \Rightarrow Q) \wedge (Q \Rightarrow P)$.

In a direct proof, we simply assume that our hypotheses P hold true, and derive the statement Q, while relying on P and previously established theorems and definitions. Here is a direct proof of a divisibility test you might have seen in your elementary school years.

Example 3.6.1
Let n be a three-digit positive integer (i.e., $n \in \mathbb{N}$ and $100 \leq n \leq 999$). We prove that if the sum of digits of n is divisible by 3, then n is divisible by 3.

Proof The statement we need to prove has the form $P \Rightarrow Q$. The hypothesis P is the assumption that n is a three-digit positive integer, whose sum of digits is divisible by 3. Denote the digits of n by x, y and z (for instance, if $n = 729$, then $x = 7$, $y = 2$ and $z = 9$). We assume that $x + y + z$ is divisible by 3.

Our task is to prove that n itself is divisible by 3, which is our Q. To do so, observe that

$$n = 100x + 10y + z,$$

as x is the hundreds digit, y is the tens, and z is the ones.

Consequently, we can write

$$n = (99x+x)+(9y+y)+z = (99x+9y)+(x+y+z) = 3\cdot(33x+3y)+(x+y+z).$$

We are told that $x + y + z$ is divisible by 3, that is, $x + y + z = 3k$ for some integer k (see Definition 1.4.1). Therefore,

$$n = 3 \cdot (33x + 3y) + 3k = 3 \cdot (33x + 3y + k),$$

which implies that n is divisible by 3, as needed. □

Proof by Contrapositive

We briefly recall the method of proof by contrapositive, that has already been mentioned on page 81.

Suppose P and Q are two statements. The implication $P \Rightarrow Q$ is logically equivalent to its contrapositive $(\neg Q) \Rightarrow (\neg P)$. Consequently, an implication can be proved by proving its contrapositive. Using this strategy, we assume that the conclusion Q is *false*, and use it to prove that the hypothesis P is *also false*.

In some cases, proving the contrapositive of an implication is much easier than proving the original implication. Here is an example.

Example 3.6.2 Consider the function $f\colon \mathbb{R} \setminus \{-1\} \to \mathbb{R}, f(x) = \frac{x}{x+1}$, and let $a, b \neq -1$. Prove that if $a \neq b$, then $f(a) \neq f(b)$.

Here, our hypothesis P is $a \neq b$, and we need to prove that $f(a) \neq f(b)$. More explicitly, we show that $\frac{a}{a+1} \neq \frac{b}{b+1}$.

In this case, constructing a direct proof would mean that we begin with the inequality $a \neq b$, and manipulate it to get $\frac{a}{a+1} \neq \frac{b}{b+1}$. This can be done, but there are some advantages to using the contrapositive method here. First, the contrapositive will involve *equalities* rather than inequalities, as the negation of $x \neq y$ is $x = y$. Second, the contrapositive will require us to *simplify an equation*, a process that we are well familiar with.

Proof We prove the contrapositive:

"If $f(a) = f(b)$, then $a = b$."

If $f(a) = f(b)$, then $\frac{a}{a+1} = \frac{b}{b+1}$. We cross multiply and simplify, to get

$$a(b+1) = b(a+1) \qquad \Rightarrow \qquad ab + a = ba + b \qquad \Rightarrow \qquad a = b,$$

and the proof is completed. □

Proof by Contradiction

Proof by contradiction is one of the most important proof methods in mathematics. It is widely used in all areas of mathematics, and at all levels. To prove a statement by contradiction, we assume that *what needs to be proved is false* and show that this assumption leads to a contradiction, that is, to a statement known to be false. In other words, to prove an implication $P \Rightarrow Q$ by contradiction, we assume that both P and $\neg Q$ are true, and try to derive a contradiction. What is it that we can contradict? Anything that is known to be true, such as a definition, a previously established theorem, or the hypothesis P.

Informally, a proof by contradiction results from asking ourselves *what would happen if the thing we need to prove is wrong?* The proof itself will sort of answer this question, by showing that if what needs to be proved is false, then some other fact, that is known to be true, is also false. As contradictions are not acceptable in mathematics, we conclude that what needs to be proved must be true.

Let us clarify by discussing some examples.

Example 3.6.3 Prove that the equation $3x^5 - 7x^2 + 1 = 0$ has no *rational* solutions.

Solution

A natural way to prove, in mathematics, that something does not exist, is to assume that it does, and derive a contradiction. It can be proved, using methods from calculus, that the equation $3x^5 - 7x^2 + 1 = 0$ has three real solutions. Our task is to show that there are *no rational numbers* that solve this equation.

Proof Assume, by contradiction, that there exists a rational solution, r, to the given equation. That is,

$$3r^5 - 7r^2 + 1 = 0.$$

As $r \in \mathbb{Q}$, we can write it as $\frac{m}{n}$, with m and n integers and $n \neq 0$. We assume that the fraction $\frac{m}{n}$ is in lowest terms, meaning that it is completely simplified (for instance, $\frac{2}{3}$ and $\frac{5}{12}$ are in lowest terms, while $\frac{14}{21}$ and $\frac{18}{51}$ are not). More formally, we assume that m and n have no common divisors other than ± 1.

As $r = \frac{m}{n}$, we can replace it with $\frac{m}{n}$ in the equation above, and multiply by n^5, to get

$$3\left(\frac{m}{n}\right)^5 - 7\left(\frac{m}{n}\right)^2 + 1 = 0 \quad \Rightarrow \quad 3m^5 - 7m^2n^3 + n^5 = 0.$$

Since both m and n are integers, each can be either an even or an odd number. Let us check all possible combinations of even and odd.

- If both m and n are odd, then the numbers $3m^5$, $7m^2n^3$ and n^5 are all odd, since every product of odd numbers is odd. Therefore, the left-hand side, $3m^5 - 7m^2n^3 + n^5$ must be an odd number, as the sum of three odd numbers is odd. This contradicts the fact that $3m^5 - 7m^3n^2 + n^5$ is zero, which is an even number.
- If m is even and n is odd, then both $3m^5$ and $7m^2n^3$ are even numbers, while n^5 is odd. Again we conclude that $3m^5 - 7m^2n^3 + n^5$ must be odd, which leads to a contradiction. A similar argument shows that if n is even and m is odd, we also get a contradiction.
- If m and n are both even, it contradicts the fact that the fraction $\frac{m}{n}$ is in lowest terms.

We see that every possible scenario leads to a contradiction, which means that our initial assumption, the existence of a rational solution to the given equation, must be false. Therefore, there is no rational solution to $3x^5 - 7x^2 + 1 = 0$, and the proof is complete. □

Next, we look at a well-known and less mathematical example.

Example 3.6.4 Is it possible to cover a regular checkerboard, with two opposite corners removed, with regular dominoes? Each domino can be placed either horizontally or vertically, and covers exactly two squares (see Figure 3.1).

You might want to try and experiment a bit before you continue reading.

There are two possible answers to this question. If it is possible to cover the checkerboard, then one can prove it by *demonstrating* such a covering. But what if it is not possible? How can one prove that? Trying for hours or days and failing to find a cover cannot serve as a proof. On the other hand, checking every possible way to place dominoes on the checkerboard may take forever.

We claim that it is indeed impossible to produce a covering for the checkerboard, and we prove it by contradiction. Namely, we start by assuming that we can cover the board with dominoes, and show that this assumption leads to a contradiction. The main observation we use in the proof is that a single domino, when placed on the board, must cover exactly one white and one black square.

Solution

No, it is not possible to cover the checkerboard with dominoes.

Figure 3.1 A checkerboard with two opposite corners removed.

Proof Assume there is a way to cover the board with dominoes. As our board has 62 squares (two corners were removed), 31 dominoes are needed for the covering. Each domino must cover a black and a white square, and so overall 31 black and 31 white squares would be covered.

But this is impossible! The two removed corner-squares have the same color, and hence the number of white squares is not the same as the number of black squares (we have 30 squares of one color, and 32 of the other). This is a contradiction, and thus there is no way to cover our board with dominoes. □

Here is another non-existence proof example.

Example 3.6.5 There are no *natural numbers* x, y for which $x^2 - y^2 = 1$.

Proof Again, we assume, by contradiction, that there are $x, y \in \mathbb{N}$ for which $x^2 - y^2 = 1$. We then factor the left-hand side, and write the equation as $(x + y)(x - y) = 1$. The numbers $x + y$ and $x - y$ are integers, whose product is 1, which means that both $x + y$ and $x - y$ are equal to 1 (note that $x + y > 0$, which rules out the possibility of $x + y = x - y = -1$).

But the only solution to $x + y = x - y = 1$ is $x = 1$ and $y = 0$, which contradicts the fact that y is a natural number. Therefore, the equation $x^2 - y^2 = 1$ has no natural solutions. □

We end this section with a proof of a famous theorem – the infinitude of prime numbers – and we show Euclid's proof of the theorem by contradiction. Recall that a prime number p is a natural number, greater than one, whose only positive divisors are 1 and p. In the proof, we rely on the following fact, which will be proved later in Chapter 4 (see Theorem 4.5.1):

Every natural number greater than 1 is divisible by a prime number.

Theorem 3.6.6 *There are infinitely many prime numbers.*

Proof Suppose, by contradiction, that there are finitely many prime numbers, and denote them by $p_1, p_2, p_3, \ldots, p_k$. That is, we assume that there are exactly k prime numbers. Define

$$M = p_1 \cdot p_2 \cdot p_3 \cdots p_k + 1.$$

The number M is greater than each of the p_i, and hence is not a prime number itself. However, every natural number greater than 1 is divisible by a prime number, and so M must be divisible by one of the p_i. Now, as

$$M - p_1 \cdot p_2 \cdot p_3 \ldots p_k = 1,$$

we get that 1 is divisible by one of the p_i, which is a contradiction: a natural number cannot be divisible by a larger number. We thus conclude that there are infinitely many prime numbers. □

3.7 Problems

3.1 Let A be the phrase "$\frac{1}{2} < x < \frac{5}{2}$," B the phrase "$x \in \mathbb{Z}$," C the phrase "$x^2 = 1$" and D the phrase "$x = 2$." Which of the following are true for all $x \in \mathbb{R}$? Explain.

a. $A \Rightarrow C$
b. $(A \wedge B) \Rightarrow C$
c. $D \Rightarrow [A \wedge B \wedge (\neg C)]$

3.2 For each statement, write the meaning in English and decide whether it is true or false (x and y represent real numbers). Explain your decision briefly.

a. $\forall x \forall y (x \geq y)$
b. $\exists x \exists y (x \geq y)$
c. $\exists y \forall x (x \geq y)$
d. $\forall x \exists y (x \geq y)$
e. $\forall x \exists y (x^2 + y^2 = 1)$
f. $\exists x \forall y (x^2 + y^2 = 1)$

3.3 Express the following statements using the logic symbols, and decide, for each, whether it is true or false. Explain your decision briefly.

a. There is a smallest positive real number.
b. Every integer is a product of two integers.
c. The equation $x^2 + y^2 = 1$ has a solution (x, y) in which both x and y are natural numbers.
d. Every real number can be written as a difference of two positive real numbers.

3.4 Let S, T and U be three sets. Then the statement $S \cap T \subseteq U$ can be written, using the logic symbols, as follows:

$$(\forall x)[((x \in S) \wedge (x \in T)) \Rightarrow (x \in U)].$$

a. Write the statement $S \cap T \nsubseteq U$ using the logic symbols (but without the symbol $\neg$).
b. Write the statement $S \subseteq T \cup U$ and its negation using the logic symbols.

3.5 Let $P(x)$ be the assertion "x is positive", and let $Q(x)$ be the assertion "$x^2 > x$."

a. Is the statement $(\forall x \in \mathbb{R})[P(x) \Rightarrow Q(x)]$ true or false? Why?
b. Is the statement $[(\forall x \in \mathbb{R})P(x)] \Rightarrow [(\forall x \in \mathbb{R})Q(x)]$ true or false? Why?

3.6 Consider the following two statements.

Statement R: "For every real number x, there is a real number y, such that $x + y < 1$."
Statement S: "There is a real number y, such that for all real numbers x, we have $x + y < 1$."

a. Write both statements using the logic symbols.
b. Write the negations of R and S. Use the logic symbols, but do not use the symbols $\neg$ or $\not<$.
c. Is R a true or a false statement? Is S a true or a false statement? Explain.

3.7 For what real number(s) $x \in \mathbb{R}$ is the following statement *false*? Why?

$$\text{"If } |x - 3| = 1, \text{ then } |x - 2| = 2."$$

3.8 Suppose P and Q are two statements. Construct the truth tables for the following compound statements.

a. $(P \wedge Q) \vee (\neg Q)$
b. $P \Rightarrow (P \Rightarrow Q)$
c. $(P \Rightarrow Q) \Rightarrow P$
d. $(P \Rightarrow Q) \Rightarrow (P \wedge Q)$
e. $(P \wedge Q) \Leftrightarrow (P \vee Q)$
f. $[P \vee (\neg Q)] \Rightarrow [Q \wedge (\neg P)]$

3.9 Find a statement R, in terms of P and Q, with the following truth table:

P	Q	R
T	T	F
T	F	T
F	T	T
F	F	T

3.10 a. Let P, Q, and R be three statements. If $P \vee (Q \Rightarrow (\neg R))$ is a *false* statement, what must be the truth values of P, Q and R? Why?
b. Let P, Q, R and S be four statements. If $[(P \wedge Q) \vee R] \Rightarrow (R \vee S)$ is a *false* statement, what must be the truth values of P, Q, R and S? Why?

3.11 Let P, Q and R be three statements.

a. If P is *false*, Q is *false*, and R is *true*, is the statement $[P \wedge (\neg Q)] \Rightarrow (R \vee Q)$ true or false?
b. If P, Q and R are *all true*, is the statement $[P \wedge (\neg Q)] \Rightarrow (R \vee Q)$ true or false?

3.12 Prove Parts 1, 2 and 4 in Proposition 3.4.5. Also, provide real-life examples that illustrate the equivalence of the statements.

3.13 a. Find a statement that is equivalent to $\neg(P \wedge (\neg Q))$, which is a disjunction (i.e., includes the symbol $\vee$).
b. Write a statement that is equivalent to $P \Rightarrow Q$, using only the connectives $\vee$ and $\neg$.

3.14 Explain why $(P \vee Q) \Rightarrow R$ and $(P \Rightarrow R) \wedge (Q \Rightarrow R)$ are logically equivalent statements. Try to avoid the use of truth tables.

3.15 a. Is the statement $P \Rightarrow Q$ logically equivalent to $(\neg P) \Rightarrow (\neg Q)$? Explain.
b. Is the statement $(P \wedge Q) \vee R$ equivalent to $P \wedge (Q \vee R)$? Explain.
c. Is the statement $(P \Rightarrow Q) \Rightarrow R$ equivalent to $P \Rightarrow (Q \Rightarrow R)$? Explain.
d. Is the statement $(\neg P) \Leftrightarrow (\neg Q)$ the negation of $P \Leftrightarrow Q$? Explain.

3.16 Write the *contrapositive* of the following sentences. Use words or the logic symbols, and simplify your answer. Try to avoid using the negation symbol, or words of negation such as "not" or "it is false that."

a. If k is a prime and $k \neq 2$, then k is odd.
b. If I do my assignments, I will get a good mark in the course.
c. If $x^2 + y^2 = 9$, then $-3 \leq x \leq 3$.
d. If $a^2 + b^2 = 0$, then $a = 0$ and $b = 0$.
e. If Anna is failing both history and psychology, then Anna is not graduating.

3.17 Suppose that R is a compound statement, made out of several elementary statements. We say that R is a *tautology* if R is true, regardless of the truth values of its elementary statements. Similarly, we say that R is a *contradiction* if R is false, regardless of the truth values of its elementary statements. Which of the following are tautologies? Which are contradictions? Explain.

$$P \wedge (\neg P) \quad P \Leftrightarrow (\neg P) \quad P \vee (\neg P) \quad P \Rightarrow (\neg P) \quad (P \wedge Q) \Rightarrow Q$$

3.18 In Figure 3.2, the mouse concludes that he is a cat. Find the flaw in his argument. Which connectives and quantifiers are used? Can you relate this to any of the truth tables discussed in this chapter?

Figure 3.2 Cat and mouse logic cartoon.

3.19 Write the statement "There is no set A, for which $A \in A$," without using words of negation such as "no" or "not," but with the symbol $\notin$.

3.20 Negate the following statements. You may use words or the logic symbols in your answers. Simplify the negations as much as you can.

a. For all $x \in A$ there is a $b \in B$ such that $b > x$.
b. For every positive real number x, there is a natural number n, for which $\frac{1}{n} < x$.

3.21 Write the negation of the following statements *without* using the negation symbol $\neg$. Also, for each statement, decide whether it is *true* or *false*. Explain your answer briefly.

a. $(\forall x \in \mathbb{R})(\exists y \in \mathbb{R})(x^2 > y^2)$
b. $(\exists x \in \mathbb{Z})\left[(x^2 = (x+1)^2) \Rightarrow (x^3 \in \mathbb{Z})\right]$
c. $(\forall n \in \mathbb{N})[(n-1)^3 + n^3 \neq (n+1)^3]$
d. $[(\forall x \in \mathbb{R})(x > 0)] \Rightarrow [(\forall x \in \mathbb{R})(x = x + 1)]$
e. $(\forall x \in \mathbb{R})[\,(x^2 \leq -1) \Rightarrow [(x+1)^2 = x^2 + 1]\,]$
f. $(\forall x \in \mathbb{R})[(x > 0) \Rightarrow (\exists n \in \mathbb{N})(n \cdot x > 1)]$
g. $(\forall x \in \mathbb{R})(\exists y \in \mathbb{R})[(x+y)^2 = x^2 + y^2]$
h. $(\exists y \in \mathbb{R})(\forall x \in \mathbb{R})(|x+y| = |x| + |y|)$
i. $(\forall x \in \mathbb{Q})(\exists n \in \mathbb{N})(n \cdot x \in \mathbb{Z})$
j. $(\forall x \in \mathbb{R})(\forall y \in \mathbb{R})\left[((x + y \leq 7) \wedge (xy = x)) \Rightarrow (x < 7)\right]$

3.22 For each statement below, write it and its negation using the logic symbols. Make sure to simplify the negation as much as you can. Also, decide whether the given statement is true or false. Explain your decision briefly.

a. There exists an integer M, such that $x^2 \leq M$ for all real numbers x.
b. There is a real number y, such that $|x - y| = |x| - |y|$ for every real number x.

c. For all real numbers x, $(x-6)^2 = 4$ implies $x = 8$.
d. For all real numbers x, y, if $x^2 - y^2 = 9$, then $|x| \geq 3$.
e. For every real number x, if $(x-1)(x-3) = 3$, then $x - 1 = 3$ or $x - 3 = 3$.

3.23 Write the negation of the following statement in words.

"For every field $\mathbb{F}$, and every $a \in \mathbb{F}$, if $a^3 = 1$ then $a = 1$."

Is this statement true or false? Explain.

3.24 Let P be the statement $(\forall x \in \mathbb{Z})[x(x-1) \geq 0]$.

a. It is a common error to believe that the negation of P is $(\forall x \notin \mathbb{Z})[x(x-1) < 0]$. Why is this wrong? What is the correct way to negate the statement?
b. Which statement is true: P or $\neg P$? Explain.

3.25 In mathematics, we often denote elements of an infinite sequence using a lowercase letter and an index. For instance, a_1, a_2 and a_3 are the first, second and third elements of a sequence (a_n). In general, for every $n \in \mathbb{N}$, a_n is the nth term of the sequence. We say that (a_n) is *increasing*, if $a_n < a_{n+1}$ for all $n \in \mathbb{N}$. Using words, write the statement "(a_n) is *not* increasing." Do not use any words of negation.

3.26 Review Example 3.6.1 before attempting this exercise.

a. Prove that a four-digit positive integer, whose sum of digits is divisible by 3, is also divisible by 3.
b. Prove that if n is a three-digit positive integer that is divisible by 3, then its sum of digits is also divisible by 3. Can you combine this observation and Example 3.6.1 into one if-and-only-if statement?
c. Generalize our previous discoveries to prove that "A natural number n is divisible by 3 if and only if its sum of digits is divisible by 3."

3.27 Prove that a natural number is divisible by 9 if and only if its sum of digits is divisible by 9.

3.28 Is the following statement true or false? Justify your answer with a proof or a counterexample.

"For every $n \in \mathbb{N}$, we have $(n-1)^2 + n^3 = (n+1)^3$."

3.29 Use *contrapositive* to prove the following statements.

a. Let x be an integer. If $x^2 - 1$ is *not* divisible by 8, then x is even.
b. Let m and n be two integers. Prove that if $m^2 + n^2$ is divisible by 4, then both m and n are even numbers.
c. Let $x, y \in \mathbb{R}$. If x and y are both positive, then $\sqrt{x+y} \neq \sqrt{x} + \sqrt{y}$.

d. Let $x, y \in \mathbb{R}$. If $x \neq y$, then $\frac{x}{\sqrt{x^2+1}} \neq \frac{y}{\sqrt{y^2+1}}$.

e. Let $x \in \mathbb{R}$. If $x^3 + 5x = 40$, then $x < 3$.

3.30 Prove that the following equations have *no rational* solutions.

a. $x^3 + x^2 = 1$

b. $x^3 + x + 1 = 0$

c. $x^5 + 3x^3 + 7 = 0$

d. $x^5 + x^4 + x^3 + x^2 + 1 = 0$

3.31 Prove that the following equations have *no natural* solutions.

a. $x^2 - 4y^2 = 7$

b. $x^2 - y^2 = 10$

3.32 Prove that the equation $x^2 + x + 1 = y^2$ has no natural solutions. (Hint: Multiply by 4 and complete the square.)

3.33 Two squares from each of two opposite corners are deleted from a checkerboard, as shown in Figure 3.3. Prove that the remaining squares cannot be fully covered using copies of the T-shape and their rotations. Each T-shape covers exactly four squares.

3.34 Read the proof of Theorem 3.6.6. The proof argues that if $p_1, \ldots, p_k$ are prime numbers, then $M = p_1 \cdots p_k + 1$ must be either a new prime number, or a number that is divisible by a prime other than $p_1, \ldots, p_k$.

a. Verify this argument by assuming that $k = 4$, $p_1 = 2$, $p_2 = 3$, $p_3 = 5$ and $p_4 = 7$. What is M? Is it a prime number?

b. Repeat with $p_1, \ldots, p_k$ the primes $7, 11, 13, 19$. And again with $2, 5, 11, 19, 23$. Use a computing device to calculate M.

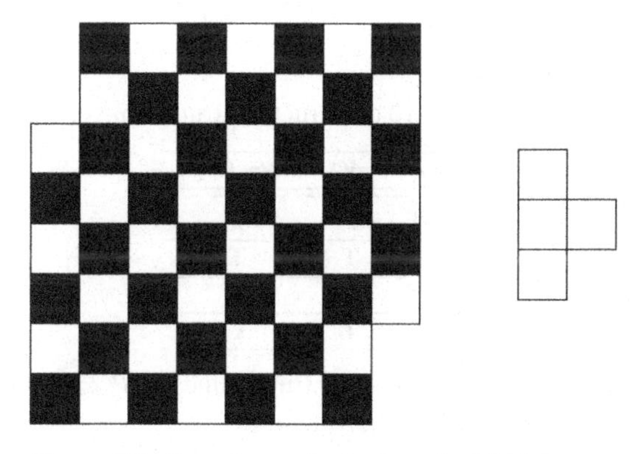

Figure 3.3 Covering a checkerboard with T-shapes.

3.8 Solutions to Exercises

Solution to Exercise 3.1.7

- "The smallest natural number is 1 *and* there is no largest natural number."
  This is a true statement.
- "For every two sets A and B, we have $A \subseteq B$ *or* $B \subseteq A$."
  This is a false statement. For instance, if $A = \{1,2\}$ and $B = \{2,3\}$, then $A \nsubseteq B$ and $B \nsubseteq A$.
- "For every function $f\colon \mathbb{R} \to \mathbb{R}$, *if* f is bounded, *then* f^2 is also bounded."
  This is a true statement. If $M > 0$ and $|f(x)| \leq M$ for all $x \in \mathbb{R}$, then $|f^2(x)| \leq M^2$ for all x, and so f^2 is bounded as well.
- "149 is a prime number *if and only if* 147 is a prime number."
  This is a false statement, as 149 is prime, while 147 is not ($147 = 3\times7\times7$).
- "*Not* every real number is rational."
  This is a true statement. As we mentioned, many real numbers, such as $\sqrt{2}$, are irrational.

Solution to Exercise 3.1.8
The quantifiers are "for every" and "there exists." The connectives are "if-then," "not" and "or."

Solution to Exercise 3.1.10

(i) Let x be an element of A. If $x \notin B$, then $x \in A \setminus B$, which is impossible as $A \setminus B = \varnothing$. The only other option is that $x \in B$, and so $A \subseteq B$.

(ii) Using the Arithmetic-Geometric-Mean Inequality, we get

$$\frac{3r + \frac{2}{r}}{2} \geq \sqrt{3r \cdot \frac{2}{r}} \qquad \Rightarrow \qquad 3r + \frac{2}{r} \geq 2 \cdot \sqrt{3 \cdot 2} = 2\sqrt{6}.$$

Solution to Exercise 3.4.4
Using $\neg Q$ as a helper column, we construct the truth table of $P \Leftrightarrow (\neg Q)$:

P	Q	$\neg Q$	$P \Leftrightarrow (\neg Q)$
T	T	F	**F**
T	F	T	**T**
F	T	F	**T**
F	F	T	**F**

We now see that R and S have the same truth values as $P \Leftrightarrow (\neg Q)$, and hence all three statements are logically equivalent.

4 Mathematical Induction

Mathematical induction is a technique which is often used to prove that a statement $P(n)$, that depends on a variable n, is valid for every natural number. Equivalently, one may think of $P(n)$ as representing an infinite sequence of statements, one for every $n \in \mathbb{N}$:

$$P(1), P(2), P(3), \ldots.$$

When dealing with infinitely many statements, there is no way we can prove them all by proving each statement individually. Induction is a tool we can often use to bypass this difficulty.

Mathematical induction is an extremely powerful proof technique. It is not restricted to specific areas of mathematics, and thus can be used to prove statements in algebra, geometry, number theory, analysis, etc. Moreover, induction is a useful tool at all levels of mathematics. It is used to prove elementary statements about numbers, as well as advanced statements in, say, topology and modern algebra.

4.1 The Principle of Mathematical Induction

We begin by discussing a few motivating examples. We then present the Principle of Mathematical Induction and demonstrate how we use it to prove various statements.

Example 4.1.1 In Chapter 1 we proved the Arithmetic-Geometric Mean Inequality (see Theorem 1.3.6). This inequality can be generalized to more than two numbers, as follows:

$$\sqrt[n]{x_1 \cdot x_2 \cdots x_n} \leq \frac{x_1 + x_2 + \cdots + x_n}{n} \qquad \text{(for } x_1, x_2, \ldots, x_n \geq 0\text{)}.$$

How would one prove such an inequality? We have already proved it for the case where $n = 2$, and in Problem 1.18, even for $n = 3$ and $n = 4$. We can spend more time and construct proofs for $n = 5$ and $n = 6$, but that would not justify the inequality for all n. Mathematical induction can be used to construct a proof for *every* natural number n (see Problem 4.34).

Example 4.1.2 The Triangle Inequality (Theorem 1.3.12) can also be generalized to more than two numbers. We will soon prove that for all $x_1, \ldots, x_n \in \mathbb{R}$, we have

$$|x_1 + x_2 + \cdots + x_n| \leq |x_1| + |x_2| + \cdots + |x_n|.$$

We have already proved the inequality for the case $n = 2$, and it is not too hard to prove it for other values of n. However, proving it for *every* number of xs (that is, for every n) can be done by induction.

Example 4.1.3 Finally, take a look at the following equalities:

$$1+3=4, \quad 1+3+5=9, \quad 1+3+5+7=16, \quad 1+3+5+7+9=25.$$

Can you see a pattern? It looks as if the sum of consecutive odd natural numbers, starting at 1, always results in a square number. But how can we be sure? We can keep checking more and more sums, and with computers, we can even check thousands of sums in a fraction of a second, but that would still not cover all cases. To prove that

$$1 + 3 + 5 + \cdots + (2n-1) = n^2$$

for *every* natural number n, we will use induction.

We are now ready to explain how mathematical induction works, and we begin with the following core principle.

The Principle of Mathematical Induction

Suppose that for every natural number n, $P(n)$ is a mathematical statement, and that the following two conditions hold.

1. $P(1)$ is a true statement.
2. For every natural number k, $P(k)$ implies $P(k+1)$.

Then $P(n)$ is true for all n.

Remarks.

- Condition 1 is often called the *base case*, and Condition 2 the *induction step*. In most cases, proving the base case is relatively easy and quick (but must be done nevertheless), and most of the work is put in proving the induction step.
- The Principle of Mathematical Induction is often taken as an axiom for the natural numbers, or is proved from other axioms. As we did not follow an axiomatic approach to defining the natural numbers, we will accept the

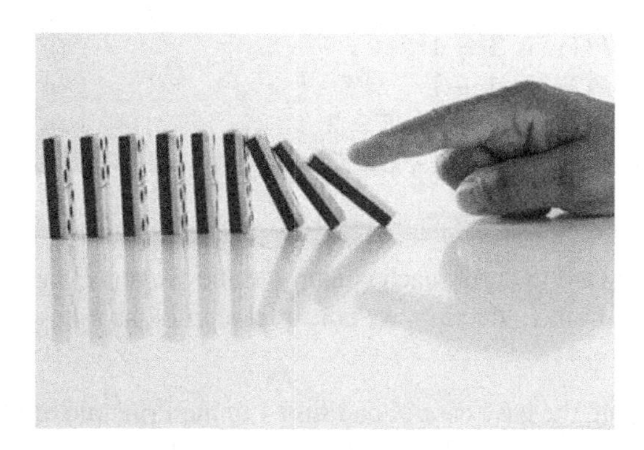

Figure 4.1 Falling dominoes.

principle without proof. In words, the principle says that if the first statement in the sequence is true, and if every statement implies the next, then all the statements are true. After all, if $P(1)$ is true, then the second condition, with $k = 1$, implies the statement $P(2)$. Applying the second condition again, this time with $k = 2$, gives that $P(2)$ implies $P(3)$. By continuing this process we obtain a chain of implications, which informally justifies the above principle:

$$P(1) \Rightarrow P(2) \Rightarrow P(3) \Rightarrow P(4) \Rightarrow \cdots \Rightarrow P(k) \Rightarrow P(k+1) \Rightarrow \cdots .$$

- Some like to compare the above principle to falling dominoes (Figure 4.1). If you line up dominoes in a straight line, then pushing the first domino creates a chain effect that eventually brings down all the dominoes. However, to make it work, two conditions must be met:

  1. someone has to push the first domino in line, and
  2. the dominoes must be close enough to each other, so that a falling domino pushes the next one in line.

  These two conditions are analogous to the conditions stated in the Principle of Mathematical Induction.

Let us take a look at a few examples.

Example 4.1.4

We begin by confirming that for every natural number n, the sum of the first n odd natural numbers is n^2.

Claim 4.1.5 *For every* $n \in \mathbb{N}$, *we have* $1 + 3 + 5 + \cdots + (2n-1) = n^2$.

Note that the equality above can be interpreted as an infinite sequence of statements $P(1), P(2), P(3), \ldots$.

$$\begin{array}{llll} P(1) & \text{is} & 1 = 1^2 & (n=1) \\ P(2) & \text{is} & 1+3 = 2^2 & (n=2) \\ P(3) & \text{is} & 1+3+5 = 3^2 & (n=3) \\ P(4) & \text{is} & 1+3+5+7 = 4^2 & (n=4) \\ \vdots & & \vdots & \vdots \end{array}$$

It is straightforward to verify each equality in the sequence, but our task is to prove them all! Using induction, we construct a proof, as follows.

Proof Verifying the base case (Condition 1 in the Principle of Mathematical Induction) is immediate. As $1 = 1^2$, $P(1)$ is a true statement, as needed.

Next, we need to prove that for every $k \in \mathbb{N}$, $P(k)$ implies $P(k+1)$. In other words, we assume that

$$1 + 3 + 5 + \cdots + (2k-1) = k^2 \qquad \text{for } \textit{some } \ k \in \mathbb{N}$$

(this is called the *induction hypothesis*), and we need to prove that

$$1 + 3 + 5 + \cdots + [2(k+1)-1] = (k+1)^2.$$

Observe that the last term in the sum $1+3+5+\cdots+[2(k+1)-1]$ is $2k+1$, and hence the preceding term must be $2k-1$. We can now prove the $(k+1)$-case, as follows:

$$\begin{aligned} 1 + 3 + 5 + \cdots + [2(k+1)-1] &= 1 + 3 + 5 + \cdots + (2k-1) + (2k+1) \\ &= [1 + 3 + 5 + \cdots + (2k-1)] + (2k+1) \\ &= k^2 + (2k+1) = (k+1)^2. \end{aligned}$$

Note how in the third step, the induction hypothesis was used to replace $1 + 3 + 5 + \cdots + (2k-1)$ with k^2. We have managed to prove $P(k+1)$ from $P(k)$, which completes the induction step of our proof. By the Principle of Mathematical Induction, the equality $1 + 3 + 5 + \cdots + (2n-1) = n^2$ is valid for all $n \in \mathbb{N}$. □

Example 4.1.6 We now prove an inequality known as *Bernoulli's Inequality*.

Claim 4.1.7 *For every $n \in \mathbb{N}$ and $x \in \mathbb{R}$ with $x \geq -1$, we have $(1+x)^n \geq 1 + nx$.*

In this case, our statement $P(n)$ is

$$\text{“For all } x \geq -1, \text{ we have } (1+x)^n \geq 1 + nx\text{”}$$

which can also be seen as representing an infinite sequence of statements, one for each n:

$n = 1$ "For all $x \geq -1$, we have $(1+x)^1 \geq 1 + 1x$"
$n = 2$ "For all $x \geq -1$, we have $(1+x)^2 \geq 1 + 2x$"
$n = 3$ "For all $x \geq -1$, we have $(1+x)^3 \geq 1 + 3x$"
and so on....

To prove that all these statements are true, we use induction.

Proof Let $x \geq -1$ be an arbitrary real number, that will be kept fixed throughout the proof. We proceed by induction on n.

For $n = 1$, the inequality becomes $(1+x)^1 \geq 1 + 1 \cdot x$, which is valid, and hence the base case holds.

Next, we assume that $(1+x)^k \geq 1+kx$ for *some* natural number k, and prove the $(k+1)$-case:

$$(1+x)^{k+1} = (1+x)^k \cdot (1+x) \geq (1+kx)(1+x) = 1+kx+x+kx^2 = 1+(k+1)x+kx^2.$$

The induction hypothesis, $(1+x)^k \geq 1+kx$, was used in the second step, together with the fact that $1+x \geq 0$. Also, as $kx^2 \geq 0$, we conclude that

$$1 + (k+1)x + kx^2 \geq 1 + (k+1)x.$$

Overall, we have that $(1+x)^{k+1} \geq 1+(k+1)x$, which is Bernoulli's Inequality for $n = k+1$. By the Principle of Mathematical Induction, Bernoulli's Inequality holds true for all $n \in \mathbb{N}$ and $x \geq -1$. □

In most future proofs by induction, we omit the explicit labelling of the statement by $P(n)$. This should not create any confusion, as long as the statement being proved is clear from the context.

Example 4.1.8
Our next example is the generalized Triangle Inequality.

Theorem 4.1.9 *For every n real numbers $x_1, x_2, \ldots, x_n$, we have*

$$|x_1 + x_2 + \cdots + x_n| \leq |x_1| + |x_2| + \cdots + |x_n|.$$

In other words, the absolute value of a sum is always less than or equal to the sum of the absolute values.

Proof We perform induction on n, the number of xs. In other words, we prove by induction that for every $n \in \mathbb{N}$, the inequality $|x_1 + x_2 + \cdots + x_n| \leq |x_1| + |x_2| + \cdots + |x_n|$ holds true for all real numbers $x_1, \ldots, x_n$.

If $n = 1$ (the base case), we get $|x_1| \leq |x_1|$, which holds true for all real numbers x_1.

Now assume that, for some $k \in \mathbb{N}$, the Triangle Inequality is valid for $n = k$ numbers (this is the induction hypothesis), and consider $k + 1$ real numbers $x_1, x_2, \ldots, x_k, x_{k+1}$.

Using Theorem 1.3.12, with $x = x_1 + \cdots + x_k$ and $y = x_{k+1}$, we have

$$|x_1+x_2+\cdots+x_k+x_{k+1}| = |(x_1+x_2+\cdots+x_k)+x_{k+1}| \leq |x_1+x_2+\cdots+x_k|+|x_{k+1}|,$$

and by the induction hypothesis,

$$\leq |x_1| + |x_2| + \cdots + |x_k| + |x_{k+1}|,$$

as needed. This proves the inequality for $n = k + 1$, and hence, by induction, for all n. □

Example 4.1.10 We use induction to prove that for every $n \in \mathbb{N}$, $2^{6n} + 3^{2n-2}$ is divisible by 5. This statement is different from the previous examples, as it does not involve an equality or an inequality. Nevertheless, it has the form "For every $n \in \mathbb{N} \ldots$," which suggests that induction may be a useful technique to use here.

Proof For $n = 1$, we get $2^6 + 3^0 = 64 + 1 = 65$, which is divisible by 5, and so the base case holds.

Assume that $2^{6k}+3^{2k-2}$ is divisible by 5 for some $k \in \mathbb{N}$ (this is the induction hypothesis). We need to prove that $2^{6(k+1)} + 3^{2(k+1)-2}$ is also divisible by 5. To do so, we write

$$\begin{aligned} 2^{6(k+1)} + 3^{2(k+1)-2} &= 2^{6k+6} + 3^{2k-2+2} = 2^6 \cdot 2^{6k} + 3^2 \cdot 3^{2k-2} \\ &= 64 \cdot 2^{6k} + 9 \cdot 3^{2k-2} = (55 + 9) \cdot 2^{6k} + 9 \cdot 3^{2k-2} \\ &= 55 \cdot 2^{6k} + 9 \cdot (2^{6k} + 3^{2k-2}). \end{aligned}$$

The term $55 \cdot 2^{6k}$ is divisible by 5, since 5 divides 55, and the term $9 \cdot (2^{6k} + 3^{2k-2})$ is divisible by 5, by the induction hypothesis. We conclude that $2^{6(k+1)} + 3^{2(k+1)-2}$ is divisible by 5. This proves the claim for $n = k + 1$, and hence, by induction, the claim is valid for all $n \in \mathbb{N}$. □

Our next example is somewhat different from the previous ones. It involves *set counting*.

Example 4.1.11 Consider the following question.

> Given a finite non-empty set S, with n elements, how many *subsets* does S have (including, of course, the empty set and the set S itself)?

Note that the question we posed is *not* a mathematical statement, and thus there is nothing we can prove (yet). We would like, however, to answer this question, and we start by looking at some special cases.

- If $n = 1$, that is S has only one element, then there are only *two* subsets: $\varnothing$ and S.
- If S has two elements, say $S = \{a, b\}$, then we get *four* subsets:

$$\varnothing, \{a\}, \{b\}, \{a, b\}.$$

- For $n = 3$, the set $S = \{a, b, c\}$ has *eight* subsets:

$$\varnothing, \{a\}, \{b\}, \{c\}, \{a, b\}, \{a, c\}, \{b, c\}, \{a, b, c\}.$$

- And for $n = 4$, we obtain *sixteen* subsets (check!).

By examining the pattern $2, 4, 8, 16, \ldots$, we conjecture that a set with n elements has 2^n subsets. This is a mathematical statement, that we now prove by induction.

Claim 4.1.12 *Let n be a natural number. A finite set, with n elements, has* 2^n *subsets.*

Proof A set S, with one element, has only two subsets, S and $\varnothing$, and thus the claim holds true for $n = 1$ (the base case).

Assume that the claim holds true for some natural number k, and consider a set S with $k + 1$ elements:

$$S = \{x_1, x_2, \ldots, x_k, x_{k+1}\}.$$

Denote by $\tilde{S}$ the set obtained from S by *removing* the element x_{k+1}:

$$\tilde{S} = \{x_1, x_2, \ldots, x_k\}.$$

By our assumption, sets with k elements have 2^k subsets, and so $\tilde{S}$ has 2^k subsets, which we denote by $A_1, A_2, \ldots, A_{2^k}$.

Remember that our task is to prove the claim for $n = k + 1$, that is, to count the number of subsets of S (and not $\tilde{S}$). However, as $\tilde{S} \subseteq S$, every subset of $\tilde{S}$ is also a subset of S. This means that $A_1, A_2, \ldots, A_{2^k}$ are also subsets of S. Are there any more? Of course there are. None of the subsets $A_1, \ldots, A_{2^k}$ contains the element x_{k+1}. We create more subsets of S by adding x_{k+1} to each of the existing subsets, as follows:

$$B_1 = A_1 \cup \{x_{k+1}\}, \quad B_2 = A_2 \cup \{x_{k+1}\}, \quad \ldots, \quad B_{2^k} = A_{2^k} \cup \{x_{k+1}\}.$$

This way, we obtain a *complete list* of all subsets of S:

$$A_1, A_2, \ldots, A_{2^k}, B_1, B_2, \ldots, B_{2^k}.$$

Every subset of S is listed precisely once. If a subset does not contain x_{k+1}, it must be one of the A_i. Otherwise, it must be one of the B_j.

So how many subsets are there in total for S? We have 2^k A_i, and 2^k B_j. Overall, there are $2^k + 2^k = 2 \cdot 2^k = 2^{k+1}$ subsets, which proves the claim for $n = k + 1$. By induction, the claim holds true for all $n \in \mathbb{N}$. □

Exercise 4.1.13
Does the above claim remain valid when $n = 0$?

4.2 Summation and Product Notation

In mathematics, we often use induction to prove identities or inequalities involving sums and products. The generalized Triangle Inequality (Theorem 4.1.9) is an example involving sums. We introduce below the commonly used Sigma and Pi notation for writing long sums and products in a more efficient way.

Definition 4.2.1 Suppose that m and n are integers with $m \leq n$, and $a_m, a_{m+1}, \ldots, a_n$ are real numbers. We define

$$\sum_{i=m}^{n} a_i = a_m + a_{m+1} + a_{m+2} + \cdots + a_n \qquad \text{(\textit{Sigma notation} for sums)}$$

$$\prod_{i=m}^{n} a_i = a_m \cdot a_{m+1} \cdot a_{m+2} \cdots a_n \qquad \text{(\textit{Pi notation} for products).}$$

The index i serves as a *counter*, and other letters may be used instead. In both cases, we evaluate the term a_i for $i = m, m+1, m+2, \ldots, n$, and then add or multiply the results.

Example 4.2.2

- $\displaystyle\sum_{i=1}^{10} \frac{1}{i} = \frac{1}{1} + \frac{1}{2} + \frac{1}{3} + \cdots + \frac{1}{10}.$
- $\displaystyle\sum_{i=3}^{10} (2i) = 6 + 8 + 10 + \cdots + 20.$
- $\displaystyle\prod_{k=5}^{15} 3^{k+2} = 3^7 \cdot 3^8 \cdot 3^9 \cdots 3^{17}.$

- $\prod_{k=1}^{n} k = 1 \cdot 2 \cdot 3 \cdots n.$

  This product is also denoted as $n!$, and reads "*n factorial.*"
- If c is a real number, and $n \in \mathbb{N}$, then

$$\sum_{j=1}^{n} c = c+c+\cdots+c = n \cdot c \qquad \text{and} \qquad \prod_{j=1}^{n} c = c \cdot c \cdots c = c^n.$$

- $$\sum_{j=3}^{78}\left(\frac{1}{j}-\frac{1}{j+1}\right) = \left(\frac{1}{3}-\frac{1}{4}\right)+\left(\frac{1}{4}-\frac{1}{5}\right)+\left(\frac{1}{5}-\frac{1}{6}\right)+\cdots+\left(\frac{1}{78}-\frac{1}{79}\right)$$
$$= \frac{1}{3}-\frac{1}{79} = \frac{76}{237}.$$

  Note how most terms in the sum cancel each other, which allowed us to simplify it to a single fraction.

In the following proposition, we restate a few known properties of sums and products using the Sigma and Pi notation.

Proposition 4.2.3 *Suppose that c, a_i and b_i represent real numbers for $i = m, m+1, \ldots, n$. Then the following identities hold.*

1. $\sum_{i=m}^{n}(c \cdot a_i) = c \cdot \sum_{i=m}^{n} a_i.$
2. $\sum_{i=m}^{n}(a_i + b_i) = \left(\sum_{i=m}^{n} a_i\right) + \left(\sum_{i=m}^{n} b_i\right).$
3. $\prod_{i=m}^{n}(c \cdot a_i) = c^{n-m+1} \cdot \prod_{i=m}^{n} a_i.$
4. $\prod_{i=m}^{n}(a_i \cdot b_i) = \left(\prod_{i=m}^{n} a_i\right) \cdot \left(\prod_{i=m}^{n} b_i\right).$

We can justify these properties by writing sums and products explicitly, without the Sigma or Pi notation. For instance, to justify the first identity, we write

$$\sum_{i=m}^{n}(c \cdot a_i) = c \cdot a_m + c \cdot a_{m+1} + \cdots + c \cdot a_n = c \cdot (a_m + a_{m+1} + \cdots + a_n) = c \cdot \sum_{i=m}^{n} a_i.$$

Exercise 4.2.4

Verify Parts 2, 3 and 4 of Proposition 4.2.3 by writing each side explicitly.

Remark. By taking a closer look at Proposition 4.2.3 we notice a connection to some of the field axioms from Section 2.3. Identity 1 is a generalization of the distributive law. Identities 2 and 4 generalize the commutativity of addition and multiplication to more than two elements. In fact, Proposition 4.2.3 remains valid in any field. A more rigorous proof of these identities can be done by induction.

The following is another version of Bernoulli's Inequality from Example 4.1.6.

Example 4.2.5 If $x_1, x_2, \ldots, x_n$ are real numbers in the interval $[-1, 0]$, then

$$\prod_{i=1}^{n}(1+x_i) \geq 1 + \sum_{i=1}^{n} x_i.$$

Note that without the Sigma and Pi notation, the inequality becomes

$$(1+x_1)\cdot(1+x_2)\cdots(1+x_n) \geq 1 + (x_1 + x_2 + \cdots + x_n).$$

Proof We perform induction on n, the number of xs.

If $n = 1$, the inequality becomes $(1+x_1) \geq (1+x_1)$, which is valid for every real number x_1.

Next, we assume that the inequality holds true for some $n = k$, and prove it for $n = k+1$. Suppose $x_1, x_2, \ldots, x_k, x_{k+1}$ are real numbers in the interval $[-1, 0]$:

$$\prod_{i=1}^{k+1}(1+x_i) = \left[\prod_{i=1}^{k}(1+x_i)\right]\cdot(1+x_{k+1}) \geq \left(1 + \sum_{i=1}^{k} x_i\right)\cdot(1+x_{k+1}).$$

Note how the induction hypothesis was used in the second step, together with the fact that $1 + x_{k+1} \geq 0$. We continue by expanding the brackets:

$$= 1 + \left(\sum_{i=1}^{k} x_i\right) + x_{k+1} + x_{k+1}\cdot\sum_{i=1}^{k} x_i = 1 + \left(\sum_{i=1}^{k+1} x_i\right) + x_{k+1}\cdot\sum_{i=1}^{k} x_i.$$

Finally, we observe that $x_{k+1}\cdot\sum_{i=1}^{k} x_i$ is a non-negative number, as $x_i \leq 0$ for all $1 \leq i \leq k+1$. We get

$$\geq 1 + \left(\sum_{i=1}^{k+1} x_i\right).$$

Putting it all together, we have shown that

$$\prod_{i=1}^{k+1}(1+x_i) \geq 1 + \sum_{i=1}^{k+1} x_i,$$

which is our inequality for $n = k+1$. This completes the proof of the proposition. □

In fact, the proof remains valid if all the xs are non-negative, which leads to the following, slightly more general claim.

Claim 4.2.6 *Suppose that $x_1, x_2, \ldots, x_n$ are real numbers in the interval $[-1, \infty)$, all having the same sign (that is, either $x_i \geq 0$ for all i or $x_i \leq 0$ for all i), then*

$$\prod_{i=1}^{n}(1+x_i) \geq 1 + \sum_{i=1}^{n} x_i.$$

Exercise 4.2.7
Explain how Claim 4.2.6 generalized Bernoulli's Inequality from Example 4.1.6.

4.3 Variations

In previous sections, we used mathematical induction to prove that a given statement is valid for *all* natural numbers. However, in some cases, we might want to show that a statement is valid for only *some* $n \in \mathbb{N}$. Can we still use induction? Yes, we can, as long as we use the appropriate variation of the original principle. Here is an example.

Example 4.3.1
For which natural numbers n is it true that $2^n \geq (n+1)^2$?

If we let $n = 1$, we get $2 \geq 4$, which is clearly false. This shows that the inequality is *not* valid for all $n \in \mathbb{N}$. Nevertheless, we know, informally, that exponential functions grow faster than quadratics, and so we expect the inequality to be valid, as long as n is large enough. But what do we mean by "large enough"? Can we make this statement precise? And moreover, can we prove it?

Let us start by trying some small values for n.

For $n = 1$:	$2^1 \geq 2^2$	False!
For $n = 2$:	$2^2 \geq 3^2$	False!
For $n = 3$:	$2^3 \geq 4^2$	False!
For $n = 4$:	$2^4 \geq 5^2$	False!
For $n = 5$:	$2^5 \geq 6^2$	False!
For $n = 6$:	$2^6 \geq 7^2$	True!
For $n = 7$:	$2^7 \geq 8^2$	True!
For $n = 8$:	$2^8 \geq 9^2$	True!

As we can see, the inequality is valid for $n = 6, 7, 8$, and it looks as if it remains valid, as long as $n \geq 6$. We now state this fact and prove it by induction. However, we use $n = 6$ as our base case, instead of $n = 1$.

Claim 4.3.2 *Let $n \in \mathbb{N}$. If $n \geq 6$, then $2^n \geq (n+1)^2$.*

Proof The base case $n = 6$ is easily verified, as $2^6 = 64$ is greater than $(6+1)^2 = 49$.

Assume that $2^k \geq (k+1)^2$ for some natural number $k \geq 6$, and compute:

$$\begin{aligned} 2^{k+1} &= 2^k \cdot 2 \geq (k+1)^2 \cdot 2 = 2k^2 + 4k + 2 = k^2 + 4k + 4 + k^2 - 2 \\ &= (k+2)^2 + (k^2 - 2) \geq (k+2)^2. \end{aligned}$$

Note how the induction hypothesis is used in the second step. In the last step, we omitted the term $k^2 - 2$, which must be positive, as $k \geq 6$. Overall, we have proved the inequality for $n = k+1$, and hence, by induction, for all $n \geq 6$. □

In the proof above, we used the following variation of the Principle of Mathematical Induction.

Variation 1

Let $\ell \in \mathbb{N}$, and suppose that for every natural number n, $P(n)$ is a mathematical statement. If $P(\ell)$ is true, and if for every $k \geq \ell$, $P(k)$ implies $P(k+1)$, then $P(n)$ is true for all $n \geq \ell$.

Other variations can be formed. For instance, Variations 2 and 3 below can be used to prove that a statement is valid for all *even* (respectively *odd*) natural numbers.

Variation 2

Suppose that for every natural number n, $P(n)$ is a mathematical statement. If $P(2)$ is true, and if for every even $k \in \mathbb{N}$, $P(k)$ implies $P(k+2)$, then $P(n)$ is true for all *even* n.

Variation 3

Suppose that for every natural number n, $P(n)$ is a mathematical statement. If $P(1)$ is true, and if for every odd $k \in \mathbb{N}$, $P(k)$ implies $P(k+2)$, then $P(n)$ is true for all *odd* n.

Here is an example in which Variation 2 is used.

Example 4.3.3 For all *even* $n \in \mathbb{N}$, $n(n^2 + 3n + 2)$ is divisible by 24.

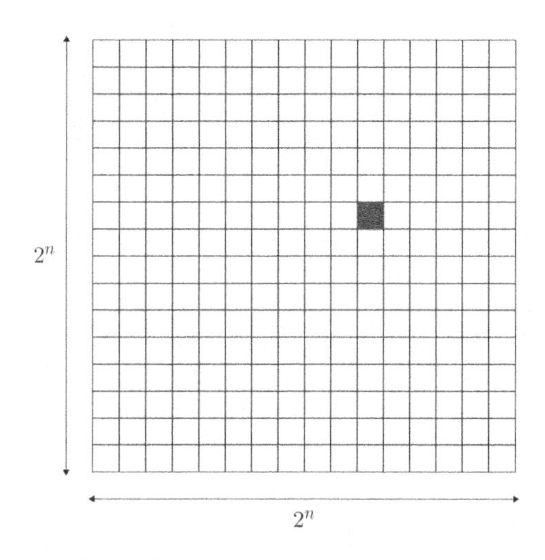

Figure 4.2 A $2^n \times 2^n$ grid with one square occupied, and an L-shape.

Proof For $n = 2$, the base case, we have $n(n^2+3n+2) = 2\cdot(4+6+2) = 24$, which is, of course, divisible by 24. Now assume that $k(k^2+3k+2)$ is divisible by 24 for some *even* $k \in \mathbb{N}$, and consider the case $n = k + 2$:

$$\begin{aligned}
&(k+2)\cdot[(k+2)^2+3(k+2)+2] \\
&\quad= (k+2)\cdot(k^2+4k+4+3k+6+2) \\
&\quad= (k+2)\cdot[(k^2+3k+2)+(4k+10)] \\
&\quad= k\cdot(k^2+3k+2)+k\cdot(4k+10)+2\cdot(k^2+3k+2+4k+10) \\
&\quad= k\cdot(k^2+3k+2)+(6k^2+24k+24).
\end{aligned}$$

The term $k \cdot (k^2 + 3k + 2)$ is divisible by 24, by the induction hypothesis. The second term, $6k^2 + 24k + 24$, is also divisible by 24, since k is even, which implies that k^2 is divisible by 4. We have proved the statement for $n = k + 2$, and by Variation 2 above, for all *even* $n \in \mathbb{N}$. □

4.4 Additional Examples

We present a few more examples of problems that can be solved by induction, and we begin with a somewhat geometric one.

Example 4.4.1 Let $n \in \mathbb{N}$ and consider a $2^n \times 2^n$ grid, with one cell occupied. We prove that the grid can be covered by L-shapes and their rotations, where each L-shape covers exactly three cells (see Figure 4.2). We often say that the grid has an *L-tiling*.

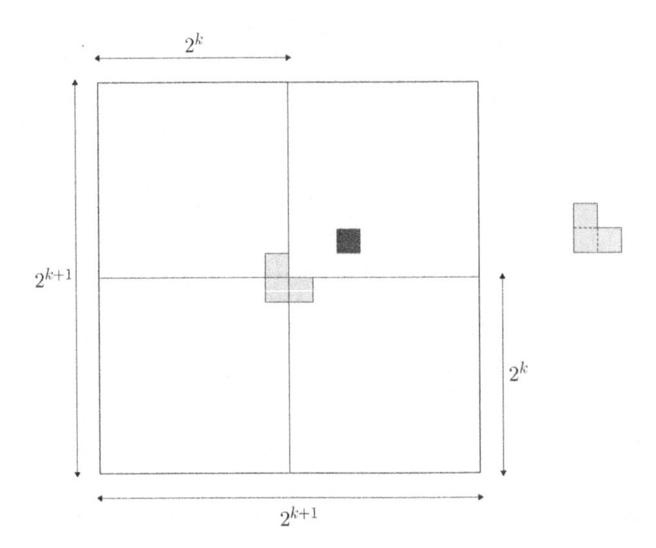

Figure 4.3 A $2^{k+1} \times 2^{k+1}$ grid divided into four $2^k \times 2^k$ grids, and with an L-shape placed at the center.

Proof For $n = 1$, we have a 2×2 grid, with one cell occupied, which can be covered with a single L-shape.

Assume that for some $k \in \mathbb{N}$, every $2^k \times 2^k$ grid, with one cell occupied, has an L-tiling. We need to show that a $2^{k+1} \times 2^{k+1}$ grid, with one cell occupied, also has an L-tiling.

To be able to use the induction hypothesis, we divide our $2^{k+1} \times 2^{k+1}$ grid into four equal parts, each having dimensions $2^k \times 2^k$. Moreover, we place one L-shape in the center, so that it occupies three cells, one in each of the parts that do not contain the occupied cell (see Figure 4.3). Each of the smaller $2^k \times 2^k$ grids has one occupied cell, so by the induction hypothesis, has an L-tiling. The original claim follows by induction. □

Recursive Definitions of Sequences

Our next example involves recursion. A recursive definition of an infinite sequence is a rule for generating its elements from previous entries.

Elements of a sequence are often denoted by a lowercase letter and subscript. For instance, the sequence (a_n) is the sequence of numbers

$$a_1, a_2, a_3, a_4, \ldots.$$

The first element of the sequence is a_1, the second is a_2, the third is a_3, and so on. In general, a_n is the nth entry of the sequence (a_n).

More precisely, an infinite sequence of real numbers can be defined as a function $f\colon \mathbb{N} \to \mathbb{R}$, as such a function gives rise to the following sequence of numbers:

$$f(1), f(2), f(3), \ldots.$$

However, we stick to the convention of labelling sequences with letters such as a, b, c, and subscripts (rather than parentheses) to denote the elements.

Example 4.4.2 In Section 4.1, we proved the following formula for all $n \in \mathbb{N}$:

$$1 + 3 + 5 + \cdots + (2n-1) = n^2.$$

Let us label the left-hand side of this identity by b_n:

$$b_n = 1 + 3 + 5 + \cdots + (2n-1).$$

That is, we have $b_1 = 1$, $b_2 = 1+3$, $b_3 = 1+3+5$, $b_4 = 1+3+5+7$, and so on. Every element of the sequence (b_n), except the first one, is obtained by adding the next odd number to the sum. Therefore, an alternative description of b_n can be given as follows:

$$b_1 = 1 \qquad \text{and} \qquad b_n = b_{n-1} + (2n-1) \ \text{ for } \ n \geq 2.$$

This is an example of a recursive definition of a sequence.

The first part $b_1 = 1$ tells us the starting value of the sequence, and the second part tells us how to calculate b_n from b_{n-1}. For instance:

$$\begin{aligned} b_2 &= b_1 + (2 \cdot 2 - 1) = 1 + 3 \\ b_3 &= b_2 + (2 \cdot 3 - 1) = 1 + 3 + 5 \\ b_4 &= b_3 + (2 \cdot 4 - 1) = 1 + 3 + 5 + 7. \end{aligned}$$

Here is another example:

$$\begin{cases} a_1 = 6 & \\ a_{n+1} = 5 \cdot a_n + 1 & \text{for } \ n \in \mathbb{N}. \end{cases}$$

The definition provides the first element of the sequence $a_1 = 6$, and then a rule for computing further elements. The equation $a_{n+1} = 5 \cdot a_n + 1$ tells us how to compute the $(n+1)$st element from the nth element. Namely, we multiply by 5 and add 1. Consequently, the first few numbers in the sequence are

$$6, 31, 156, 781, 3906, \ldots.$$

In some cases, recursive definitions are more elegant or natural than an explicit formula in terms of n, but they can also be less convenient. In the sequence (a_n) above, we cannot find the one-hundredth element a_{100} unless we first compute all the preceding 99 entries $a_1, a_2, \ldots, a_{99}$.

Are there methods for converting a recursive definition of a sequence into an explicit formula? Well, there is no general algorithm, but there are methods (and some are quite advanced) that can be applied in special cases. However, checking that a particular formula works can often be done by induction.

Example 4.4.3 Consider the recursive sequence

$$\begin{cases} a_1 = 6 \\ a_{n+1} = 5 \cdot a_n + 1 & \text{for } n \in \mathbb{N}. \end{cases}$$

We prove that $a_n = \frac{5^{n+1}-1}{4}$ for all $n \in \mathbb{N}$.

Proof Note that the recursive definition is given, and we may use it throughout the proof. It is the explicit formula that we now prove by induction.

If $n = 1$, we get

$$\frac{5^{n+1}-1}{4} = \frac{5^2-1}{4} = 6,$$

which equals a_1, and so the base case holds true.

Assume that $a_k = \frac{5^{k+1}-1}{4}$ for some $k \in \mathbb{N}$. To show that the explicit formula is valid for $n = k + 1$, we use the recursive definition, and the induction hypothesis, as follows:

$$a_{k+1} = 5 \cdot a_k + 1 = 5 \cdot \frac{5^{k+1}-1}{4} + 1 = \frac{5 \cdot 5^{k+1} - 5 + 4}{4} = \frac{5^{k+2}-1}{4}.$$

We have proved the case $n = k + 1$, and hence, by induction, our proof is completed. □

Note that induction and recursion are fundamentally different notions. Induction is used to *prove* the truth of a statement that depends on a natural number n, and recursion is a way of *defining* an infinite sequence. However, one cannot ignore some similarities between the two.

The Principle of Mathematical Induction requires that two conditions are verified: the base case $n = 1$, and the implication $P(k) \Rightarrow P(k+1)$ for all $k \in \mathbb{N}$. Analogously, a typical recursive definition of a sequence (a_n) has two parts: the initial value of the sequence a_1, and the recursive part stating a_{n+1} in terms of a_n.

For these reasons, mathematical induction is a natural tool used for proving and verifying various statements involving recursion.

A Fallacy

Induction must be used carefully. Both the base case and the induction step need to be properly checked, and should not be treated as mechanical procedures. A careless application of the Principle of Mathematical Induction can lead to flawed arguments. For instance, the following claim is clearly false, but can you find the mistake in its proof?

"**Claim**" In every group of n people (where $n \in \mathbb{N}$), all must have the same height.

"**Proof**" For $n = 1$, the claim holds true, as in a group with one person we can trivially assert that "all have the same height."

Now assume that the claim is valid for some $k \in \mathbb{N}$, and consider a group with $k+1$ people. We call that group S, and denote its elements by $x_1, \ldots, x_{k+1}$:

$$S = \{x_1, x_2, \ldots, x_k, x_{k+1}\}.$$

To prove that all people in S have the same height, we define the following subsets of S:

$$A = \{x_1, x_2, \ldots, x_k\}, \qquad B = \{x_2, \ldots, x_k, x_{k+1}\}.$$

In other words, A is obtained from S by removing the element x_{k+1}, and B is obtained by removing x_1. Both A and B are sets with k people, and so by the induction hypothesis, in each of the groups, all have the same height. Moreover, A and B overlap, as $A \cap B = \{x_2, \ldots, x_k\}$, and hence the height of those in A and B must be the same. We therefore conclude that all people in $S = A \cup B$ have the same height, which proves the claim for $n = k + 1$. By induction, the claim holds true for all $n \in \mathbb{N}$. □

What is going on here? Where is the mistake in the proof?

The base case seems fine. The sets A and B indeed have k elements each, so applying the induction hypothesis on both is legitimate (the hypothesis can be applied as many times as needed). The problem arises when we claim that "A and B overlap." It is true that when S has three elements or more, the intersection $A \cap B$ is non-empty. However, if $S = \{x_1, x_2\}$, then $A = \{x_1\}$ and $B = \{x_2\}$, which implies that $A \cap B = \varnothing$, and the argument breaks.

The fact that an argument, that is supposed to hold for all $k \in \mathbb{N}$, fails for a single value of k, is enough to invalidate the whole proof!

4.5 Strong Mathematical Induction

We devote the last section of this chapter to another variation of the Principle of Mathematical Induction.

The Principle of Strong Mathematical Induction

Suppose that for every natural number n, $P(n)$ is a mathematical statement, and that the following two conditions hold.

1. $P(1)$ is a true statement.
2. For every natural number k, the statements $P(1), P(2), \ldots, P(k)$ imply $P(k+1)$.

Then $P(n)$ is true for all n.

As we can see, the only difference between the original principle and this one, is in the second condition, the induction step. In the ordinary Principle of Mathematical Induction, we had to show that for every $k \in \mathbb{N}$, $P(k) \Rightarrow P(k+1)$. That is, that the kth statement implies the $(k+1)$st. With strong induction, we assume, in the induction step, that all the statements $P(1), P(2), \ldots, P(k)$ are true, and prove from that assumption the statement $P(k+1)$. In other words, our induction hypothesis seems *stronger*, and allows us to use any of the statements for $n = 1, \ldots, k$ in our proof.

As an example, we prove the following theorem.

Theorem 4.5.1 *Every natural number $n \geq 2$ can be written as a product of prime numbers.*

Note that if n is itself a prime number, we still view it as a product with a single factor.

Proof We use strong induction in our proof, and $n = 2$ as our base case.

If $n = 2$, the theorem is valid, as 2 is a prime number. Assume that the theorem holds true for $n = 2, 3, 4, \ldots, k$ for some natural number $k \geq 2$, and consider $n = k+1$. If $k+1$ happens to be a prime number, the theorem applies. Otherwise, $k+1$ is composite, and is divisible by some natural number $2 \leq m \leq k$. Equivalently, $k+1 = m \cdot \ell$, where m, ℓ are natural numbers between 2 and k.

Both m and ℓ are natural numbers, greater than 1 and smaller than $k+1$, and hence covered by the induction hypothesis. Therefore, m and ℓ are products of prime numbers, and consequently, so is $k+1 = m \cdot \ell$.

By the Principle of Strong Mathematical Induction, the theorem is valid for all natural numbers $n \geq 2$. □

Pay close attention to the proof of Theorem 4.5.1. Why was strong induction needed here? Could we carry the above argument with ordinary induction?

Here is another example.

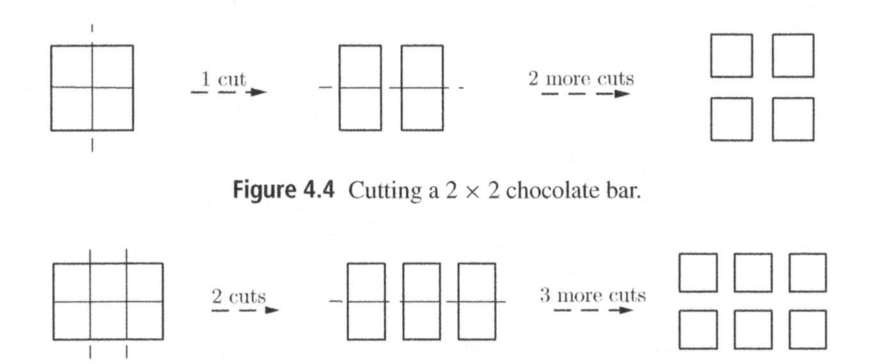

Figure 4.4 Cutting a 2×2 chocolate bar.

Figure 4.5 Cutting a 2×3 chocolate bar.

Example 4.5.2 Let n be a natural number, and consider a rectangular chocolate bar with n squares.

How many cuts are needed to break the bar into 1×1 squares? In this context, cuts are performed either horizontally or vertically, and a cut can be applied to one piece at a time.

Let us start by checking a couple of special cases. If we have a square chocolate bar, with $n = 4$ squares, then 3 cuts are needed (Figure 4.4).

For a 2×3 bar with 6 squares, we need 5 cuts altogether (Figure 4.5). After experimenting a little more, it seems that the number of cuts needed is one less than the number of squares. We can now formulate a claim, and prove it by induction.

Claim 4.5.3 *Let $n \in \mathbb{N}$. Then $n-1$ cuts are needed to break a rectangular chocolate bar with n squares into 1×1 pieces.*

Proof If $n = 1$, then our bar consists of a single square, and no cuts are needed. As $n-1 = 1-1 = 0$, the claim holds true, and the base case is verified.

Assume that the claim is valid for $n = 1, 2, \ldots, k$ for some $k \in \mathbb{N}$, and consider a chocolate bar with $k + 1$ squares. To be able to use the induction hypothesis, we perform an arbitrary first cut (Figure 4.6), which breaks the bar into two separate pieces, say with ℓ and m squares (and so $\ell + m = k + 1$). Note that both ℓ and m must be smaller than $k + 1$, and thus are covered by the induction hypothesis. Therefore, $\ell-1$ cuts are needed for the part with ℓ squares, and $m-1$ cuts are needed for the part with m squares. Overall, the number of cuts needed to break our original bar with $k + 1$ squares into 1×1 pieces is

$$1 + (\ell-1) + (m-1) = \ell + m-1 = (k + 1)-1 = k.$$

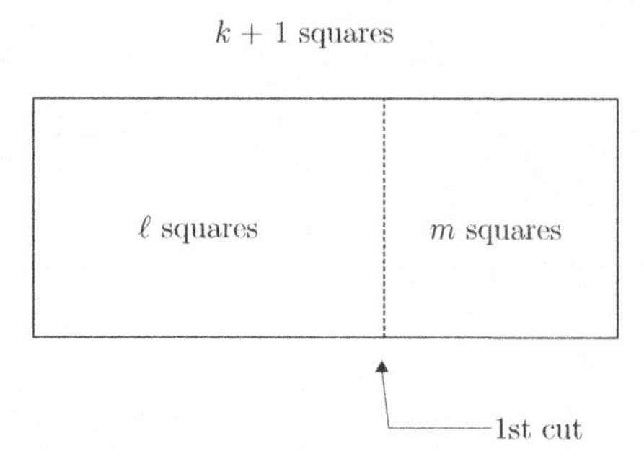

Figure 4.6 The first cut of a bar with $k+1$ squares.

The first term represents the initial cut. As we obtained k cuts for a bar with $k+1$ squares, we have proved the claim for $n=k+1$. By strong induction, the claim is true for all $n \in \mathbb{N}$. □

Our last example is a theorem, related to *binary representations*. This important notion has many applications in mathematics, computer science and digital electronics.

A binary representation of a natural number n is a representation in which only two digits, 0 and 1, are used. Other numbers, such as negative integers and non-integers can also be represented in such a way, but we will focus on natural numbers only.

The core idea is based on the following theorem, which we will prove shortly.

Theorem 4.5.4 *Every natural number n can be expressed as a sum of distinct non-negative integer powers of 2. This representation is unique up to the order of the summands.*

A non-negative integer is either 0 or a natural number, and so the non-negative integer powers of 2 are

$$2^0 = 1,\ 2^1 = 2,\ 2^2 = 4,\ 2^3 = 8,\ 2^4 = 16, \ldots.$$

The theorem says that every natural number can be written as a sum of such powers, *without repetitions* (hence "distinct") and in a unique way. These sums can have one or more summands, for instance:

$$\begin{aligned} 17 &= 2^4 + 2^0 \\ 128 &= 2^7 \end{aligned}$$

$$\begin{aligned} 42 &= 2^5 + 2^3 + 2^1 \\ 65 &= 2^6 + 2^0 \\ 312 &= 2^8 + 2^5 + 2^4 + 2^3. \end{aligned}$$

To represent such numbers in a binary form, we first add a coefficient of 1 in front of each power of 2 that appears in the sum, and a coefficient of 0 in front of the intermediate powers that do not appear:

$$\begin{aligned} 17 &= 1 \cdot 2^4 + 0 \cdot 2^3 + 0 \cdot 2^2 + 0 \cdot 2^1 + 1 \cdot 2^0 \\ 128 &= 1 \cdot 2^7 + 0 \cdot 2^6 + 0 \cdot 2^5 + 0 \cdot 2^4 + 0 \cdot 2^3 + 0 \cdot 2^2 + 0 \cdot 2^1 + 0 \cdot 2^0 \\ 42 &= 1 \cdot 2^5 + 0 \cdot 2^4 + 1 \cdot 2^3 + 0 \cdot 2^2 + 1 \cdot 2^1 + 0 \cdot 2^0 \\ 65 &= 1 \cdot 2^6 + 0 \cdot 2^5 + 0 \cdot 2^4 + 0 \cdot 2^3 + 0 \cdot 2^2 + 0 \cdot 2^1 + 1 \cdot 2^0 \\ 312 &= 1 \cdot 2^8 + 0 \cdot 2^7 + 0 \cdot 2^6 + 1 \cdot 2^5 + 1 \cdot 2^4 + 1 \cdot 2^3 + 0 \cdot 2^2 + 0 \cdot 2^1 + 0 \cdot 2^0. \end{aligned}$$

The ones and zeros are used to form the binary representation of a given natural number.

Ordinary decimal representation	**Binary representation**
17	10001
128	10000000
42	101010
65	1000001
312	100111000

Here is the general definition.

Definition 4.5.5 Let n be a natural number. Suppose that $k \in \mathbb{N} \cup \{0\}$, and that $a_0, a_1, \ldots, a_k \in \{0, 1\}$, such that

$$n = a_k \cdot 2^k + a_{k-1} \cdot 2^{k-1} + \cdots + a_2 \cdot 2^2 + a_1 \cdot 2^1 + a_0 \cdot 2^0.$$

The finite sequence $a_k a_{k-1} \ldots a_2 a_1 a_0$ is called the *binary representation* of n.

Theorem 4.5.4 guarantees the existence and uniqueness of binary representations.

Proof We prove existence and uniqueness separately, and in both parts, we use strong induction.

Proof of Existence

The base case is quickly verified, as $1 = 2^0$.

Let $k \in \mathbb{N}$, and assume that the theorem holds true for $n = 1, 2, 3, \ldots, k$. We need to prove that $k + 1$ can be expressed as a sum of distinct non-negative integer powers of 2, and we do that by looking at the following two possible cases.

- **Case 1: k is even.** By assumption, the theorem applies to $n = k$, and so we can write

$$k = 2^{a_1} + 2^{a_2} + \cdots + 2^{a_m},$$

where $a_1, \ldots, a_m$ are distinct *positive* integers. As k is even, the term 2^0 does not appear in the sum. Consequently,

$$k + 1 = 2^0 + 2^{a_1} + 2^{a_2} + \cdots + 2^{a_m},$$

and we have expressed $k + 1$ in the required form.
- **Case 2: k is odd.** The argument used in Case 1 will not work here, so we use a different approach (see Problem 4.28). As k is odd, $k + 1$ is even, and we can write $k + 1 = 2m$, for some $m \in \mathbb{N}$. As m is smaller than $k + 1$, the induction hypothesis applies, and we have

$$m = 2^{a_1} + 2^{a_2} + \cdots + 2^{a_m},$$

for some non-negative distinct integers $a_1, \ldots, a_m$. This time, one of the a_i may be zero!

We conclude that

$$k + 1 = 2m = 2^{a_1+1} + 2^{a_2+1} + \cdots + 2^{a_m+1},$$

as needed. Note that since $a_1, a_2, \ldots, a_m$ are distinct, so are $a_1 + 1, a_2 + 1, \ldots, a_m + 1$.

We have proved the existence part of the theorem for $n = k + 1$, and hence, by strong induction, for every $n \in \mathbb{N}$.

Proof of Uniqueness

The only way to represent $n = 1$ in the required form is to write $n = 1 = 2^0$, as all higher powers of 2 are greater than one. This proves uniqueness of representation for $n = 1$, which is our base case.

Let $k \in \mathbb{N}$, and suppose that all integers $n = 1, 2, \ldots, k$ can be expressed uniquely as a sum of non-negative integer powers of 2. Consider $n = k + 1$.

- First, we claim that if 2^m is the highest power of 2 such that $2^m \leq k+1$, then 2^m must participate in every representation of $k+1$ as a sum of non-negative integer powers of 2.

  This follows from the fact that

$$1 + 2 + 2^2 + \cdots + 2^{m-1} = 2^m - 1 < k + 1$$

(see Problem 4.13).

- Next, suppose that $k+1$ can be represented in potentially two different ways as sums of non-negative integer powers of 2:

$$k+1 = 2^{a_1} + 2^{a_2} + \cdots + 2^{a_\ell} + 2^m = 2^{b_1} + 2^{b_2} + \cdots + 2^{b_r} + 2^m.$$

Here, the a_i and b_j are integers, satisfying

$$0 \leq a_1 < a_2 < \cdots < a_\ell < m \quad \text{and} \quad 0 \leq b_1 < b_2 < \cdots < b_r < m.$$

By subtracting 2^m from the above equalities, we get

$$k+1-2^m = 2^{a_1} + 2^{a_2} + \cdots + 2^{a_\ell} = 2^{b_1} + 2^{b_2} + \cdots + 2^{b_r}.$$

Now, as $k+1-2^m$ is a natural number smaller than $k+1$, it is covered by our induction hypothesis, and hence has a unique representation as a sum of non-negative integer powers of 2. Therefore, we conclude that $\ell = r$ and $a_i = b_i$ for all i.

That is, the two potentially different representations of $k+1$ are, in fact, the same, and so only one representation exists. We have proved the uniqueness statement for $n = k+1$, and hence, by strong induction, for all $n \in \mathbb{N}$.

□

4.6 Problems

4.1 Prove the following equalities for all $n \in \mathbb{N}$.

a. $1^2 + 3^2 + 5^2 + \cdots + (2n-1)^2 = \frac{4n^3-n}{3}$.
b. $1^2 + 2^2 + 3^2 + \cdots + n^2 = \frac{1}{6}n(1+n)(1+2n)$.

4.2 Prove the following inequalities by induction for all $n \in \mathbb{N}$.

a. $5^n + 5 < 5^{n+1}$.
b. $1 + 2 + 3 + \cdots + n \leq n^2$.
c. $\frac{1}{\sqrt{1}} + \frac{1}{\sqrt{2}} + \frac{1}{\sqrt{3}} + \cdots + \frac{1}{\sqrt{n}} \geq \sqrt{n}$.
d. $\frac{1}{\sqrt{1}} + \frac{1}{\sqrt{2}} + \frac{1}{\sqrt{3}} + \cdots + \frac{1}{\sqrt{n}} \leq 2\sqrt{n}$.

4.3 (**Harder!**) Let $0 < a < 1$. Prove that for every $n \in \mathbb{N}$, $(1-a)^n < \frac{1}{1+n \cdot a}$.

4.4 Prove that for every $n \in \mathbb{N}$, $3^{4n+2} + 1$ is divisible by 10.

4.5 Prove that for every $n \in \mathbb{N}$, $n^3 + 2n$ is divisible by 3.

4.6 Prove that for all $n \in \mathbb{N}$, $4^{2n} - 1$ is divisible by 5.

4.7 Let a be an integer different than 1. Prove, *by induction*, that for every $n \in \mathbb{N}$, $a^n - 1$ is divisible by $a - 1$.

4.8 Prove that $\frac{1}{3}n^3 + \frac{1}{2}n^2 + \frac{1}{6}n$ is *an integer* for every $n \in \mathbb{N}$.

4.9 Compute the following expressions (obtain a single number).

a. $\displaystyle\sum_{k=1}^{100}\left[k \cdot (-1)^k\right]$

b. $\displaystyle\sum_{k=2}^{200}\left(\frac{1}{k} - \frac{1}{k+1}\right)$

c. $\displaystyle\prod_{k=1}^{69} 2^{k-35}$

d. $\displaystyle\prod_{i=10}^{99} \frac{i}{i+1}$

4.10 As mentioned in Section 4.2, Proposition 4.2.3 remains valid in any field. Rigorous proofs of these properties can be done by induction. To get a feeling for such proofs, prove the following statement by induction.

Let $\mathbb{F}$ be a field, $n \in \mathbb{N}$, and $a_1, \ldots, a_n, b_1, \ldots, b_n \in \mathbb{F}$.
Prove that $\displaystyle\sum_{i=1}^{n}(a_i + b_i) = \sum_{i=1}^{n} a_i + \sum_{i=1}^{n} b_i$.

Note that the induction will be done on n (the number of a_i and b_i). Use Example 4.1.8 as a model.

4.11 Which of the following is true for every $a_1, \ldots, a_n, b_1, \ldots, b_n \in \mathbb{R}$? Justify your answer briefly. Provide counterexamples for the false statements.

a. $\displaystyle\left(\sum_{k=1}^{n} a_k\right) \cdot \left(\sum_{k=1}^{n} b_k\right) = \sum_{k=1}^{n}(a_k \cdot b_k)$.

b. $\displaystyle\sum_{k=1}^{n} a_k - \sum_{k=1}^{n} b_k = \sum_{k=1}^{n}(a_k - b_k)$.

c. $\displaystyle\prod_{k=1}^{n} a_k - \prod_{k=1}^{n} b_k = \prod_{k=1}^{n}(a_k - b_k)$.

d. $\displaystyle\left(\prod_{k=1}^{n} a_k\right) / \left(\prod_{k=1}^{n} b_k\right) = \prod_{k=1}^{n} \frac{a_k}{b_k}$, assuming that $b_1, b_2, \ldots, b_n$ are all non-zero.

4.12 Prove the following identities.

a. $\displaystyle\sum_{i=1}^{n} \frac{i}{2^i} = 2 - \frac{n+2}{2^n}$ (for $n \geq 1$).

b. $\displaystyle\sum_{i=1}^{n}(3i - 2) = \frac{n(3n-1)}{2}$ (for $n \geq 1$).

c. $\displaystyle\prod_{i=2}^{n}\left(1 - \frac{1}{i^2}\right) = \frac{n+1}{2n}$ (for $n \geq 2$).

4.13 Let $q \neq 1$. Prove that for every $n \in \mathbb{N}$, we have

$$\sum_{i=0}^{n-1} q^i = \frac{q^n - 1}{q - 1}.$$

4.14 Let $P(1), P(2), \ldots$ be a sequence of statements. Write down variations (in the style of those on page 109), for proving that $P(n)$ is true for the following.

a. All n which are a multiple of 3.
b. $n = 2, 5, 8, 11, \ldots$.
c. $n = 11, 13, 15, 17, \ldots$.

4.15 Prove, by induction, that $10^n - 1$ is divisible by 11 for every *even* natural number n.

4.16 a. Show that for every $k \in \mathbb{N}$, if $2^{3k-1} + 5 \cdot 3^k$ is divisible by 11, then $2^{3(k+2)-1} + 5 \cdot 3^{k+2}$ is also divisible by 11.
b. Which of the following statements is true? Explain.

(i) For every *odd* number $n \in \mathbb{N}$, $2^{3n-1} + 5 \cdot 3^n$ is divisible by 11.
(ii) For every *even* number $n \in \mathbb{N}$, $2^{3n-1} + 5 \cdot 3^n$ is divisible by 11.

4.17 Let (a_n) be a sequence such that $a_1 = 1$ and $a_{n+1} = a_n + 3n(n+1)$ for $n \in \mathbb{N}$. Prove that $a_n = n^3 - n + 1$ for $n \in \mathbb{N}$.

4.18 Let (a_n) be a sequence given by

$$\begin{cases} a_1 = 3 \\ a_{n+1} = a_n + 6n(n+1) \qquad (\text{for } n \in \mathbb{N}). \end{cases}$$

Prove, by induction, that $a_n = 2n^3 - 2n + 3$ for every $n \in \mathbb{N}$.

4.19 a. Prove that $3x^2 + 3 \geq (x+1)^2 + 1$ for all real numbers x (can this be done by induction?).
b. Consider the following recursively defined sequence:

$$\begin{cases} a_1 = 2 \\ a_{n+1} = 3 \cdot a_n \qquad (\text{for } n = 1, 2, \ldots). \end{cases}$$

Use Part a to prove, by induction, that $a_n \geq n^2 + 1$ for all $n \in \mathbb{N}$.

4.20 Justify Parts 2, 3 and 4 of Proposition 4.2.3, by writing sums and products explicitly (i.e., without sigmas or pis).

4.21 Let (x_n) be a sequence given by the following recursion formula:

$$x_1 = 3, \quad x_2 = 7 \qquad \text{and} \qquad x_{n+1} = 5 \cdot x_n - 6 \cdot x_{n-1} \qquad \text{for} \quad n \geq 2.$$

Prove that for all $n \in \mathbb{N}$, $x_n = 2^n + 3^{n-1}$.

4.22 Consider the following sequence defined recursively:

$$\begin{cases} a_1 = 5, \qquad a_2 = 8 \\ a_{n+1} = 2a_n - a_{n-1} + 2 \qquad \text{for} \quad n \geq 2, \end{cases}$$

Prove that $a_n = n^2 + 4$ for all $n \in \mathbb{N}$.

4.23 The sequence (a_n) is defined recursively by

$$\begin{cases} a_1 = 6, \quad a_2 = 8 \\ a_n = 4 \cdot a_{n-1} - 4 \cdot a_{n-2} \quad \text{for } n > 2. \end{cases}$$

Prove that $a_n = (4 - n) \cdot 2^n$ for all $n \in \mathbb{N}$.

4.24 Let (a_n) be a sequence satisfying $a_1 = a_2 = 1$ and $a_n = \frac{1}{2}\left(a_{n-1} + \frac{2}{a_{n-2}}\right)$ for $n \geq 3$. Prove that $1 \leq a_n \leq 2$ for all $n \in \mathbb{N}$.

4.25 Consider the following recursively defined sequence:

$$a_1 = \frac{5}{2}, \qquad a_{n+1} = \frac{1}{2} \cdot (a_n + 2) \qquad (\text{for } n \in \mathbb{N}).$$

Prove that $a_n > \frac{1}{2^n}$ for all $n \in \mathbb{N}$.

4.26 Let (a_n) be a sequence satisfying $a_n = 2a_{n-1} + 3a_{n-2}$ for $n \geq 3$. Given that a_1, a_2 are odd, prove that a_n is odd for $n \in \mathbb{N}$.

4.27 Consider a regular triangular board of side length 2^n. The board consists of 4^n equilateral triangles of side length 1. Show that if one of the corner triangles is removed, then the remaining board can be tiled by isosceles trapezoids (as in Figure 4.7), each covering exactly three triangles.

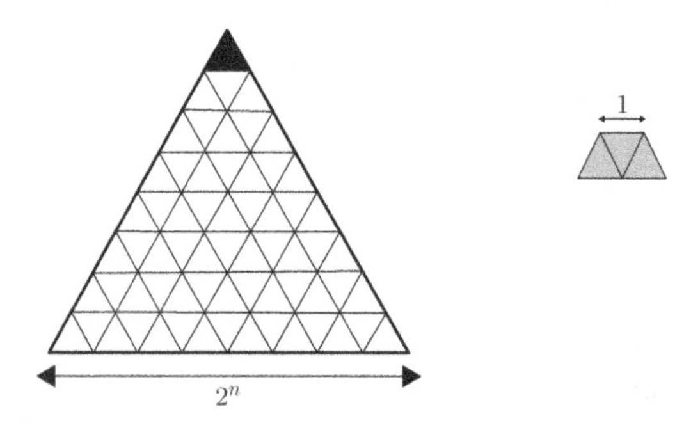

Figure 4.7 A regular triangular board, and an isosceles trapezoid.

4.28 In the proof of Theorem 4.5.4, why were we not able to use the argument in Case 1 for the case where k is odd? Also, why did we have to use strong induction? Which part of the argument cannot be carried out with ordinary induction?

4.29 Prove that every $n \in \mathbb{N}$ can be written as a product of an odd integer and a non-negative integer power of 2. For instance: $36 = 9 \cdot 2^2$, $80 = 5 \cdot 2^4$, $17 = 17 \cdot 2^0$, etc.

Hint: Use strong induction on n. In the induction step, treat the cases "k even" and "k odd" separately.

4.30 Let x be a non-zero real number, such that $x + \frac{1}{x}$ is an *integer*. Prove that for all $n \in \mathbb{N}$, the number $x^n + \frac{1}{x^n}$ is also an integer.

4.31 Find the mistake in the following "proof."

> "*Claim*" The numbers $0, 1, 2, 3, \ldots$ are all even.
> "*Proof*" We use strong induction to prove the statement "n is even" for $n = 0, 1, 2, 3, \ldots$.
> Base case: $n = 0$ is an even number, hence the statement is true for $n = 0$.
> Assume that the statement is true for $n = 0, 1, 2, \ldots, k$, and consider $n = k + 1$.
> By assumption, both 1 and k are even numbers, and hence so is their sum $k + 1$. It thus follows that the statement holds for all $n = 0, 1, 2, 3, \ldots$.

4.32 Find the mistake in the following "proof."

> "*Claim*" For all $n \in \mathbb{N}$, we have $5^n = 5$.
> "*Proof*" For $n = 1$, we have $5^1 = 5$, which proves the base case. Now assume that the claim holds true for $n = 1, 2, \ldots, k$. Then we have
>
> $$5^{k+1} = \frac{5^k \cdot 5^k}{5^{k-1}} = \frac{5 \cdot 5}{5} = 5$$
>
> (we used the induction hypothesis in the second equality). This proves the $n = k + 1$ case. By strong induction, the claim follows.

4.33 What do you think about the following proof of the statement "*Every person is bald*"?

> A person with a single hair is clearly bald, which confirms our base case. Now, assume that a person with k hairs is bald. Then obviously, adding one more hair to a bald person will leave that person bald. By induction, it follows that every person with n hairs is bald. In other words, everyone is bald!

4.34 (**Harder!**) **The General Arithmetic-Geometric Mean Inequality**. In this exercise, we prove the general version of the inequality, stating that for every $x_1, x_2, \ldots, x_n \geq 0$, we have

$$\sqrt[n]{x_1 \cdot x_2 \cdots x_n} \leq \frac{x_1 + x_2 + \cdots + x_n}{n}.$$

The proof uses induction, but is a bit tricky. We break it into several steps.

a. Prove that if x and y are real numbers, satisfying $0 \leq x \leq 1 \leq y$, then $x + y \geq xy + 1$.
(Hint: Note that $x-1 \leq 0$ and $y-1 \geq 0$.)

b. Assume that $a_1, a_2, \ldots, a_n$ are non-negative real numbers (with $n \geq 2$), whose product is 1:

$$a_1 \cdot a_2 \cdots a_n = 1.$$

Explain why $a_i \leq 1 \leq a_j$ for some $i \neq j$.

c. Prove, by induction on n, the following claim:

If $a_1, \ldots, a_n$ are non-negative real numbers, with $a_1 \cdots a_n = 1$, then $a_1 + \cdots + a_n \geq n$.

(Hints: To carry the induction step, note that a product of $k + 1$ numbers can be thought of as a product of k numbers:

$$a_1 \cdot a_2 \cdot a_3 \cdots a_{k+1} = (a_1 \cdot a_2) \cdot a_3 \cdots a_{k+1}.$$

You will also need to use Parts a and b in your proof.)

d. Finally, prove the General Arithmetic-Geometric Mean Inequality. To do so, note that if one of the x_i is zero, the inequality follows immediately (why?). If all the x_i are positive, use Part c with

$$a_1 = \frac{x_1}{\sqrt[n]{x_1 \cdot x_2 \cdots x_n}}, \quad a_2 = \frac{x_2}{\sqrt[n]{x_1 \cdot x_2 \cdots x_n}}, \quad a_3 = \frac{x_3}{\sqrt[n]{x_1 \cdot x_2 \cdots x_n}}, \text{ etc.}$$

4.7 Solutions to Exercises

Solution to Exercise 4.1.13

Yes, the claim remains valid for $n = 0$. A set with zero elements is simply the empty set. In this case, we have only one subset, the empty set itself. This is consistent with the claim, as $2^0 = 1$.

Solution to Exercise 4.2.4

2.

$$\sum_{i=m}^{n} (a_i + b_i) = (a_m + b_m) + (a_{m+1} + b_{m+1}) + \cdots + (a_n + b_n)$$

$$
\begin{aligned}
&= (a_m + a_{m+1} + \cdots + a_n) + (b_m + b_{m+1} + \cdots + b_n) \\
&= \sum_{i=m}^{n} a_i + \sum_{i=m}^{n} b_i.
\end{aligned}
$$

3.

$$
\begin{aligned}
\prod_{i=m}^{n}(c \cdot a_i) &= (c \cdot a_m) \cdot (c \cdot a_{m+1}) \cdots (c \cdot a_n) \\
&= (c \cdot c \cdots c) \cdot (a_m \cdot a_{m+1} \cdots a_n) = c^{n-m+1} \cdot \prod_{i=m}^{n} a_i.
\end{aligned}
$$

4.

$$
\begin{aligned}
\prod_{i=m}^{n}(a_i \cdot b_i) &= (a_m \cdot b_m) \cdot (a_{m+1} \cdot b_{m+1}) \cdots (a_n \cdot b_n) \\
&= (a_m \cdot a_{m+1} \cdots a_n) \cdot (b_m \cdot b_{m+1} \cdots b_n) \\
&= \prod_{i=m}^{n} a_i \cdot \prod_{i=m}^{n} b_i.
\end{aligned}
$$

Solution to Exercise 4.2.7

If all the x_i in Claim 4.2.6 are equal to each other, say $x_1 = x_2 = \cdots = x_n = x$, then we get $(1+x)^n \geq 1 + nx$, which is Bernoulli's Inequality.

5 Bijections and Cardinality

Given two infinite sets, is there a sensible way to decide which one is larger? For instance, if A is the set of even integers, and B is the interval $[0, 1]$, is there a way to compare their size?

In this section, we focus on such questions, and introduce the notion of *cardinality*, which is used to describe "how many elements" a (potentially infinite) set has. This leads to some interesting and counter-intuitive consequences.

However, we must first prepare the ground by discussing injections, surjections, bijections and related results.

5.1 Injections, Surjections and Bijections

Consider a function $f\colon \{a, b, c, d\} \to \{\alpha, \beta, \gamma, \delta, \epsilon\}$, given by the diagram in Figure 5.1. The symbols $\alpha, \beta, \gamma, \delta, \epsilon$ represent the first five letters of the Greek alphabet: alpha, beta, gamma, delta and epsilon.

Indeed, the diagram defines a function from the set $\{a, b, c, d\}$ to the set $\{\alpha, \beta, \gamma, \delta, \epsilon\}$ (see Definition 2.2.1). Suppose now, that we decide to reverse the arrows, and obtain the diagram in Figure 5.2. Look at this diagram and attempt the following exercise.

Exercise 5.1.1
Does the new diagram define a function from $\{\alpha, \beta, \gamma, \delta, \epsilon\}$ to $\{a, b, c, d\}$? Explain.

To be able to reverse the arrows and obtain a function, the original function must satisfy two conditions, injectivity and surjectivity, which we define below.

Definition 5.1.2 Let $f\colon A \to B$ be a function.

1. f is *injective* (or an *injection*, or a *one-to-one function*), if for every $y \in B$, there is *at most* one $x \in A$, for which $f(x) = y$.
2. f is *surjective* (or a *surjection*, or an *onto function*), if for every $y \in B$, there is *at least* one $x \in A$, for which $f(x) = y$.
3. f is *bijective* (or a *bijection*), if it is *both* injective and surjective.

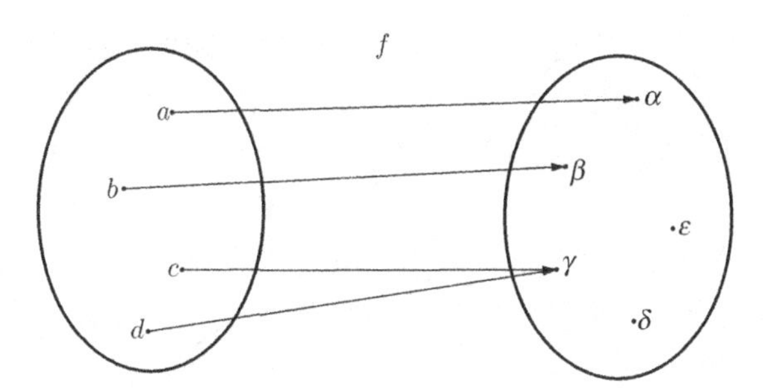

Figure 5.1 A function $f\colon \{a, b, c, d\} \to \{\alpha, \beta, \gamma, \delta, \epsilon\}$.

Remarks.

- In the previous example, f is *neither injective nor surjective.* We have $f(c) = f(d) = \gamma$, and hence f is *not* an injection, and as δ and ϵ are not images of elements in the domain, f is also *not* a surjection.
- There are many ways to state the definition of injectivity and surjectivity. Here are a few alternatives.
  - f is *injective*, if for every $x_1, x_2 \in A$, $x_1 \neq x_2$ implies $f(x_1) \neq f(x_2)$. That is, f maps distinct elements in the domain to distinct elements in the codomain.
  - f is *injective*, if for every $x_1, x_2 \in A$, $f(x_1) = f(x_2)$ implies $x_1 = x_2$. This is simply the contrapositive of the implication $x_1 \neq x_2 \Rightarrow f(x_1) \neq f(x_2)$.
  - f is *surjective*, if $f(A) = B$ (i.e., if the image of f is the whole codomain B).
  - f is a *bijection*, if for *every* $y \in B$, there is *exactly one* $x \in A$, for which $f(x) = y$.

  Convince yourself that these definitions are equivalent to the ones given in Definition 5.1.2. We will freely use the most convenient formulation in our proofs and examples.

In order to be able to reverse the arrows and obtain a function that goes in the opposite direction, our initial function must be both injective and surjective. In other words, it must be a bijection. Let us formalize this idea.

Definition 5.1.3 Let $f\colon A \to B$ be a bijection. The *inverse* of f is a function $g\colon B \to A$, that assigns to every $y \in B$, the only $x \in A$ for which $f(x) = y$. We denote the inverse function by f^{-1}.

The definition of the inverse may sound confusing. All it says is that if f maps an element $a \in A$ to some $b \in B$, then f^{-1} maps b back to a:

$$f(a) = b \qquad \Leftrightarrow \qquad f^{-1}(b) = a.$$

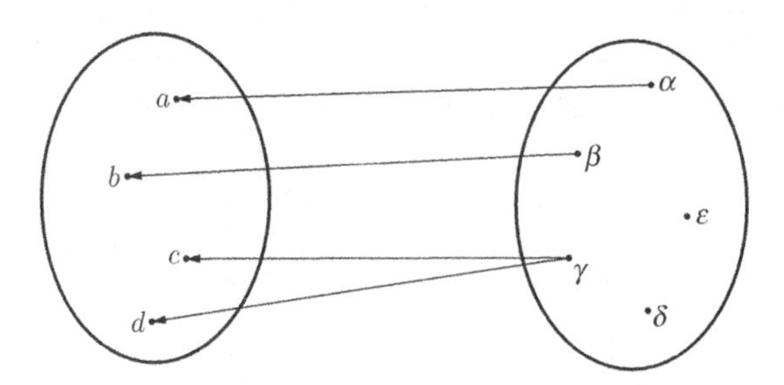

Figure 5.2 Reversing the arrows in the diagram of f.

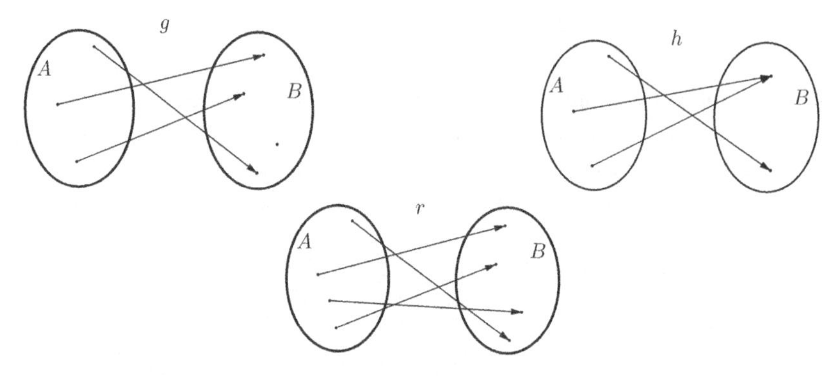

Figure 5.3 Three functions.

In other words, f^{-1} undoes the effect of f.

Example 5.1.4 g, h and r are functions from sets A to B, as shown in Figure 5.3. The function g is *injective*, as there are no two elements in A which are mapped to the same element in B. In other words, there are no two arrows pointing to the same element in B. However, g is *not surjective* as there is a "lonely" element in B, which is not the image of an element in A.

The function h is *onto*, as every element in B is an image, but it is *not one-to-one*, since there are two arrows pointing to the same element in B.

Finally, r is both an *injection* and a *surjection*, as every element in B is the image of *exactly one* element in A. Consequently, r is a *bijection*, and hence can be inverted.

Example 5.1.5 $f\colon \mathbb{R} \to \mathbb{R}, f(x) = \dfrac{1}{1+x^2}$.

f is *not an injection*, as we can find two x with the same image. For instance, $f(2) = f(-2) = \frac{1}{5}$. f is *not a surjection* either, since $f(x) > 0$ for all $x \in \mathbb{R}$, and hence there is no $x \in \mathbb{R}$ for which $f(x) = -1$.

Example 5.1.6 $g\colon \mathbb{N} \to \mathbb{N}, g(n) = n^3 + 1$.

Note that the domain of g is the set of natural numbers, and so for every n in the domain, $g(n) = n^3 + 1 \geq 1 + 1 = 2$. This means that for all $n \in \mathbb{N}$, $g(n) \neq 1$, which shows that g is *not surjective*. However, g is an *injection*. We prove this by showing that distinct elements in the domain are mapped to distinct images. Indeed, if $n_1 \neq n_2$ are two natural numbers, then

$$n_1^3 \neq n_2^3 \quad \Rightarrow \quad n_1^3 + 1 \neq n_2^3 + 1 \quad \Rightarrow \quad g(n_1) \neq g(n_2).$$

Example 5.1.7 We proceed with a less standard example. Let $\mathcal{P}$ denote the set of all *non-zero polynomials with real coefficients*. Recall that a polynomial is a sum of terms of the form $a \cdot x^n$, where a is a real number, and n is a non-negative integer. For instance, $5x^3 + 8x^7 - 4$ and $x - \frac{1}{2}x^4$ are polynomials, while $\sqrt{x}$ and $\frac{1}{x} + 2x$ are not.

Every polynomial has a *degree*, which is the highest exponent appearing in the polynomial, with a non-zero coefficient. We often denote the degree of a polynomial by deg(). For instance,

$$\begin{aligned} \deg(5x^3 + 8x^7 - 4) &= 7 \\ \deg\left(x - \frac{1}{2}x^4\right) &= 4 \\ \deg(2x + 7) &= 1 \\ \deg((2 + x^2)^{13}) &= 26. \end{aligned}$$

Now, define a function D from the set of non-zero polynomials to the set of non-negative integers, that assigns to each polynomial $p(x)$, its degree:

$$D\colon \mathcal{P} \to \mathbb{N} \cup \{0\}, \qquad D(p(x)) = \deg(p(x)).$$

Is D injective? Is it surjective? Clearly, there are different polynomials of the same degree. For example, x^3+x+2 and $5x^3-4x^2$ are both cubic polynomials, and so have degree 3:

$$D(x^3 + x + 2) = D(5x^3 - 4x^2) = 3.$$

As D assigns the same image to two distinct elements, it is *not injective*. What about surjectivity? If n is a non-negative integer, can we always find a polynomial with degree n? Of course we can. $p(x) = x^n$ is a polynomial of degree n. In other words, $D(x^n) = n$ for $n = 0, 1, 2, \ldots$, which shows that *D is a surjective function*.

When a function happens to be a bijection, it has an inverse. In the case where the function is given by a formula, we can find a formula for the inverse function by expressing x in terms of y.

Example 5.1.8 $h\colon \mathbb{R} \setminus \{-1\} \to \mathbb{R} \setminus \{1\},\ h(x) = \dfrac{x}{x+1}.$

To find out whether h is injective, surjective, neither or both, we often draw the graph of the function, using calculus or a graphing software, and try to get some intuition regarding surjectivity and injectivity. Then, we can proceed to proving our conjectures.

In this example, *h is a bijection*. Here is the proof.

To show that h is one-to-one, we start by assuming that $h(x_1) = h(x_2)$ for some $x_1, x_2 \neq -1$. We argue that $x_1 = x_2$ as follows:

$$h(x_1) = h(x_2) \quad \Rightarrow \quad \frac{x_1}{x_1+1} = \frac{x_2}{x_2+1}$$

$$\Rightarrow \quad x_1x_2 + x_1 = x_2x_1 + x_2 \quad \Rightarrow \quad x_1 = x_2.$$

Now we show that h is surjective. Let $y \neq 1$. Our task is to show that $h(x) = y$ for some x in the domain of h. We do that by solving the equation $h(x) = y$ for x:

$$\frac{x}{x+1} = y \quad \Rightarrow \quad x = xy + y \quad \Rightarrow \quad x(1-y) = y \quad \Rightarrow \quad x = \frac{y}{1-y}.$$

Note that, in the final step, we could divide by $1-y$ since $y \neq 1$. Also note that $\frac{y}{1-y} \neq -1$ (why?), and so x lies in the domain of h. Overall, we have shown that every y in the codomain of h has an x, which implies that h is surjective.

The bijectivity of h implies that h can be inverted. For the inverse function, the domain and codomain are switched, and so $h^{-1}\colon \mathbb{R}\setminus\{1\} \to \mathbb{R}\setminus\{-1\}$. How can we find a formula for h?

Remember that inverting a function means, informally, switching the roles of x and y, and so our goal is to express x as a function of y. In fact, we have already done that, when we showed that h is surjective:

$$x = \frac{y}{1-y}.$$

We therefore conclude that $h^{-1}(y) = \frac{y}{1-y}$, or $h^{-1}(x) = \frac{x}{1-x}$, after switching back to x.

Example 5.1.9 One of the commonly used trigonometric functions is the sine function:

$$f\colon \mathbb{R} \to \mathbb{R}, \qquad f(x) = \sin x.$$

The graph of f looks like a wave, which clearly shows that f is neither injective nor surjective. More formally, $\sin(0) = \sin(\pi) = 0$, and thus f is *not one-to-one*, and $\sin(x) \neq 2$ for every x, which implies that f is *not onto either*.

However, we can easily "turn" f into a bijection, by restricting its domain and codomain to the intervals $\left[-\frac{\pi}{2}, \frac{\pi}{2}\right]$ and $[-1, 1]$, respectively. Consequently, the function

$$g\colon \left[-\frac{\pi}{2}, \frac{\pi}{2}\right] \to [-1, 1], \qquad g(x) = \sin x$$

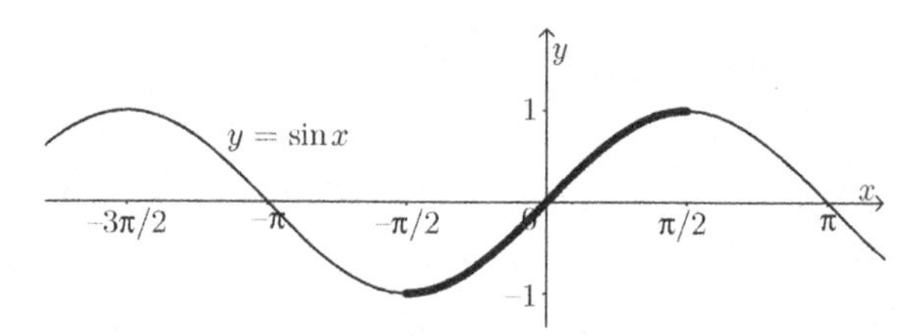

Figure 5.4 The restriction of $y = \sin x$ to $\left[-\frac{\pi}{2}, \frac{\pi}{2}\right]$ is one-to-one.

is a bijection (this can be derived from the geometric definition of the sine function, or seen by simply looking at the restricted graph, see Figure 5.4). The inverse function, g^{-1}, is commonly called the *inverse sine* or the *arcsine function*:

$$g^{-1}: [-1, 1] \to \left[-\frac{\pi}{2}, \frac{\pi}{2}\right], \qquad g^{-1}(x) = \sin^{-1} x = \arcsin x.$$

Example 5.1.10 A similar discussion can be carried out for the function $f(x) = x^2$. As a function from $\mathbb{R}$ to $\mathbb{R}$, f is neither injective nor surjective. However, if we view f as a function from $[0, \infty)$ to $[0, \infty)$, then it is a bijection, and its inverse is the square-root function. In other words, the function

$$f: [0, \infty) \to [0, \infty), \qquad f(x) = x^2$$

is a bijection, and its inverse is the function

$$f^{-1}: [0, \infty) \to [0, \infty), \qquad f^{-1}(x) = \sqrt{x}.$$

Sometimes, it may be difficult to prove directly that a function is injective or surjective. In these cases, we may use indirect methods for doing so. For instance, monotone functions are one-to-one. Here is a definition you might have seen in your calculus class.

Definition 5.1.11 Let $A \subseteq \mathbb{R}$, and $f: A \to \mathbb{R}$ a function. We say that f is a *strictly increasing* (respectively *strictly decreasing*) *function*, if for every $x_1 < x_2$ in A, we have $f(x_1) < f(x_2)$ (respectively $f(x_1) > f(x_2)$).

A function f is *strictly monotone*, if it is either strictly increasing or strictly decreasing.

Remarks.

- The wording of this definition might seem weird at first, if you are not used to the word "respectively," which allows us to compress definitions and mathematical statements. If we read the first sentence of Definition 5.1.11

without the parentheses, we obtain the definition of a *strictly increasing* function:

"f is a *strictly increasing function*, if for every $x_1 < x_2$ in A, we have $f(x_1) < f(x_2)$."

However, if we replace "increasing" and "$f(x_1) < f(x_2)$" with the content in parentheses, we get the definition of a *strictly decreasing* function:

"f is a *strictly decreasing function*, if for every $x_1 < x_2$ in A, we have $f(x_1) > f(x_2)$."

- Informally, the definition says that for a strictly increasing function, the y-values *increase* as we increase the x-values. For a strictly decreasing function, the y-values *decrease* as the x-values increase (see Figure 5.5).

We are now ready to prove the following proposition.

Proposition 5.1.12 *Let $A \subseteq \mathbb{R}$. If $f : A \to \mathbb{R}$ is a strictly monotone function, then f is injective.*

Proof We assume that f is a strictly monotone function, and we prove that f is one-to-one, by showing that $x_1 \neq x_2$ implies $f(x_1) \neq f(x_2)$ for arbitrary $x_1, x_2 \in A$.

As $x_1 \neq x_2$, either $x_1 < x_2$ or $x_2 < x_1$. Since f is strictly monotone, either $f(x_1) < f(x_2)$ or $f(x_1) > f(x_2)$. In either case, we see that $f(x_1) \neq f(x_2)$, and hence f is injective, as needed. □

This last proof is extremely short. Do not let that fool you! Make sure you understand the argument. What did we assume? What did we prove? Which definitions did we use? Did we cover all cases?

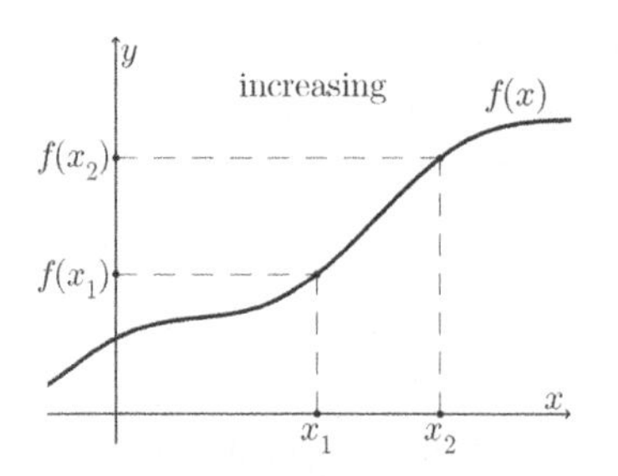

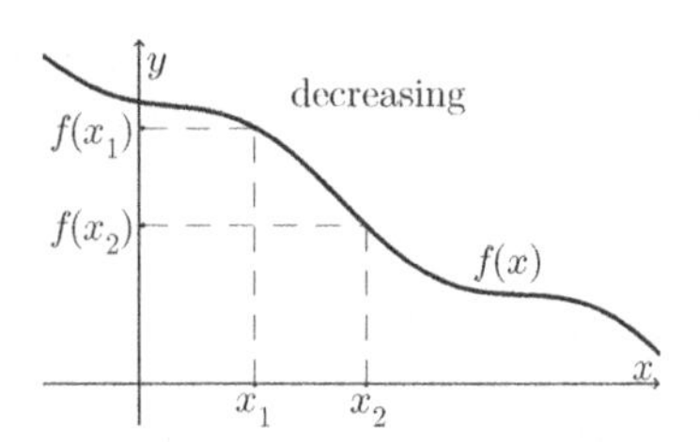

Figure 5.5 Strictly monotone functions.

Example 5.1.13 The function $f: (-1, 1) \to \mathbb{R}$, $f(x) = \frac{2x}{1-x^2}$ is injective, but it may require some work to prove this fact directly from Definition 5.1.2. However, using calculus, we see that the derivative is always positive:

$$f'(x) = \frac{2(1+x^2)}{(1-x^2)^2} > 0 \qquad \text{for } -1 < x < 1.$$

We conclude that f is strictly increasing, and hence, by Proposition 5.1.12, f is one-to-one.

Note that in this book, we normally try to avoid using calculus. This example is an exception.

5.2 Compositions

Every two functions f and g, with codomain $\mathbb{R}$, can be added, subtracted, multiplied and divided to create new functions. There is, however, one more important operation that can be performed on functions – we can compose them! Namely, we can apply one after the other, and we can do that even when their domain and codomain are *not* sets of numbers.

Definition 5.2.1 Let $f: A \to B$ and $g: B \to C$ be two functions (see Figure 5.6). The *composition* of g with f, denoted as $g \circ f$, is the function from A to C, given by

$$g \circ f(x) = g(f(x)) \qquad \text{for } x \in A.$$

Note that the only requirement for composition is that the codomain of f is equal to the domain of g.

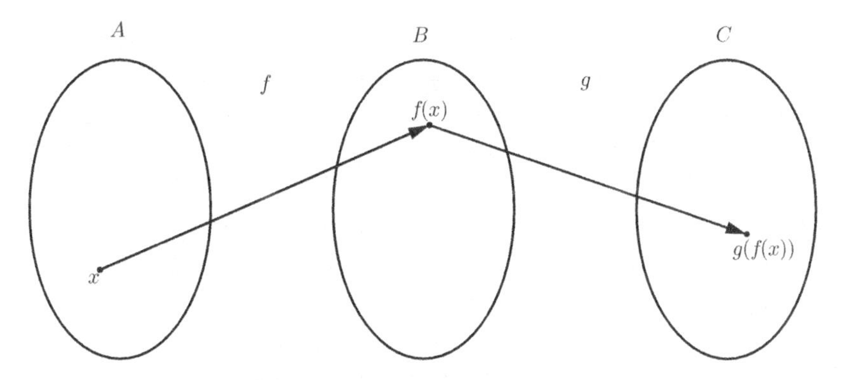

Figure 5.6 Composition of functions.

Example 5.2.2 Consider the functions $f, g\colon \mathbb{R} \to \mathbb{R}$, given by

$$f(x) = \frac{1}{1+x^2} \qquad \text{and} \qquad g(x) = e^x.$$

If we compose g with f, we obtain the function

$$g \circ f(x) = g(f(x)) = g\left(\frac{1}{1+x^2}\right) = e^{\frac{1}{1+x^2}}.$$

We can also apply g first and compose f with g, to get

$$f \circ g(x) = f(g(x)) = f(e^x) = \frac{1}{1+(e^x)^2}.$$

As we can see, $g \circ f \neq f \circ g$, which means that when we compose two functions, *the order of composing matters*.

Example 5.2.3 Let f and g be the following two functions:

$$\begin{cases} f\colon \mathbb{N} \times \mathbb{N} \to \mathbb{Z} \\ f(m,n) = m - n \end{cases} \qquad \qquad \begin{cases} g\colon \mathbb{Z} \to \mathbb{R} \\ g(k) = \sqrt{|k|}. \end{cases}$$

We can compose g with f, as the codomain of f is the domain of g. We get the function $g \circ f\colon \mathbb{N} \times \mathbb{N} \to \mathbb{R}$, given by

$$g \circ f(m,n) = g(f(m,n)) = g(m-n) = \sqrt{|m-n|}.$$

Note that the composition $f \circ g$ is *undefined*, as the codomain of g, which is $\mathbb{R}$, is different than the domain of f, which is $\mathbb{N} \times \mathbb{N}$.

Exercise 5.2.4

Consider the function $f\colon \mathbb{R} \times \mathbb{R} \to \mathbb{R} \times \mathbb{R}$ given by

$$f(x,y) = (x+2y, xy).$$

Find an expression for the composition $f \circ f$.

What is the relation between compositions, surjectivity and injectivity? Does injectivity of functions imply the injectivity of their composition? What if a composition is known to be surjective? Does that mean that the functions themselves must be surjective? These sort of questions arise frequently in mathematical proofs. Our next example deals with one such question.

Example 5.2.5 Let $f\colon A \to B$ and $g\colon B \to C$ be two functions. Prove that if $g \circ f$ is injective, then so is f.

Solution

Note that we are given that $g \circ f$ is one-to-one. Our task is to prove that the inner function, f, is one-to-one as well. We do so by showing that $f(x_1) = f(x_2)$ implies $x_1 = x_2$ for all $x_1, x_2 \in A$.

If $f(x_1) = f(x_2)$, then, by applying the function g on both sides of this equality, we get

$$g(f(x_1)) = g(f(x_2)) \qquad \text{or} \qquad g \circ f(x_1) = g \circ f(x_2).$$

However, $g \circ f$ is known to be injective, and hence we conclude, from $g \circ f(x_1) = g \circ f(x_2)$, that $x_1 = x_2$. This completes the proof of injectivity of f, as needed.

The last example raises the following question.

Question: If the composition $g \circ f$ is injective, must also g be injective?

The answer can be either *yes* or *no*. If g must be injective, we could probably prove it using an argument similar to the one we used for showing that f is injective. Otherwise, we should look for a counterexample showing that g need not be injective.

Indeed, it is not hard to construct functions f and g, such that $g \circ f$ is injective, while g is not. Figure 5.7 describes such a counterexample. Can you see that $g \circ f$ is injective, while g is not? We therefore conclude that g *need not* be injective.

We end this section with a proposition stating that injectivity, surjectivity and bijectivity are properties which are preserved under composition. The last part of the proposition gives a formula for computing the inverse of a composed function.

The proposition is important for two reasons. First, once proved, it can be used in other proofs and arguments. Second, proving such propositions is an excellent opportunity to practice the definitions of composition, injectivity and surjectivity, and to build logical arguments.

Proposition 5.2.6 *1. The composition of two injections is an injection.*

2. *The composition of two surjections is a surjection.*
3. *The composition of two bijections is a bijection.*
4. *If $f: A \to B$ and $g: B \to C$ are two bijections, then $(g \circ f)^{-1} = f^{-1} \circ g^{-1}$. In other words, the inverse of a composition of two bijections, is the composition of their inverses, in reverse order.*

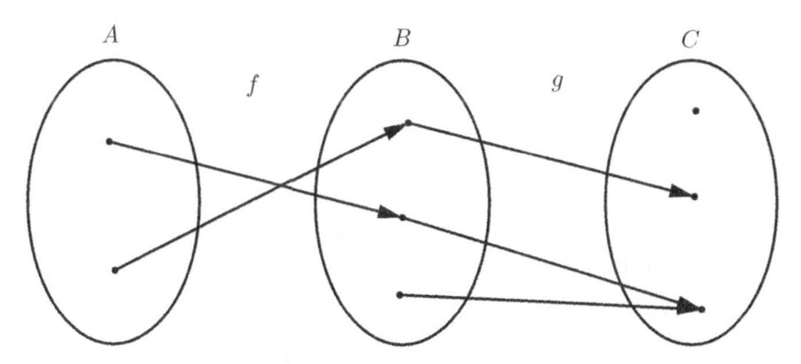

Figure 5.7 $g \circ f$ is injective, while g is not.

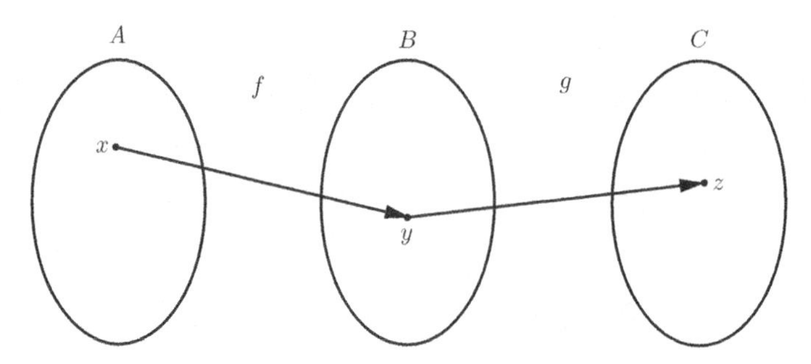

Figure 5.8 Every $z \in C$ is the image of some $x \in A$ under $g \circ f$.

Exercise 5.2.7
Prove Part 1 of Proposition 5.2.6.

We now prove the remaining parts of Proposition 5.2.6.

Proof 2. Let $f: A \to B$ and $g: B \to C$ be two surjections. Our task is to prove that $g \circ f: A \to C$ is also surjective (see Figure 5.8).

Let $z \in C$ be an arbitrary element. As g is surjective, $z = g(y)$ for some $y \in B$. Similarly, f is surjective, and so $y = f(x)$ for some $x \in A$. Over all, we have

$$z = g(y) = g(f(x)) = g \circ f(x),$$

which shows that $g \circ f$ is surjective (every $z \in C$ is the image of some $x \in A$ under $g \circ f$).

3. This part follows right away from Parts 1 and 2, as bijections are functions which are both injective and surjective.
4. Note that both functions $(g \circ f)^{-1}$ and $f^{-1} \circ g^{-1}$ are functions from C to A. Our task is to show that they are equal to each other, i.e., we need to prove that $(g \circ f)^{-1}(z) = f^{-1} \circ g^{-1}(z)$ for every $z \in C$.

Fix $z \in C$, and denote $y = g^{-1}(z)$ and $x = f^{-1}(y)$. Therefore,

$$f^{-1} \circ g^{-1}(z) = f^{-1}(g^{-1}(z)) = f^{-1}(y) = x.$$

On the other hand, as $g(y) = z$ and $f(x) = y$, we have

$$g \circ f(x) = g(f(x)) = g(y) = z \qquad \Rightarrow \qquad (g \circ f)^{-1}(z) = x.$$

We have proved that $(g \circ f)^{-1}(z) = f^{-1} \circ g^{-1}(z) = x$, which completes the proof of Part 4. □

5.3 Cardinality

In everyday life, we are often required to compare the number of objects of two sets, and decide whether they are of equal size, or whether perhaps one is larger than the other. For instance, when scheduling classes at university, we need to make sure that the number of students, in a particular class, does not exceed the number of seats available in the classroom. In a parking lot, the attendant needs to ensure that the number of cars entering is not larger than the number of available parking spots.

Comparing the size, that is, the number of elements, of two sets, seems to be a straightforward task. Nevertheless, in order to fully understand the notion of cardinality, it is crucial that we take a closer look at the process of comparing the number of objects of two sets.

Question: How do we compare the sizes of two finite sets A and B?

The quick answer is, of course, *by counting*. All we have to do is count the number of elements in A and in B, and compare the two resulting numbers. If the numbers are equal, the sets have the same size. Otherwise, one set is larger than the other.

There is, however, another method for comparing sets, which is more important and useful in mathematics.

Imagine, for instance, that in a particular university classroom, every student occupies exactly one seat. If some of the seats in the room are unoccupied, we can immediately conclude that there are more seats than students in the room. There is no need to do any counting at all!

Here is another example. In a parent and tot swimming class in a community center, parents are required to supervise and hold their children throughout the lesson. A lifeguard, watching the class, notices that each parent is holding exactly one child, and that each child is held by exactly one parent. The lifeguard concludes right away, that the number of children equals the number of parents.

Again, no counting is needed. The participants in class form pairs, each of which consists of exactly one parent and one child. This complete matching implies that there is an equal number of parents and children.

The matching method is interesting, for several reasons. Anthropologists have found that in some primitive cultures, where the notions of numbers and counting are not developed, people often use matching. That is, counting is done against objects, such as maize kernels or against body parts, such as head, arms, fingers, etc. It is believed that the matching method preceded the use of abstract numbers for counting.

Moreover, it is also known that children use the matching method before they learn how to count. For instance, a young child can help set up the table for dinner, by placing one spoon next to each plate, and then realize that he is short of spoons, or has too many. This task does not require the ability to count and use numbers.

Our interest in the matching method is for a different reason though. In mathematics, we often need to deal with *infinite sets* such as intervals on the number line and regions in the plane. However, for infinite sets, the counting method cannot be applied. Which set is larger, the integers $\mathbb{Z}$ or the closed interval $[0, 1]$? Both sets are infinite, and counting the number of elements in each is not possible. Fortunately, the matching method can be used, and that is how we proceed.

Remark. In fact, even the counting method can be seen as matching. Counting objects in a set is nothing but pairing them with the natural numbers: we assign the number 1 to an element, the number 2 to another, and so on. This observation leads to the following characterization of finite sets:

> A set A is *finite* if either it is empty, or there exists a bijection between A and the set $\{1, 2, 3, \ldots, n\}$ for some $n \in \mathbb{N}$.

We are now ready to present our main definition.

Definition 5.3.1 Two sets A and B are said to *have the same cardinality*, if there is a bijection between them.

Remarks.

- *Cardinality* is the accurate and more general term used in mathematics for *the number of elements* in a set. It is more precise than *size*, which may refer to length or area, and not necessarily to the number of objects.
- By *a bijection between them*, we mean either a bijection from A to B, or a bijection from B to A. In fact, if there is a bijection going from A to B, then its inverse is a bijection from B to A.

As you will soon realize, it is extremely important to work closely with the definition of cardinality when facing the problem of comparing two sets, and especially when dealing with infinite sets. We have very little experience in working with infinite sets in our everyday life, and relying on our intuition can be risky.

Example 5.3.2 The sets

$$A = \{\text{Sunday, Monday, Tuesday, Wednesday, Thursday, Friday, Saturday}\}$$

and

$$B = \{1, 2, 3, 4, 5, 6, 7\}$$

have the same cardinality, as they both contain seven elements. More formally, we can easily construct a bijection $f\colon A \to B$ by defining

$$f(\text{Sunday}) = 1, \ f(\text{Monday}) = 2, \ \ldots, \ f(\text{Saturday}) = 7,$$

and so by Definition 5.3.1, the two sets *have the same cardinality.*

Example 5.3.3 The sets $A = \{x, y, z\}$ and $B = \{\varnothing\}$ do not have the same cardinality. This is clear if the counting method is used: A has three elements, and B has only one. However, we should develop the habit of relying on Definition 5.3.1 in our arguments, and not on counting.

We claim that there is no way of forming a bijection between A and B, as every function $f\colon A \to B$ must assign the element $\varnothing$ to each element in A, and so $f(x) = f(y) = f(z) = \varnothing$. This shows that there are no *injective* functions from A to B, and hence these sets *do not have the same cardinality.*

Example 5.3.4 Consider the sets $A = \{1, 2, \ldots, 100\}$ and $B = \mathbb{R}$ (all real numbers). It is quite clear that A and B do not have the same cardinality, as A is finite and B is infinite. Can we use Definition 5.3.1 to confirm our intuition? If $f\colon A \to B$ is a function, then its image, $f(A)$, consists of at most 100 elements. As $\mathbb{R}$ contains more than 100 numbers, f cannot be surjective. Therefore, there is no bijection between A and B, and the two sets *do not have the same cardinality.*

Example 5.3.5 Let $A = \mathbb{N} = \{1, 2, 3, \ldots\}$ and $B = \{-1, -2, -3, \ldots\}$. Note that both sets are infinite, and so counting the number of elements is not an option. However, it is quite easy to pair elements of one set with elements from the other set. Define a function $f\colon A \to B$ by $f(n) = -n$.

This function is a bijection from A to B, which implies that the two sets *have the same cardinality.*

Example 5.3.6 This example, though similar to the previous one, may surprise you at first. Let A be the set of natural numbers, and B the set of *even* positive integers:

$$A = \{1, 2, 3, \ldots\} \qquad \text{and} \qquad B = \{2, 4, 6, \ldots\}.$$

Do these sets have the same cardinality? Well, intuitively, it may seem that B is "smaller" than A, as it is a proper subset of A. This intuition, however, is based on our everyday experience with finite sets, and so we must be very careful in applying it to infinite sets. Referring to Definition 5.3.1, we should ask ourselves: *Can we form a bijection from* $\boldsymbol{A}$ *to* $\boldsymbol{B}$ *(or from* $\boldsymbol{B}$ *to* $\boldsymbol{A}$*)*? And the answer is yes, we can. If we define

$$g\colon A \to B \qquad \text{by} \qquad g(n) = 2n,$$

then g is a bijection from A to B, which proves that A and B *do have the same cardinality.*

In other words, *there are as many natural numbers as even positive integers*! Again, this may be, at first, quite counter-intuitive. With time and practice, your intuition will adjust to align with the formal definition of having the same cardinality. Meanwhile, consider your intuitions about counting with suspicion, as they come from your interactions with finite collections of objects. Work closely with Definition 5.3.1 and you will be safe.

Exercise 5.3.7
Consider the following sets:

$$A = \{1,2\}, \quad B = \{\{1,2\}\} \quad \text{and} \quad C = \{\{1\},\{2\}\}.$$

Do A and B have the same cardinality? How about A and C? Justify your answers.

We continue with a few more examples.

Example 5.3.8 This example is more geometric in nature. Consider the two line segments AB and CD in Figure 5.9, and let us think of them as sets of points in the plane. Do AB and CD, as sets of points, have the same cardinality? Clearly, AB is shorter than CD. Does that imply that the line segments do not have the same cardinality? One way to construct a function $f: AB \to CD$ is to assign, to each point in AB the point right below it on CD (see Figure 5.10). The function f is a one-to-one function, but not onto (why?). Does that mean that AB and CD do not have the same cardinality? No, it does not. Read carefully Definition 5.3.1 one more time. Two sets have the same cardinality if

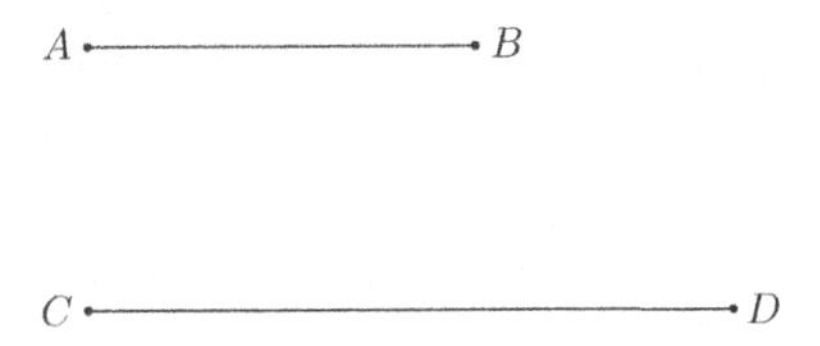

Figure 5.9 Two line segments of different length.

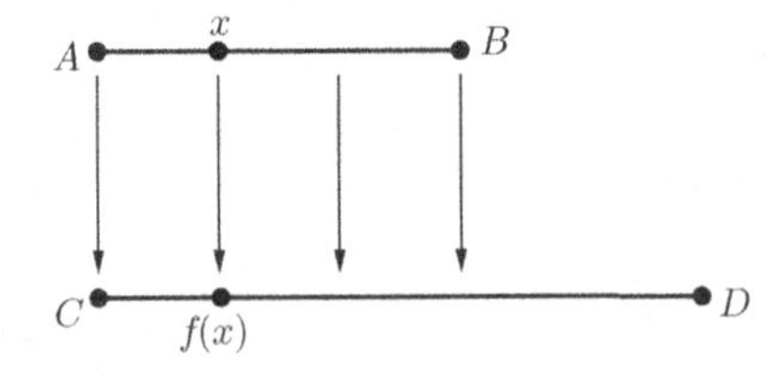

Figure 5.10 An injection from AB to CD.

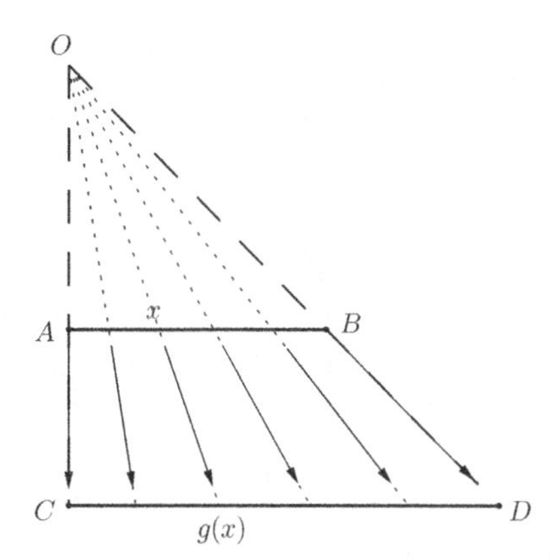

Figure 5.11 A bijection from AB to CD.

there exists a bijection between them. The fact that we were able to construct a function which is not bijective does not imply that a bijection does not exist.

In fact, there is a way to form a bijection between the two line segments! First, we draw line segments AC and BD, then extend them until they meet at some point O, as in Figure 5.11. Then, for every point x on AB we assign the point $g(x)$ on CD, obtained by extending the line segment from O to x until it hits CD. This way, we obtain a bijection from AB to CD, which shows that *the two segments have the same cardinality*.

This example shows that the notion of cardinality is different from the notion of length (which we do not define precisely in this book). Two segments may have the same cardinality (as sets of points), even though they are of different length.

Example 5.3.9 Do the open interval $(0, 1)$ and the set of real numbers $\mathbb{R}$ have the same cardinality? Namely, can we form a bijection between the two sets?

Here, our knowledge from calculus and pre-calculus may be handy. A bijection from $(0, 1)$ to $\mathbb{R}$ will need to "go to plus and minus infinity," and this can be achieved by using vertical asymptotes. For instance, the graph in Figure 5.12 represents a bijection from $(0, 1)$ to $\mathbb{R}$. We can also provide an explicit formula for such a bijection. Functions such as $f(x) = \frac{1}{1-x} - \frac{1}{x}$ and $g(x) = \tan[\pi(x - 0.5)]$ have a graph that is similar, when restricted to the interval $(0, 1)$, to the one in the diagram. Proving that f and g are indeed bijections can be done, for instance, by using calculus and properties of the tangent function (which we omit).

In conclusion, the open interval $(0, 1)$ and the whole real line *have the same cardinality*. That is, they have the "same number of elements."

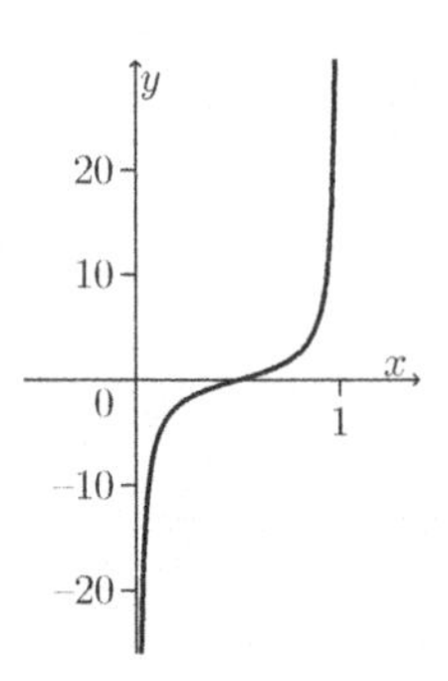

Figure 5.12 A bijection from (0, 1) to $\mathbb{R}$.

5.4 Cardinality Theorems

In this section, we prove a few fundamental theorems involving cardinality. First, we prove that the rational and natural numbers are sets of the same cardinality. Then, we show that the real and the natural numbers *do not* have the same cardinality. Finally, we discuss Cantor's Theorem on the cardinality of power sets.

As you will see, the proofs of these theorems are more sophisticated than previous proofs discussed in this book, and involve clever ideas and unusual constructions. Moreover, the statements themselves are deep, often counter-intuitive, and with far-reaching consequences. The power of mathematical proofs will be fully felt and demonstrated in this section.

Theorem 5.4.1 *The set of natural numbers,* $\mathbb{N}$*, and the set of rational numbers,* $\mathbb{Q}$*, have the same cardinality.*

In other words, there are as many rational numbers as natural numbers. Geometrically, the natural numbers can be thought of as an infinite sequence of dots on the number line, with consecutive dots being one unit apart. On the other hand, the rational numbers are occupying every region on the number line (there are infinitely many rationals in every interval). We say that $\mathbb{Q}$ is *dense* in $\mathbb{R}$.

Looking at Figure 5.13, it may by quite difficult to believe that the two sets of numbers are of the same size. You might wonder whether maybe every two infinite sets have the same cardinality? This question will be answered by the next theorem. Nevertheless, leaving intuition aside, if indeed $\mathbb{N}$ and $\mathbb{Q}$ have the same cardinality, how would one go about proving it? Namely, how can we form a bijection between the rational and the natural numbers. There is no "first rational number" which can be assigned to the number 1 (and then "a second" for 2, etc.). The proof below uses a surprising and clever strategy to construct a bijection. Here it is.

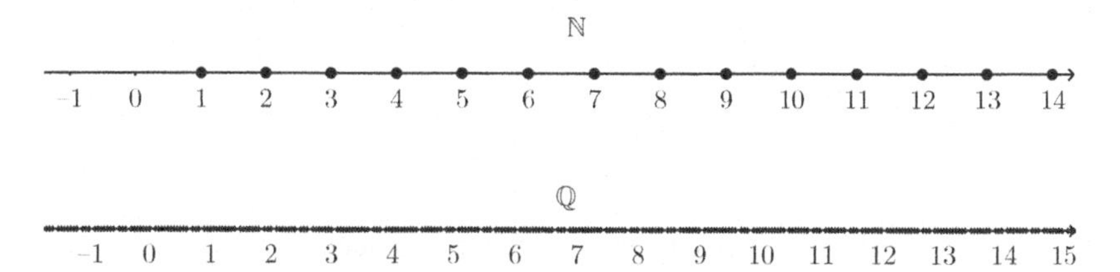

Figure 5.13 The natural numbers and the rational numbers on the number line.

Proof The proof involves several steps.

Step 1: Defining A_k. We define, for each $k \in \mathbb{N}$, the set

$$A_k = \left\{\frac{a}{b} : a, b \in \mathbb{N} \quad \text{and} \quad a + b = k\right\}.$$

For instance, if $k = 4$ then A_k is the set of all fractions with positive numerator and denominator, that add up to 4:

$$A_4 = \left\{\frac{1}{3}, \frac{2}{2}, \frac{3}{1}\right\}.$$

If $k = 7$, we get the set

$$A_7 = \left\{\frac{1}{6}, \frac{2}{5}, \frac{3}{4}, \frac{4}{3}, \frac{5}{2}, \frac{6}{1}\right\},$$

and so on (what is A_1?). Looking carefully at a few more special cases, we conclude that each A_k contains exactly $k-1$ elements, with numerators going from 1 to $k-1$, and corresponding denominators, going from $k-1$ to 1:

$$A_k = \left\{\frac{1}{k-1}, \frac{2}{k-2}, \frac{3}{k-3}, \ldots, \frac{k-1}{1}\right\}.$$

Step 2: Forming a sequence. We arrange the elements of the sets A_k in an infinite sequence, as follows:

$$\underbrace{\frac{1}{1}}_{A_2}, \underbrace{\frac{1}{2}, \frac{2}{1}}_{A_3}, \underbrace{\frac{1}{3}, \frac{2}{2}, \frac{3}{1}}_{A_4}, \underbrace{\frac{1}{4}, \frac{2}{3}, \frac{3}{2}, \frac{4}{1}}_{A_5}, \underbrace{\frac{1}{5}, \frac{2}{4}, \frac{3}{3}, \frac{4}{2}, \frac{5}{1}}_{A_6}, \underbrace{\frac{1}{6}, \frac{2}{5}, \frac{3}{4}, \frac{4}{3}, \frac{5}{2}, \frac{6}{1}}_{A_7}, \ldots.$$

Step 3: Removing repeated terms. Note that this sequence contains many repeated elements. For instance, $\frac{1}{2}$, from A_3, is repeated as $\frac{2}{4}$ in A_6, and $\frac{1}{1}$, which is simply the number 1, is repeated many times, as $\frac{2}{2}, \frac{3}{3}, \frac{4}{4}$, etc. We remove, from our sequence, every rational number that has already appeared:

$$\frac{1}{1}, \frac{1}{2}, \frac{2}{1}, \frac{1}{3}, \cancel{\frac{2}{2}}, \frac{3}{1}, \frac{1}{4}, \frac{2}{3}, \frac{3}{2}, \frac{4}{1}, \frac{1}{5}, \cancel{\frac{2}{4}}, \cancel{\frac{3}{3}}, \cancel{\frac{4}{2}}, \frac{5}{1}, \frac{1}{6}, \frac{2}{5}, \frac{3}{4}, \frac{4}{3}, \frac{5}{2},$$

$$\frac{6}{1}, \frac{1}{7}, \cancel{\frac{2}{6}}, \frac{3}{5}, \cancel{\frac{4}{4}}, \frac{5}{3}, \cancel{\frac{6}{2}}, \frac{7}{1} \cdots$$

We denote the elements of the resulting sequence by $a_1, a_2, a_3, \ldots$. That is, $a_1 = \frac{1}{1}, a_4 = \frac{1}{3}, a_6 = \frac{1}{4}, a_{11} = \frac{5}{1}$, etc.
This sequence has *two* important features.

- There are no repeated numbers.
- Every *positive* rational number appears as an element of the sequence. For instance, $\frac{17}{35}$ appears in A_{52}, as $17 + 35 = 52$. In general, the rational number $\frac{a}{b}$, with $a, b \in \mathbb{N}$, appears in A_{a+b}.

Step 4: Adding the remaining rational numbers. We now extend the sequence (a_n) to include all rational numbers. We do so by inserting 0 as its first element, and negative rational numbers after each positive element, as follows:

$$0, a_1, -a_1, a_2, -a_2, a_3, -a_3, a_4, -a_4, a_5, -a_5, a_6, -a_6, \ldots.$$

Or, more explicitly,

$$0, \frac{1}{1}, -\frac{1}{1}, \frac{1}{2}, -\frac{1}{2}, \frac{2}{1}, -\frac{2}{1}, \frac{1}{3}, -\frac{1}{3}, \frac{3}{1}, -\frac{3}{1}, \ldots.$$

This modified sequence contains *all the rational numbers*: positive, negative and zero. We can now finally construct the desired bijection.

Step 5: Forming the bijection. As we now have a sequence containing *all the rational numbers*, and *without repetitions*, we can define a function $f : \mathbb{N} \to \mathbb{Q}$ according to the following diagram:

$$\begin{array}{cccccccccccc} \mathbb{N} & 1 & 2 & 3 & 4 & 5 & 6 & 7 & 8 & 9 & \cdots \\ & \downarrow & \downarrow & \downarrow & \downarrow & \downarrow & \downarrow & \downarrow & \downarrow & \downarrow & \\ \mathbb{Q} & 0 & a_1 & -a_1 & a_2 & -a_2 & a_3 & -a_3 & a_4 & -a_4 & \cdots \end{array}$$

The function f sends 1 to 0, the even natural numbers $2, 4, 6, \ldots$ to the positive rational numbers $a_1, a_2, a_3, \ldots$, and the odd natural numbers $3, 5, 7, \ldots$ to the negative rational numbers $-a_1, -a_2, -a_3, \ldots$. All the rational numbers are covered, which implies that f is surjective. f is also injective, as our sequence $a_1, a_2, a_3, \ldots$ from Step 3 has no repetition.

In conclusion, we have constructed a bijection from $\mathbb{N}$ to $\mathbb{Q}$, and so these two sets have the same cardinality. □

The approach used in the proof of the theorem can be generalized. Whenever elements of a set can be arranged in an infinite sequence, without repetitions, it has the same cardinality as $\mathbb{N}$. Such sets are said to be *countable*.

Definition 5.4.2 A set that has the same cardinality as $\mathbb{N}$ is called a *countable set*. An infinite set that is *not* countable, is called an *uncountable set*.

Remarks.

1. In some books, the term "countable" refers to sets that either are finite or have the same cardinality as $\mathbb{N}$. In this textbook, we follow the above definition, and so countable sets are infinite sets.
2. It follows that every set falls into *exactly one* of the following categories:
   - a finite set,
   - a countable set,
   - an uncountable set.

Example 5.4.3 The following sets are countable.

- $\mathbb{N}$ itself (why?).
- The rational numbers $\mathbb{Q}$ (this is Theorem 5.4.1).
- The negative integers (see Example 5.3.5 from page 141).
- The even positive integers (see Example 5.3.6 from page 141).
- The set of all integers $\mathbb{Z}$, as we can arrange them in an infinite sequence: $0, 1, -1, 2, -2, 3, -3, \ldots$.
- $\mathbb{N} \times \mathbb{N}$ (see Problem 5.32).

Do there exist infinite sets which are *not* countable? In general, is it possible for two infinite sets *not to have the same cardinality*? Are some infinite sets "larger" than others?

The following theorem answers these questions.

Theorem 5.4.4 *The set of real numbers $\mathbb{R}$ is uncountable.*

In other words, there is no way to arrange all the real numbers in one infinite sequence. How would one prove that "there is no way of doing something"? One common approach is to use the method of *proof by contradiction*. Namely, we assume that there is a bijection between $\mathbb{N}$ and $\mathbb{R}$, and show that this assumption leads to a contradiction. Again, the proof of the theorem is far from being straightforward, and involves an extremely clever construction, known as *Cantor's Diagonalization Argument*.

Proof In Example 5.3.9 (page 143), we proved that $\mathbb{R}$ and the interval $(0, 1)$ have the same cardinality. Hence, it is enough to prove that $(0, 1)$ is uncountable.

Assume, by contradiction, that $(0, 1)$ is countable, and suppose that $f: \mathbb{N} \to (0, 1)$ is a bijection. Then, for each $n \in \mathbb{N}$, $f(n)$ is some real number between 0 and 1, whose decimal expansion must start as 0.___.... If the decimal expansion happens to be finite, such as 0.25, zeros can be added to make it infinite: 0.250000....

We can therefore picture the sequence $f(1), f(2), f(3), \ldots$ as an infinite array of digits:

$$
\begin{aligned}
f(1) &= 0.\_\_\_\_\_\_\_\_\_\_\_\_\_\_\_\_\_\_ \quad \cdots \\
f(2) &= 0.\_\_\_\_\_\_\_\_\_\_\_\_\_\_\_\_\_\_ \quad \cdots \\
f(3) &= 0.\_\_\_\_\_\_\_\_\_\_\_\_\_\_\_\_\_\_ \quad \cdots \\
f(4) &= 0.\_\_\_\_\_\_\_\_\_\_\_\_\_\_\_\_\_\_ \quad \cdots \\
f(5) &= 0.\_\_\_\_\_\_\_\_\_\_\_\_\_\_\_\_\_\_ \quad \cdots \\
f(6) &= 0.\_\_\_\_\_\_\_\_\_\_\_\_\_\_\_\_\_\_ \quad \cdots \\
&\vdots
\end{aligned}
$$

To prove that f is not a surjection, we now create a real number, in the interval $(0, 1)$, that is not the image of any $n \in \mathbb{N}$ under f.

Define a number $x \in (0, 1)$, with decimal expansion $0.a_1a_2a_3a_4a_5 \ldots$, as follows:

$$
a_k = \begin{cases} 5 & \text{if the } k\text{th digit of } f(k) \text{ is } 4 \\ 4 & \text{otherwise.} \end{cases}
$$

For instance, if the first few images of f are

$$
\begin{aligned}
f(1) &= 0.\mathbf{3}9283434832\ldots \\
f(2) &= 0.1\mathbf{4}563203947\ldots \\
f(3) &= 0.59\mathbf{5}52551920\ldots \\
f(4) &= 0.120\mathbf{4}9271927\ldots \\
f(5) &= 0.1052\mathbf{2}961203\ldots \\
f(6) &= 0.81024\mathbf{0}14459\ldots \\
&\vdots
\end{aligned}
$$

then the number x begins as 0.454544....

In other words, we look at the digits on the *diagonal* of the above array, and pick the kth digit of x to be 5, if the corresponding digit on the diagonal is 4, and 4 in all other cases.

The number x is defined in such a way that its kth digit is *different* than the kth digit of $f(k)$, and therefore $x \neq f(k)$ for all $k \in \mathbb{N}$. Consequently, x is a real number in $(0, 1)$ that does not appear in the list $f(1), f(2), f(3), \ldots$, and so f is *not surjective*. This contradicts the fact that f is a bijection, and hence $\mathbb{R}$ must be uncountable. This concludes the proof of the theorem. □

Remarks.

- There is a little technical issue in the representation of the images of f. Some numbers have two different decimal representations, such as $0.5000\ldots = 0.4999\ldots$, and so we have to make a choice. We follow the convention of choosing the one ending with zeros rather than the one ending with nines.
- In the proof we used the digits 4 and 5 to construct x. There is nothing special about 4 and 5. Any two digits can be used.

Our last theorem is a general cardinality statement, involving *power sets*.

Definition 5.4.5 Given a set X, we define its *power set*, $P(X)$, to be the set of all subsets of X:

$$P(X) = \{A : A \subseteq X\}.$$

For instance, if $X = \{a, b\}$, then $P(X) = \{\varnothing, \{a\}, \{b\}, \{a, b\}\}$. In Chapter 4, we proved that a set with n elements has 2^n subsets (see Claim 4.1.12 on page 105). However, power sets also exist for infinite sets. For example, the power set of the natural numbers, $P(\mathbb{N})$, is the collection of all subsets of $\mathbb{N}$. Sets such as $\{1, 10, 22, 114\}$, $\{2, 4, 6, 8, \ldots\}$ and $\{1, 10, 100, 1000, 10000, \ldots\}$ are *elements* of $P(\mathbb{N})$, while the sets $\{-1, 0, 1, 2, 3\}$ and $\{1, \frac{1}{2}, \frac{1}{3}, \frac{1}{4}, \ldots\}$ are not. We write:

$$\begin{aligned}
\{1, 10, 22, 114\} &\in P(\mathbb{N}) \\
\{2, 4, 6, 8, \ldots\} &\in P(\mathbb{N}) \\
\{1, 10, 100, 1000, 10000, \ldots\} &\in P(\mathbb{N}) \\
\{-1, 0, 1, 2, 3\} &\notin P(\mathbb{N}) \\
\{1, \tfrac{1}{2}, \tfrac{1}{3}, \tfrac{1}{4}, \ldots\} &\notin P(\mathbb{N}).
\end{aligned}$$

We can now state Cantor's famous theorem on power sets.

Theorem 5.4.6 (Cantor's Theorem) *Let X be a set. Then the sets X and $P(X)$ do not have the same cardinality.*

This theorem has a straightforward proof if we assume that X is finite. If X has n elements, where n is either 0 or a natural number, then, according to

Claim 4.1.12, $P(X)$ has 2^n elements. As $n < 2^n$ for every $n \in \mathbb{N} \cup \{0\}$, which can be shown by induction, the theorem is confirmed. This strategy, however, is not applicable to infinite sets. The proof, done by contradiction, uses a clever construction and pure logic.

Proof We assume, by contradiction, that X and $P(X)$ do have the same cardinality, and so there is a bijection $f\colon X \to P(X)$. We define the following subset of X:

$$D = \{a \in X : a \notin f(a)\} \subseteq X.$$

That is, D is the set of all elements a of X, which are *not* elements of their image $f(a)$.

As D is a subset of X, it is *one of the elements* of the power set $P(X)$ (i.e., $D \in P(X)$). However, since f is a bijection, D must be the image of some element in X:

$$D = f(y) \qquad \text{for some } y \in X.$$

Now we are getting closer to our contradiction. As y is an element of X, and D is a subset of X, either y is or is not an element of D. We take a close look at each of these options.

Case 1: $y \in D$. According to the definition of D, if $y \in D$, then $y \notin f(y)$. But this implies that $y \notin D$, as $f(y) = D$, which is impossible.

Case 2: $y \notin D$. If $y \notin D$, then $y \notin f(y)$ (again, as $f(y) = D$), which means that y satisfies the requirement $a \notin f(a)$ for being in the set D. We thus conclude that $y \in D$, which is also impossible.

We therefore have a contradiction, as one of the two cases above must occur. We have no choice but to abandon our initial assumption that X and $P(X)$ have the same cardinality, which concludes the proof of the theorem. □

Informally, the power set of a given set X has "more elements" or a "larger cardinality" than X (a notion to be made precise in the next section). Cantor's Theorem implies that for every set, no matter how large, there is an even larger set – its power set. This means, for instance, that in the following sequence, no two sets have the same cardinality (this can be proved using tools from the next section, see Problem 5.51):

$$\mathbb{N}\,,\ P(\mathbb{N})\,,\ P(P(\mathbb{N}))\,,\ P(P(P(\mathbb{N})))\,,\ \ldots.$$

Roughly speaking, we may say that there is "no largest infinity" in the sense of cardinality.

5.5 More Cardinality and the Schröder–Bernstein Theorem

In Section 5.3 we introduced the notion of "having the same cardinality." However, we did not define what it means for one set to have a *larger* or *smaller cardinality* than another set (a notion that does exist, and is widely used, for finite sets). The following definition is a natural extension of Definition 5.3.1.

Definition 5.5.1 Let A and B be two sets. We say the following.

1. A and B have the *same cardinality*, and we write $|A| = |B|$, if there is a *bijection* from A to B.
2. A has cardinality *less than or equal to* the cardinality of B, and we write $|A| \leq |B|$, if there is an *injection* from A to B.
3. A has cardinality *greater than or equal to* the cardinality of B, and we write $|A| \geq |B|$, if there is an *injection* from B to A.

We assign natural meaning to statements such as $|A| < |B|$, $|A| \neq |B|$, etc. For instance, $|A| < |B|$ means that A has cardinality less than, but not the same as, the cardinality of B.

Example 5.5.2 The set $\{-4, -2, 0, 2, 4\}$ has cardinality less than the set $\{1, 2, 3, 4, 5, 6, 7, 8, 9, 10\}$. We write

$$|\{-4, -2, 0, 2, 4\}| \leq |\{1, 2, 3, 4, 5, 6, 7, 8, 9, 10\}|$$

or

$$|\{-4, -2, 0, 2, 4\}| < |\{1, 2, 3, 4, 5, 6, 7, 8, 9, 10\}|.$$

Example 5.5.3 $\mathbb{N}$ and $\mathbb{Q}$ are both countable sets by Theorem 5.4.1, so we write $|\mathbb{N}| = |\mathbb{Q}|$. Similarly, we have $|(0, 1)| = |\mathbb{R}|$.

Example 5.5.4 The function $f \colon \mathbb{Z} \to \mathbb{R}$, that sends every integer to itself, is an injection, and so $|\mathbb{Z}| \leq |\mathbb{R}|$. However, $\mathbb{Z}$ is countable, while $\mathbb{R}$ is not (by Theorem 5.4.4), and thus $|\mathbb{Z}| < |\mathbb{R}|$.

Example 5.5.5 For every set X, the function from X to $P(X)$, sending $a \in X$ to $\{a\} \in P(X)$, is injective. Moreover, X and $P(X)$ never have the same cardinality by Cantor's Theorem. We can summarize these facts by saying that for every set X, $|X| < |P(X)|$.

A word of caution is in place. We have been using symbols such as $\leq$, $=$ and $<$ ever since elementary school, to compare real numbers, and order them: $5 < 19$, $-5 > -7.5$, $\frac{1}{2} = \frac{5}{10}$, etc. Basic properties of these relations became natural, and we use them freely without having any doubts. For instance, if

$a < b$ and $b < c$, then $a < c$ for every $a, b, c \in \mathbb{R}$. Now that we have extended the use of these symbols to comparing cardinalities of both finite and infinite sets, we need to review these "obvious" properties of $<$, $\leq$ and $=$, and check whether they still apply in the context of cardinality. We should ask ourselves the following.

1. If $|A| \leq |B|$ and $|B| \leq |C|$ for sets A, B, C, is it necessarily true that $|A| \leq |C|$?
2. If A and B are two sets, is it necessarily true that either $|A| \leq |B|$ or $|A| \geq |B|$. That is, given two sets, is there always an injection going from one of them to the other?
3. If $|A| \leq |B|$ and $|A| \geq |B|$, can we conclude that $|A| = |B|$?

Fortunately, the answer to all the three questions is affirmative. However, the proofs can be non-trivial. Question 1 is left as an exercise (see Problem 5.49). To fully answer Question 2, a more advanced set-theoretic tool is needed which is beyond the scope of this book. Question 3 is answered by the famous Schröder–Bernstein Theorem.

Theorem 5.5.6 (Schröder–Bernstein Theorem) *Let A and B be two sets. If* $|A| \leq |B|$ *and* $|A| \geq |B|$, *then* $|A| = |B|$.

In other words, if there are injections $f : A \to B$ *and* $g : B \to A$, *then there is a bijection* $h : A \to B$.

The proof of this theorem is outlined in Problem 5.52 , and is quite technical. The main difficulty here is that f and g are *known to be injective*, but *neither of them is necessarily a surjection* (see Figure 5.14). A bijection from A to B needs to be constructed from f and g, and doing so requires some effort. Nevertheless, this theorem provides a powerful tool for proving that two sets have the same cardinality, as illustrated in the following example.

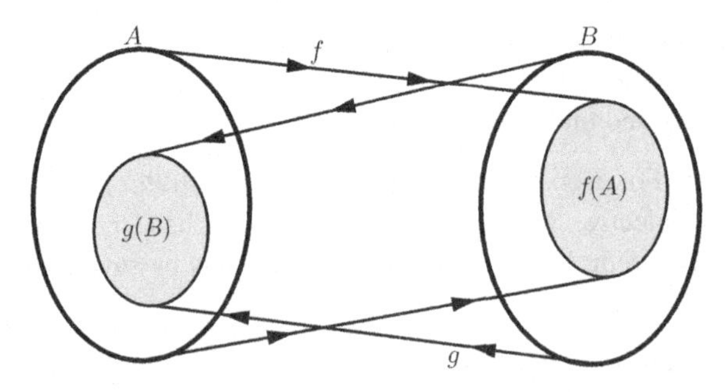

Figure 5.14 $f : A \to B$ and $g : B \to A$ are injections, but not necessarily bijections.

Figure 5.15 Injections between (0, 1) and [0, 1].

Example 5.5.7 $|[0,1]| = |(0,1)|$. Namely, the intervals [0, 1] and (0, 1) have the same cardinality (see Figure 5.15).

Proof Define functions $f\colon (0,1) \to [0,1]$ and $g\colon [0,1] \to (0,1)$ as follows:

$$f(x) = x \quad \text{for} \quad 0 < x < 1, \qquad \text{and} \qquad g(x) = \frac{x}{2} + \frac{1}{4} \quad \text{for} \quad 0 \le x \le 1.$$

The functions f and g are both injective (why?). f is simply the inclusion of the open interval (0, 1) into the closed interval [0, 1]. The function g shrinks the interval [0, 1] by a factor of 2, and translates it a quarter unit to the right (i.e., g sends the interval [0, 1] to [0.25, 0.75]). Since we have exhibited the two required injections, according to the Schröder–Bernstein Theorem, (0, 1) and [0, 1] have the same cardinality. □

5.6 Problems

5.1 Figure 5.16 shows (part of) the graph of a function $f\colon \mathbb{R} \to \mathbb{R}$. The domain and the codomain of f can be restricted to various intervals. In each case, decide whether the given restriction is *an injection, a surjection, a bijection, or neither*. Explain your answer briefly.

a. $f\colon [-3,3] \to [-6,6]$
b. $f\colon [-4,0] \to [0,6]$
c. $f\colon [1,4] \to [-7,2]$
d. $f\colon [-2,2] \to [-7,7]$
e. $f\colon [-4,-1] \to [-2,7]$
f. $f\colon [0,4] \to [-6,0]$

5.2 For each of the following functions, decide whether it is *an injection, a surjection, a bijection, or neither*. Justify your answer.

a. $p\colon \mathbb{N} \times \mathbb{N} \to \mathbb{N}$, $p(a,b) = \frac{ab(b+1)}{2}$.
b. $f\colon [0,\infty) \to \mathbb{R}$, $f(x) = \sqrt{x}$.
c. $g\colon \mathbb{R}^2 \to \mathbb{R}$, $g(x,y) = |x+y|$.
d. $h\colon \mathbb{R} \to \mathbb{R}$, $h(x) = x^3$.
e. $r\colon \mathbb{R} \to \mathbb{R}$, $r(x) = \frac{x}{1+x^2}$.

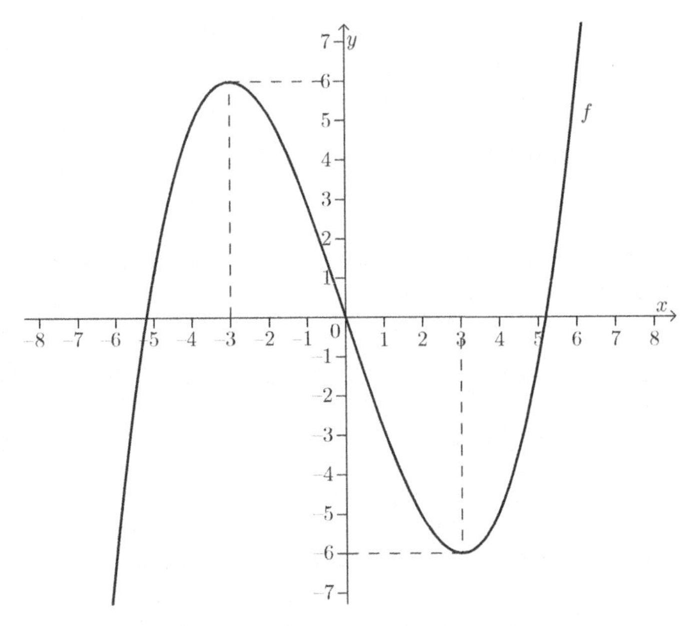

Figure 5.16 The graph of a function $f\colon \mathbb{R} \to \mathbb{R}$.

f. $q\colon \mathbb{Z} \to \mathbb{Z}$, $\quad q(n) = n + 1$.
g. $t\colon \mathbb{N} \times \mathbb{N} \to \mathbb{N}$, $\quad t(a,b) = a \cdot b$.

5.3 For each of the following functions, decide whether it is *an injection, a surjection, a bijection, or neither*. Justify your answer.

a. $p\colon \mathbb{N} \times \mathbb{N} \to \mathbb{N}$, $\quad p(a,b) = \frac{(a+1)b(b+1)}{2}$.
b. $f\colon \mathbb{R} \to \mathbb{R}$, $\quad f(x) = 2x + 1$.
c. $g\colon \mathbb{R}^2 \to \mathbb{R}$, $\quad g(x,y) = x + y$.
d. $h\colon \mathbb{R} \to \mathbb{R}$, $\quad h(x) = \frac{x^2}{1+x^2}$.
e. $r\colon \mathbb{R} \to [0,\infty)$, $\quad r(x) = |x|$.
f. $q\colon \mathbb{N} \to \mathbb{N}$, $\quad q(n) = n + 1$.
g. $t\colon \mathbb{N} \times \mathbb{N} \to \mathbb{N}$, $\quad t(a,b) = a + b$.

5.4 Let $f\colon \mathbb{R} \to \mathbb{R}$ given by $f(x) = \begin{cases} x & \text{if } x \leq 0 \\ \frac{1}{x} & \text{if } x > 0. \end{cases}$

Is f an *injection*? Is it a *surjection*? Explain.

5.5 Which function is one-to-one and *not* onto? Explain.

- $f\colon \mathbb{Z} \to \mathbb{Z}$, $f(x) = x^3$
- $g\colon \mathbb{R} \to \mathbb{R}$, $g(x) = x^3$
- $h\colon \mathbb{R} \to \mathbb{R}$, $h(x) = x^2$
- $r\colon \mathbb{R} \times \mathbb{R} \to \mathbb{R}$, $r(x,y) = x^2 + y$

5.6 Which function is an *injection*? Explain.

- $f\colon \mathbb{R} \to \mathbb{R}$, $f(x) = x^2$
- $g\colon \mathbb{Q} \to \mathbb{Q}$, $g(x) = x^2$
- $h\colon \mathbb{Z} \to \mathbb{Z}$, $h(x) = x^2$
- $p\colon \mathbb{N} \to \mathbb{N}$, $p(x) = x^2$

5.7 Which function is a *bijection*? Explain.

- $f\colon \mathbb{R} \to \mathbb{R}$, $f(x) = x^6$
- $g\colon [0,\infty) \to [0,\infty)$, $g(x) = x^6$
- $h\colon \mathbb{Z} \to \mathbb{Z}$, $h(x) = x^6$
- $p\colon \mathbb{Q} \to [0,\infty)$, $p(x) = x^6$

5.8 Let $f\colon \mathbb{R} \to \mathbb{R}$ be an arbitrary function. Each of the following statements, written using the logic symbols, describes a property of f (e.g., injectivity, monotonicity, boundedness, etc.). Identify the property described by each statement.

a. $(\forall y \in \mathbb{R})(\exists x \in \mathbb{R})(f(x) = y)$
b. $(\exists M \in \mathbb{R})(\forall x \in \mathbb{R})(|f(x)| \leq M)$
c. $(\forall x_1 \in \mathbb{R})(\forall x_2 \in \mathbb{R})[(x_1 \neq x_2) \Rightarrow (f(x_1) \neq f(x_2))]$

5.9 Prove that the function $f\colon \mathbb{R} \to \mathbb{R},\ f(x) = x \cdot |x|$ is a bijection.

5.10 Consider the function $f\colon \mathbb{N} \to \mathbb{N},\ f(n) = n + (-1)^{n+1}$.

a. Compute $f(1), f(2), f(3)$ and $f(4)$.
b. Is f an *injection*?
c. Is f a *surjection*?

5.11 Consider the sets

$$A = \{1, 2, 3, \ldots, 365\},\quad B = \{\text{Jan, Feb, } \ldots, \text{ Dec}\},\quad C = \{1, 2, 3, \ldots, 31\}$$

and the function $f{:}A \to B \times C$, defined by

$$f(x) = (\text{month for day } x, \text{ day of the month for day } x),$$

in a regular year (not leap year). For example, $f(365) = (\text{Dec}, 31)$, since the date of the last day of the year is December 31st.

a. What are $f(1), f(32)$, and $f(359)$?
b. Is f *injective*? Explain.
c. Is f *surjective*? Explain.

5.12 Let $f\colon [0, 2] \to [5, 6]$ be a function.

a. If the restrictions of f to $[0, 1]$ and to $[1, 2]$ are injective functions, must f be an injection? Explain.
b. If the restrictions of f to $[0, 1]$ and to $[1, 2]$ are surjective functions, must f be a surjection? Explain.

5.13 Let $f{:}\mathbb{R} \to \mathbb{R}$ be a function, such that $|f(x) - f(y)| \geq 5|x - y|$ for all $x, y \in \mathbb{R}$. Show that f is injective.

5.14 **(Harder!)** Show that the function $f\colon \mathbb{N} \to \mathbb{R}$, $f(n) = \sin n$ is one-to-one.
(Hint: You may use, without proof, the fact that π is an irrational number.)

5.15 Let $f\colon A \to B$ be an arbitrary function.

a. Prove that if f is a bijection (and hence invertible), then $f^{-1}(f(x)) = x$ for all $x \in A$, and $f(f^{-1}(x)) = x$ for all $x \in B$.
b. Conversely, show that if there is a function $g\colon B \to A$, satisfying $g(f(x)) = x$ for all $x \in A$, and $f(g(x)) = x$ for all $x \in B$, then f is a bijection, and $f^{-1} = g$.

5.16 a. Is it true that $\sin^{-1}(\sin x) = x$ for all $x \in \mathbb{R}$? Explain.
b. The *inverse tangent function* is defined as the inverse of the function

$$f\colon \left(-\frac{\pi}{2}, \frac{\pi}{2}\right) \to \mathbb{R}, \qquad f(x) = \tan x.$$

We write $f^{-1}(x) = \tan^{-1} x$. Is it true that $\tan(\tan^{-1} x) = x$ for all $x \in \mathbb{R}$? Explain.
c. What is the inverse function of $g\colon (-\infty, 0] \to [0, \infty)$, $g(x) = x^2$? Explain.

5.17 Consider the functions $\begin{cases} f\colon \mathbb{N} \times \mathbb{N} \to \mathbb{R} \\ f(m,n) = \frac{m+1}{n+1} \end{cases}$ and $\begin{cases} g\colon \mathbb{R} \to \mathbb{R} \\ g(x) = \sqrt{x^2+5}. \end{cases}$
Compute $g \circ f(9,4)$ and $f(g(2), g(\sqrt{20}))$.

5.18 Consider the following functions:

$$\begin{cases} f\colon \mathbb{Z} \to \mathbb{Z} \times \mathbb{Z} \\ f(m) = (2m, m-1) \end{cases} \qquad \begin{cases} g\colon \mathbb{Z} \times \mathbb{Z} \to \mathbb{Z} \\ g(m,n) = |m \cdot n|. \end{cases}$$

a. Is f *injective*? Explain.
b. Is g *surjective*? Explain.
c. State the *domain* and the *codomain* of $g \circ f$, and write a *formula* for this composition.
d. State the *domain* and the *codomain* of $f \circ g$, and write a *formula* for this composition.

5.19 Let $g\colon \mathbb{Z} \to \mathbb{Z}$ be given by $g(x) = 2x + 1$.

a. Does g have an inverse? Explain.
b. Find an expression for the function $g \circ g \circ g$. Simplify your answer.

5.20 Let $f\colon A \to \mathbb{R}$ be a function (where A is a set). Define a new function $g\colon A \to \mathbb{R}$ by $g(x) = 3 \cdot [f(x)]^2 + 1$. Prove that if g is injective, then f is injective.

5.21 Consider the function $f\colon [0,\infty) \to [0,\infty)$, $f(x) = \frac{x}{x+1}$. Prove, by induction, that for every $n \in \mathbb{N}$, $f^n(x) = \frac{x}{1+n\cdot x}$.

Note: f^n is the function obtained by composing n copies of f: $f^n = \underbrace{f \circ f \circ \cdots \circ f}_{n \text{ times}}$.

5.22 Consider the function $g\colon \mathbb{R} \to \mathbb{R}$, $g(x) = 2x+1$. Prove, by induction, that for every $n \in \mathbb{N}$, $g^n(x) = 2^n \cdot x + 2^n - 1$.
Note: g^n is the function obtained by composing n copies of g: $g^n = \underbrace{g \circ g \circ \cdots \circ g}_{n \text{ times}}$.

5.23 For each of the following statements, decide whether it is *true* or *false*. Justify your answer with a proof or a counterexample.

a. Every surjective function $f\colon \mathbb{R} \to \mathbb{R}$ is unbounded (i.e., not bounded).
b. Every unbounded function $f\colon \mathbb{R} \to \mathbb{R}$ is surjective.
c. Every injective function $f\colon \mathbb{R} \to \mathbb{R}$ is strictly monotone.
d. The composition of two strictly monotone functions $f, g\colon \mathbb{R} \to \mathbb{R}$ is also strictly monotone.

5.24 Is the following statement necessarily true? Provide a proof or a counterexample:

"If $h\colon A \to B$, $g\colon B \to C$ and $f\colon B \to C$ are three functions, and $g \circ h = f \circ h$, then $g = f$."

5.25 Let $f\colon A \to B$ be a bijection, where A and B are subsets of $\mathbb{R}$. Prove that if f is strictly increasing (respectively decreasing) on A, then f^{-1} is strictly increasing (respectively decreasing) on B.

5.26 Find examples of functions f, g from $\mathbb{R}$ to $\mathbb{R}$, satisfying the following conditions, or prove that such examples do not exist.

a. $g \circ f$ is injective, but g is not injective.
b. $g \circ f$ is surjective, but g is not surjective.
c. $g \circ f$ is surjective, but f is not surjective.
d. f and g are not injective, but $g \circ f$ is injective.

5.27 (**Harder!**) Find two functions $f, g\colon \mathbb{R} \to \mathbb{R}$, such that $f \circ g$ is bijective, while $g \circ f$ is not a bijection.

5.28 Show that the following pairs of sets have the same cardinality.

a. Integers divisible by 3 and the *even* positive integers.
b. $\mathbb{R}$ and the interval $(0,\infty)$.

c. The interval $[0,2)$ and the set $[5,6) \cup [7,8)$.
d. The intervals $(-\infty,-1)$ and $(-1,0)$.

5.29 **Understanding the Proof of Theorem 5.4.1.** Read the proof of Theorem 5.4.1, and answer the following questions.

a. Write down explicitly the sets A_1, A_5 and A_{10}, defined in Step 1.
b. Which elements of A_{10} appeared in one of the previous A_k?
c. Find three A_k that contain the number 0.28.
d. Why was it necessary to remove, in Step 3, repeated terms from our initial sequence?
e. Referring to the sequence (a_n) from Step 3, what are a_{19}, a_{22} and a_{27}?
f. The bijection $f\colon \mathbb{N} \to \mathbb{Q}$, from Step 5, can be described using algebraic expressions (instead of a diagram). Complete the following alternative definition of f:

$$f(n) = \begin{cases} \underline{\hspace{6cm}} & \text{if } \quad n = 1 \\ \underline{\hspace{6cm}} & \text{if } n \text{ is even} \\ \underline{\hspace{6cm}} & \text{if } n \text{ is odd and greater than 1.} \end{cases}$$

5.30 **An Alternative Proof of Theorem 5.4.1.** The heart of the proof of Theorem 5.4.1 is to arrange all the rational numbers in one infinite sequence. However, there is more than one way to do so. The following pattern can be used to arrange all the rational numbers *in the open interval* $(0,1)$ in an infinite sequence (with repetitions):

$$\frac{1}{2}, \frac{1}{3}, \frac{2}{3}, \frac{1}{4}, \frac{2}{4}, \frac{3}{4}, \frac{1}{5}, \frac{2}{5}, \frac{3}{5}, \frac{4}{5}, \frac{1}{6}, \ldots.$$

We can then add reciprocals and negatives to include all rational numbers. Use these ideas to construct an alternative proof of Theorem 5.4.1.

5.31 a. Let A and B be disjoint sets, which are both countable. Prove that $A \cup B$ is also countable.

b. Use Part a to show that the set of all *irrational* real numbers is uncountable.

5.32 Prove that $\mathbb{N} \times \mathbb{N}$ (the set of all pairs of natural numbers) is a countable set. To do so, define, for each $k \in \mathbb{N}$, the set

$$A_k = \{(m,n) \colon m, n \in \mathbb{N} \text{ and } m + n = k\},$$

and follow the outline of the proof of Theorem 5.4.1.

5.33 a. Find an example of a set A and a *non-empty* subset $B \subseteq A$, such that A and $A \setminus B$ have the same cardinality.

b. Explain why if A is a *finite* set, and $B \subseteq A$ is a subset for which A and $A \setminus B$ have the same cardinality, then $B = \varnothing$.

5.34 Prove that the set of all *finite* subsets of $\mathbb{N}$ is a countable set.
Hint: For each $k \in \mathbb{N} \cup \{0\}$, define A_k to be the collection of all finite subsets of $\mathbb{N}$ whose elements add up to k. For example:

$$A_5 = \{\{1,4\},\{2,3\},\{5\}\},\ A_6 = \{\{1,2,3\},\{1,5\},\{2,4\},\{6\}\}.$$

5.35 Let A be the set of all infinite sequences consisting of 0s and 1s (i.e., sequences such as $010101010\ldots$, $1010010001000\ldots$, etc.). Prove that A is *uncountable*.
Hint: Assume that A is countable (i.e., its elements can be arranged in a list), and construct a sequence of zeros and ones which is not on that list. Use Cantor's Diagonalization Argument.

5.36 For each set, decide whether it is finite, countable, or uncountable. Explain your answer briefly.

$$P(\mathbb{N}), \quad \left\{1, \frac{1}{2}, \frac{1}{3}, \frac{1}{4}, \ldots\right\} \cap [0.03, 1], \quad \mathbb{Q}, \quad \mathbb{Z}, \quad (0, \infty).$$

5.37 For each set, decide whether it is finite, countable, or uncountable. Explain your answer briefly.

$P(\mathbb{Z})$, $(2,3)$, all the prime numbers, $\mathbb{Q} \cap [0,1]$, $\mathbb{N} \cap (-\infty, 1000)$.

5.38 Let $A = \left\{2, \frac{1}{2}, 3, \frac{1}{3}, 4, \frac{1}{4}, 5, \frac{1}{5}, \ldots\right\}$. Which set is *infinite* and *uncountable*? Explain.

$$A, \quad A \cap \mathbb{N}, \quad P(A \cap [0,1]), \quad P(A \cap [1,2]).$$

5.39 Let A and B be two *infinite* sets that *do not* have the same cardinality.

a. If A is countable, must B be uncountable? Explain.
b. If A is uncountable, must B be countable? Explain.

5.40 Let $A = \{4,5,6,7\}$. Fill in the blanks with either $\in$ or $\subseteq$.

$\{4\}$ ____ A $\quad \{5,6\}$ ____ $P(A)$ $\quad \{\varnothing\}$ ____ $P(A)$ $\quad 5$ ____ A

A ____ $P(A)$ $\quad \{A, \varnothing\}$ ____ $P(A)$

5.41 Fill in the blanks with either $\in$ or $\subseteq$.

$\varnothing$ ____ $\mathbb{Z}$ $\quad \mathbb{N}$ ____ $P(\mathbb{Z})$ $\quad P(\mathbb{N})$ ____ $P(\mathbb{Q})$

$\{\sqrt{2}, 4.5\}$ ____ $\mathbb{R}$ $\quad \frac{3}{4}$ ____ $\mathbb{Q}$ $\quad \mathbb{R}$ ____ $P(\mathbb{R})$

5.42 Let $A = \{1, 2, \{1, 2\}\}$.

a. Fill in the blanks with either $\in$ or $\subseteq$.

$\{2\}$ ____ A $\quad$ $\{\{1\}\}$ ____ $P(A)$ $\quad$ $\{1,2\}$ ____ $P(A)$

$\{\{1,2\}\}$ ____ A $\quad$ $\{1,\{1,2\}\}$ ____ $P(A)$

b. How many elements are in the set $P(A) \setminus A$?

5.43 a. Let $X = \{0,1\}$. List all the elements in $P(P(X))$.
b. Let $A = \{1,2\}$ and $B = \{2,3\}$. Find the sets $P(A \cap B)$ and $P(A) \cup P(B)$.

5.44 For each statement, decide whether it is *true* or *false*. Justify with a proof or a counterexample.

a. For every two sets A, B we have $P(A \cap B) = P(A) \cap P(B)$.
b. For every two sets A, B we have $P(A \cup B) = P(A) \cup P(B)$.
c. For every two sets A, B, if $A \cap B = \varnothing$, then $P(A) \cap P(B) = \{\varnothing\}$.

5.45 Consider the function $f\colon P(\mathbb{Z}) \to P(\mathbb{N})$, $f(A) = A \cap \mathbb{N}$.

a. What are $f(\{-2,-1,0,1,2\})$, $f(\{-1,-2,-3,\dots\})$ and $f(\mathbb{N})$?
b. Is f *surjective*? Explain.
c. Is f *injective*? Explain.

5.46 Let $f\colon \mathbb{N} \to P(\mathbb{N})$ be given by $f(n) = \{n+1, n+2, n+3, \dots\}$.

a. Find the set $f(3) \cap [-8, 8]$.
b. Is f an *injection*? Explain.
c. Is f a *surjection*? Explain.

5.47 Let X be a non-empty set, and define B to be the set of all functions from X to the two-element set $\{0, 1\}$:

$$B = \{f\colon X \to \{0,1\}\}.$$

Construct a *bijection* $h\colon P(X) \to B$.

5.48 Let A and B be two sets. Prove that if $P(A \cup B) = P(A) \cup P(B)$, then $A \subseteq B$ or $B \subseteq A$.

5.49 Let A and B be two sets. Prove that if $|A| \le |B|$ and $|B| \le |C|$, then $|A| \le |C|$.

5.50 Use the Schröder–Bernstein Theorem to prove the following.

a. $|[0,1]| = |[0,1)|$
b. $|[0,\infty)| = |(0,\infty)|$
c. $|[0,1]| = |\mathbb{R}|$
d. $|\mathbb{R} \setminus \mathbb{Z}| = |\mathbb{R}|$

5.51 Use the Schröder–Bernstein Theorem to show that for every set A, A and $P(P(A))$ do not have the same cardinality.

5.52 **Proof of the Schröder–Bernstein Theorem.** Recall that the Schröder–Bernstein Theorem says that if $f\colon A \to B$ and $g\colon B \to A$ are two

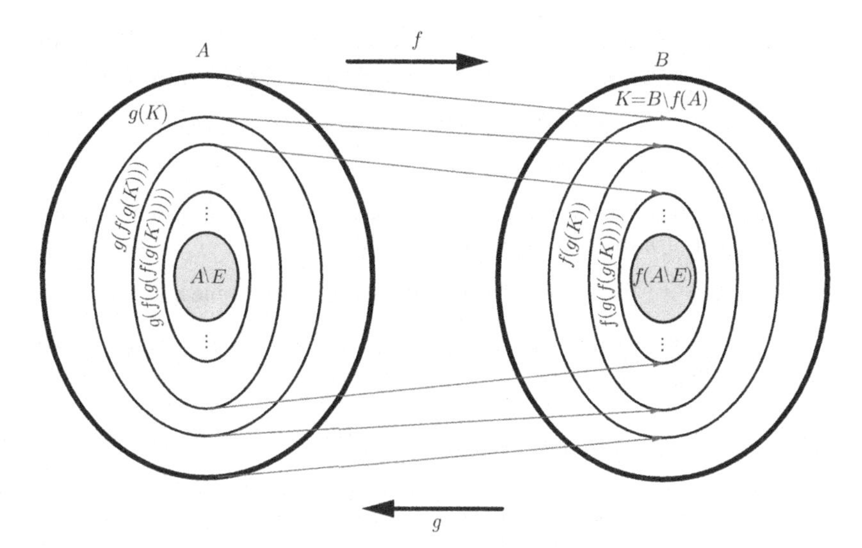

Figure 5.17 The proof of the Schröder–Bernstein Theorem.

injections, then there is a *bijection* $h\colon A \to B$. To prove the theorem, we need to construct the function h from the functions f and g.

It is important to realize that since we have no information whatsoever about the sets A, B, and the functions f, g (other than the fact that they are one-to-one), we have little choice in constructing h. The function h must send every $a \in A$ to some $b \in B$ via either f or g.

The key to understanding the proof is making sense of the diagram in Figure 5.17. Here is what is going on. Denote by K the set of all the elements in B which are not in the image of f (i.e., $K = B \setminus f(A)$). As h is supposed to be a surjection, it must send some elements of A to K. The function f will not be able to do that, and so g has to be used. In other words, h will send elements in $g(K)$ to B using g:

$$h(a) = g^{-1}(a) \quad \text{if} \quad a \in g(K).$$

Now that h has been defined on $g(K)$, we must take care of the rest of the elements of A, namely $A \setminus g(K)$. In other words, we need to define a bijection from $A \setminus g(K)$ to $B \setminus K = f(A)$. But we now face the exact same problem: the set $f(A \setminus g(K))$ is a subset of $f(A)$. Elements in $f(A)$ which are not in $f(A \setminus g(K))$ cannot be reached through f, so g must be used again. That is,

$$h(a) = g^{-1}(a) \quad \text{if} \quad a \in g(f(g(K))).$$

This process can be repeated indefinitely. To formalize it, we denote, for every $n = 0, 1, 2, \ldots$ the composition of n copies of $f \circ g$ by $(f \circ g)^n$:

$$(f \circ g)^n = \underbrace{(f \circ g) \circ (f \circ g) \circ \cdots \circ (f \circ g)}_{n \text{ times}}$$

(if $n = 0$, we interpret $(f \circ g)^n$ as the identity function, sending each element of B to itself). We define

$$E = \{a \in A : a = g \circ (f \circ g)^n(b) \text{ for some } b \in K \text{ and some } n = 0, 1, 2, \ldots\}$$

(note that E is just the union of $g(K), g(f(g(K))), g(f(g(f(g(K))))), \ldots$). Finally, we define:

$$h: A \to B, \qquad h(a) = \begin{cases} g^{-1}(a) & \text{if } a \in E \\ f(a) & \text{if } a \notin E. \end{cases}$$

The rest of the work is left for you. Answer the following questions.

a. Why do we use the notation $g^{-1}(a)$, when g is not necessarily invertible (as a function from A to B)?
b. The identity $f(A) \setminus f(A \setminus g(K)) = f(g(K))$ was used in the proof (where?). Prove that it is indeed a valid identity.
c. Prove that h is one-to-one. Namely, show that if $h(a_1) = h(a_2)$ for some $a_1, a_2 \in A$, then $a_1 = a_2$.
   Hint: If a_1, a_2 are both in E or both not in E, then clearly $a_1 = a_2$.
d. Prove that h is onto.
   Hint: If $b \in B$, then either $b \in f(A \setminus E)$ or $b \notin f(A \setminus E)$.

5.7 Solutions to Exercises

Solution to Exercise 5.1.1

No. The diagram does not represent a function, for the following reasons.

- In the new diagram, both c and d are assigned to the element γ. Functions *cannot* assign two images to an element in the domain.
- In the new diagram, no elements are assigned to δ and ϵ. However, a function must assign an image to *every* element in its domain.

Solution to Exercise 5.2.4

$$f \circ f(x, y) = f(f(x, y)) = f(x + 2y, xy) = ((x + 2y) + 2xy, (x + 2y)xy).$$

Once simplified, we get $f \circ f(x, y) = (x + 2y + 2xy, x^2y + 2y^2x)$.

Solution to Exercise 5.2.7

Suppose that $f: A \to B$ and $g: B \to C$ are two injections. We show that $g \circ f: A \to C$ is also injective.

If $x_1, x_2 \in A$ such that $g \circ f(x_1) = g \circ f(x_2)$, it follows that $g(f(x_1)) = g(f(x_2))$. From the injectivity of g we conclude that $f(x_1) = f(x_2)$, and from the injectivity of f we conclude that $x_1 = x_2$. Consequently, $g \circ f$ is an injection.

Solution to Exercise 5.3.7

A and B do not have the same cardinality. A has two elements, and B has a single element. The only function from A to B maps both 1 and 2 to $\{1, 2\}$, and hence is not a bijection.

A and C do have the same cardinality, as the map $f: A \to C$, given by $f(1) = \{1\}$ and $f(2) = \{2\}$ is a bijection.

6 Integers and Divisibility

This chapter is devoted to studying, in more depth, the set of integers $\mathbb{Z}$, its structure and properties. The integers play a fundamental role in many areas of mathematics, science, and beyond. The integers are closely related to the set of natural numbers, and thus are often used in problems involving counting, sequences, and structures with finitely many elements (such as finite fields).

You might wonder why can we not simply study the real numbers instead? After all, the integers are a subset of the real numbers, so once we understand the reals, we ought to have a solid understanding of the integers as well? Well, unfortunately, in mathematics, studying a certain object does not always reveal properties of its sub-objects.

Consider, for instance, the notion of *divisibility*. Given two integers a and b, with $b \neq 0$, either a is or is not divisible by b. However, it makes no sense to apply this notion to the set of real numbers: is 72.5 divisible by $\sqrt{5}$? Yes it is, and every real number is divisible by any other non-zero real number. Divisibility is a notion that becomes interesting and relevant when dealing with integers, and gives rise to other important notions, such as *even* and *prime* numbers.

6.1 Divisibility and the Division Algorithm

We begin by recalling Definitions 1.4.1 and 1.4.3 from Chapter 1, and introduce a common and useful notation.

Definition 6.1.1 Let a be an integer, and b a *non-zero* integer.

1. We say that a is *divisible* by b, or that b *divides* a, and write $b|a$, if there exists an integer m for which $a = m \cdot b$. We write $b \nmid a$ to indicate that a is *not* divisible by b.
2. A natural number $p > 1$ is called a *prime number* if the only natural numbers that divide p are 1 and p.

Example 6.1.2 40 is divisible by 8, which can be written as $8|40$.

- 13 divides 52, which we can write as $13|52$.
- 72 is not a multiple of 7, and hence $7\nmid 72$.
- Every non-zero integer is divisible by itself and by 1. Thus, for every $k \in \mathbb{Z}$, with $k \neq 0$, we have $1|k$ and $k|k$.
- Similarly, for every non-zero $k \in \mathbb{Z}$, we have

$$k|5k, \quad k|k^2, \quad k|0, \quad k|(k^3 - 7k), \quad \text{etc.}$$

Exercise 6.1.3
If a and b are non-zero integers, such that $a|b$ and $b|a$, what can we conclude about a and b?

When dividing an integer by another, say 35 by 8, the result may not be an integer. However, we know from our elementary school years, that we can perform *division with remainder*: 35 divided by 8 gives 4, with a reminder of 3. In other words, we can break 35 into four groups of 8 (and not more), after which we are left with three unused units:

$$35 = 4 \cdot 8 + 3.$$

Division with remainder allows us to perform division without leaving the world of integers (i.e., without referring to fractions or decimals). The following theorem gives a general description of this operation.

Theorem 6.1.4 (The Division Algorithm) *For every $a, b \in \mathbb{N}$, there is a unique pair of integers q and r, with $q \geq 0$ and $0 \leq r < b$, such that*

$$a = q \cdot b + r$$

(q is called the quotient, and r the remainder).

Remark. Note that the theorem simply claims the existence of quotient and remainder, without giving an explicit way for calculating them. Thus, the term "algorithm" is misleading, as there in no algorithm here at all. Nevertheless, it is quite common to refer to this theorem as the *Division Algorithm*, as we will do in this book.

There are, however, algorithms for computing q and r, such as long division that you may have encountered at school.

Example 6.1.5
- Dividing $a = 20$ by $b = 3$ gives a quotient of $q = 6$ and a remainder of $r = 2$, as $20 = 6 \cdot 3 + 2$.
- If $a = 47$ and $b = 10$, then the quotient is $q = 4$, and the remainder is 7, as $47 = 4 \cdot 10 + 7$.
- 28 is divisible by 7. In this case, $q = 4$ and $r = 0$: $28 = 4 \cdot 7 + 0$.

- What if $a < b$? For instance, say $a = 12$ and $b = 34$? Then we cannot squeeze even one group of 34 into 12, and so the quotient must be zero. The remainder, in this case, is equal to a:

$$12 = 0 \cdot 34 + 12.$$

In general, whenever $a < b$, the quotient q is zero, and $r = a$.

Proof of Theorem 6.1.4 The theorem states that *there are* numbers q, r, and that they are *unique*. Therefore, our proof has two parts: *the existence part*, showing that there exist such q and r, and *the uniqueness part*, showing that q and r are unique.

Proof of Existence

We prove existence by *strong induction* on a. That is, we treat b, throughout the proof, as a fixed unknown integer.

For $a = 1$, we distinguish between two cases.

- If $b = 1$, we can write $1 = 1 \cdot b + 0$, and so $q = 1$ and $r = 0$ satisfy the conclusion of the theorem (dividing 1 by 1 gives a quotient of 1 and a remainder of 0).
- If $b > 1$, we write $1 = 0 \cdot b + 1$, and hence $q = 0$ and $r = 1$ are appropriate choices.

We have proved the theorem for $a = 1$, regardless of the value of b, which confirms the base case.

Now assume that the theorem holds true for $a = 1, 2, \ldots, k$, for some $k \in \mathbb{N}$, and consider $a = k + 1$. We need to show that $k + 1 = q \cdot b + r$, for some integers q, r with $q \geq 0$ and $0 \leq r < b$. Again, we proceed by cases.

- ***k*+1 < *b*.** In this case we attempt to divide a natural number by a large number, which means that the quotient must be zero. Indeed, we can write

$$k + 1 = 0 \cdot b + (k + 1),$$

proving the theorem in this case with $q = 0$ and $r = k + 1$.
- ***k*+1 = *b*.** Here our a and b are the same, and thus the quotient is 1 and the remainder is 0:

$$k + 1 = 1 \cdot b + 0.$$

- ***k*+1 > *b*.** In this case, we finally use our induction hypothesis. From the fact that $k + 1 > b$ and $b \in \mathbb{N}$, we conclude that $k + 1 - b$ is a natural number, smaller than $k + 1$, and hence covered by our induction hypothesis. We have

$$k + 1 - b = q \cdot b + r$$

for some integers q, r, with $q \geq 0$ and $0 \leq r < b$. Rewriting this equality as

$$k + 1 = (q + 1) \cdot b + r$$

shows that the theorem holds for $a = k + 1$ as well, with quotient $q + 1$ and remainder r.

In conclusion, we have proved the existence part of the theorem for $a = k + 1$ and an arbitrary $b \in \mathbb{N}$, and hence for all a and b.

Proof of Uniqueness

A common way to prove uniqueness in mathematics is to assume that there are two elements satisfying the conclusion of a theorem, and then show that these two elements are, in fact, equal to each other.

To apply this strategy in our case, we assume that there are two pairs of q and r as required by the theorem. Namely, suppose that

$$a = q_1 \cdot b + r_1 \qquad \text{and} \qquad a = q_2 \cdot b + r_2$$

for some integers q_1, q_2, r_1, r_2, with $q_1, q_2 \geq 0$ and $0 \leq r_1, r_2 < b$. We show uniqueness by proving that $q_1 = q_2$ and $r_1 = r_2$.

By equating the two right-hand sides, we get

$$q_1 \cdot b + r_1 = q_2 \cdot b + r_2 \qquad \Rightarrow \qquad (q_1 - q_2) \cdot b = r_2 - r_1.$$

Note that since both r_1 and r_2 are between 0 and b, their difference must be between $-b$ and b:

$$-b < r_2 - r_1 < b.$$

However, from $(q_1 - q_2) \cdot b = r_2 - r_1$ we see that $r_2 - r_1$ must be an integer multiple of b. Well, the only multiple of b that is strictly between $-b$ and b is 0, and thus

$$r_2 - r_1 = 0 \qquad \Rightarrow \qquad r_1 = r_2.$$

Now, it follows that $(q_1 - q_2) \cdot b = r_2 - r_1 = 0$, which implies $q_1 = q_2$ (as $b > 0$). This completes the proof of the uniqueness part of the theorem. □

Example 6.1.6 Let n be an integer. Show that if n is *not divisible* by 5, then $n^4 - 1$ *is divisible* by 5.

We can easily confirm the claim for a few special cases.

- The number $n = 2$ is not divisible by 5, but $2^4 - 1 = 15$ is a multiple of 5.
- Similarly, if we take $n = 9$, then $9^4 - 1 = 6560$ is divisible by 5.

Our task is to prove the claim for every $n \in \mathbb{N}$, and we can do so by using the Division Algorithm.

Proof According to the Division Algorithm, there exist integers q and r, with $q \geq 0$ and $0 \leq r < 5$, such that

$$n = q \cdot 5 + r.$$

We know that n is not divisible by 5, and hence r cannot be zero. That is, $r = 1, 2, 3$ or 4.

Note that $n^4 - 1 = (n^2 - 1)(n^2 + 1)$. To prove that $n^4 - 1$ is divisible by 5, it is enough to show that either $n^2 - 1$ or $n^2 + 1$ is a multiple of 5.

If $r = 1$ or 4, then $n = 5q + 1$ or $n = 5q + 4$, and hence

$$n^2 - 1 = (5q + 1)^2 - 1 = 25q^2 + 10q = 5 \cdot (5q^2 + 2q)$$

or

$$n^2 - 1 = (5q + 4)^2 - 1 = 25q^2 + 40q + 15 = 5 \cdot (5q^2 + 8q + 3).$$

If $r = 2$ or 3, then $n = 5q + 2$ or $n = 5q + 3$, and we get

$$n^2 + 1 = (5q + 2)^2 + 1 = 5 \cdot (5q^2 + 4q + 1)$$

or

$$n^2 + 1 = (5q + 3)^2 + 1 = 5 \cdot (5q^2 + 6q + 2).$$

This shows that for $r = 1, 2, 3, 4$ either $n^2 - 1$ or $n^2 + 1$ is divisible by 5, which implies that $n^4 - 1$ is divisible by 5, as needed. □

6.2 Greatest Common Divisors and the Euclidean Algorithm

The greatest common divisor of two integers is a useful notion in number theory, with many applications in mathematics and science. It is defined as follows.

Definition 6.2.1 Let a and b be two integers, not both zero. The *greatest common divisor* (or, in short, the *GCD*) of a and b, denoted as $\gcd(a, b)$, is the largest integer that divides both numbers. If $\gcd(a, b) = 1$, we say that a and b are *relatively prime* or *coprime*.

Remark. The term "relatively prime" may seem strange at first, but it will become clearer once we present the Fundamental Theorem of Arithmetic in the next section.

Note that as 1 divides every integer, the GCD of two integers must be at least 1 and so, in particular, a natural number. Also note that the GCD of two numbers is a *symmetric operation*: $\gcd(a, b) = \gcd(b, a)$ for every $a, b \in \mathbb{Z}$ not both zero.

Example 6.2.2 Take $a = 36$ and $b = 42$. The positive divisors of a are 1, 2, 3, 4, 6, 9, 12, 18 and 36, and the divisors of b are 1, 2, 3, 6, 7, 14, 21 and 42. By comparing these lists, we can see that the largest number dividing both 36 and 42 is 6, and so $\gcd(36, 42) = 6$.

Example 6.2.3 The positive divisors of $a = 4$ are $1, 2$ and 4, and the divisors of $b = 9$ are $1, 3, 9$. Therefore, $\gcd(4, 9) = 1$. Namely, 4 and 9 are *relatively prime*.

Example 6.2.4 What if a and b are larger? Listing all their divisors, and making sure we have not missed any of them can be a long and tedious process. For instance, take $a = 23814$ and $b = 8232$. To find their greatest common divisor more efficiently, we can decompose the numbers as products of prime numbers, as guaranteed to be possible by Theorem 4.5.1 (in fact, we will shortly prove that the decomposition of a number as a product of primes is unique):

$$\begin{aligned} a &= 11907 \cdot 2 = 1323 \cdot 9 \cdot 2 = 147 \cdot 9 \cdot 9 \cdot 2 = 49 \cdot 3 \cdot 9 \cdot 9 \cdot 2 \\ &= 7 \cdot 7 \cdot 3 \cdot 9 \cdot 9 \cdot 2 = 7^2 \cdot 3^5 \cdot 2 \end{aligned}$$

$$b = 2058 \cdot 4 = 1029 \cdot 2 \cdot 4 = 343 \cdot 3 \cdot 2 \cdot 4 = 7 \cdot 49 \cdot 3 \cdot 2 \cdot 4 = 7^3 \cdot 3 \cdot 2^3.$$

Numbers, other than 1, which divide a and b, must be products of 2s, 3s and 7s. The longest such product, that divides both a and b, is $7 \cdot 7 \cdot 3 \cdot 2 = 294$, and so $\gcd(23814, 8232) = 294$.

Example 6.2.5 If $a, b \in \mathbb{N}$, and $a|b$, then $\gcd(a, b) = a$, as a clearly divides both a and b, and every number larger than a cannot divide a.

Exercise 6.2.6
Let n be a non-zero integer. What is $\gcd(n, 0)$? Explain.

For really large numbers, with dozens or even hundreds of digits, prime factorization becomes extremely difficult, and in practice impossible, *even with the aid of computers*. For instance, in 2009, a number known as RSA-768, with 232 decimal digits was factored, after *two years* of utilizing *hundreds of powerful computers*.

There is, however, an alternative method, known as the *Euclidean Algorithm*, for computing the greatest common divisor of two numbers, that *does not* involve prime factorization, or the listing of all divisors. It relies on the Division Algorithm, which makes it much more efficient from a computational point of view.

We first prove a proposition, on which the algorithm relies, and then describe the algorithm in detail.

Proposition 6.2.7 *Let a, b and k be integers, with a and b not both zero. Then* $\gcd(a, b) = \gcd(a - kb, b)$.

In other words, subtracting from a an integer multiple of b does not change the GCD.

Proof To prove the proposition, we show that the pairs (a, b) and $(a - kb, b)$ have the same set of (positive) divisors.

- If $d \in \mathbb{N}$ divides both a and b, then both are multiples of d. That is, $a = md$ and $b = ld$ for some $m, l \in \mathbb{Z}$. But then $a - kb = md - kld = (m - kl)d$, and hence $d | a - kb$.
- Conversely, if $d \in \mathbb{N}$ divides both $a - kb$ and b, then $a - kb = td$ and $b = ld$ for some $t, l \in \mathbb{Z}$. We can then write

  $$a = kb + td = kld + td = (kl + t)d,$$

  which shows that $d | a$.

We conclude that the pairs (a, b) and $(a - kb, b)$ have the same positive divisors, and hence the same greatest common divisor, as needed. □

The Euclidean Algorithm

Suppose that a and b are two natural numbers, with $a \geq b$. To compute the GCD of a and b, we apply the Division Algorithm repeatedly, as follows.

- Divide a by b, obtaining a quotient q_1 and a remainder r_1:

  $$a = q_1 \cdot b + r_1.$$

- If $r_1 = 0$ (that is, if a is divisible by b), then stop. Otherwise, divide b by r_1. Denote the new quotient and remainder by q_2 and r_2:

  $$b = q_2 \cdot r_1 + r_2.$$

- If $r_2 = 0$, then stop. Otherwise, divide r_1 by r_2. Denote the new quotient and remainder by q_3 and r_3:

  $$r_1 = q_3 \cdot r_2 + r_3.$$

- Keep repeating this process, as long as the remainder is not zero:

  $$r_2 = q_4 \cdot r_3 + r_4, \qquad r_3 = q_5 \cdot r_4 + r_5, \qquad r_4 = q_6 \cdot r_5 + r_6, \qquad \text{etc.}$$

The key observation here, is that the remainders form a *strictly decreasing list of non-negative integers*, as the remainder is always smaller than the divisor (Theorem 6.1.4):

$$r_1 > r_2 > r_3 > r_4 > \cdots .$$

On the other hand, every decreasing list of non-negative integers must terminate (can you see why?), and so at some point, a remainder of zero will show up. Using Proposition 6.2.7, we show that *the last positive remainder* must be the greatest common divisor of a and b.

Claim 6.2.8 *Let $a, b \in \mathbb{N}$ with $a \geq b$, and consider the remainders $r_1, r_2, r_3, \ldots$ obtained from the process described above. If $r_n = 0$ for some $n \in \mathbb{N}$, then $r_{n-1} = \gcd(a, b)$.*

Proof According to Proposition 6.2.7, subtracting a multiple of one number from another does not change the GCD. We can therefore apply the proposition repeatedly, and get:

$$\begin{aligned}
\gcd(a, b) &= \gcd(a - q_1 b, b) = \gcd(r_1, b) \\
&= \gcd(r_1, b - q_2 r_1) = \gcd(r_1, r_2) \\
&= \gcd(r_1 - q_3 r_2, r_2) = \gcd(r_3, r_2) \\
&= \gcd(r_3, r_2 - q_4 r_3) = \gcd(r_3, r_4) \cdots .
\end{aligned}$$

This process terminates once we get to the nth remainder, $r_n = 0$, and we conclude that

$$\begin{aligned}
\gcd(a, b) = \gcd(r_1, r_2) = \gcd(r_2, r_3) = \cdots &= \gcd(r_{n-1}, r_n) \\
&= \gcd(r_{n-1}, 0) = r_{n-1},
\end{aligned}$$

as needed (recall that $\gcd(n, 0) = n$ for every natural number n). □

The procedure of repeatedly applying the Division Algorithm, as described above, and eventually finding the GCD of two given numbers, is called the *Euclidean Algorithm*. This is one of the oldest algorithms in mathematics which is in common use. The starting point of the algorithm is a pair of natural numbers (the original a and b), and in each step, one of the numbers in the pair is replaced by a smaller number, without affecting the GCD. Thus, as we saw, the pairs

$$(a, b),\ (b, r_1),\ (r_1, r_2),\ (r_2, r_3),\ (r_3, r_4), \ldots$$

all have the same greatest common divisor. Once we get a pair with zero as one of the arguments, the other (positive) argument is the GCD of all the pairs, and in particular, of a and b.

Note that the algorithm can be used even when a and b are not both positive, since changing the sign of either a or b does not change their GCD. In other words, the following pairs all have the same GCD as the pair $(|a|, |b|)$:

$$(a, b), \qquad (-a, b), \qquad (a, -b), \qquad (-a, -b).$$

Example 6.2.9 Suppose we want to find the GCD of 154 and 35. That is, $a = 154$ and $b = 35$. We start by dividing a by b, with remainder:

$$\mathbf{154} = 4 \cdot \mathbf{35} + 14.$$

As the remainder is *not* zero, we proceed with dividing 35 by 14, with remainder:

$$\mathbf{35} = 2 \cdot \mathbf{14} + 7.$$

The remainder is still not zero, so we continue, and divide 14 by 7:

$$\mathbf{14} = 2 \cdot \mathbf{7} + 0.$$

Now, the remainder is zero (as 14 is divisible by 7), and so the GCD of the two given numbers is equal to our last non-zero remainder, which is 7:

$$\gcd(154, 35) = 7.$$

Another way to describe this process is by writing the pairs (typed above in **boldface** font) obtained in each step:

$$(154, 35) \quad \to \quad (35, 14) \quad \to \quad (14, 7) \quad \to \quad (7, 0).$$

Example 6.2.10 It is quite easy to execute the Euclidean Algorithm by hand without a calculator, even when the initial numbers are larger. For instance, here we perform the algorithm on the numbers 1533 and 150:

$$\begin{aligned} \mathbf{1533} &= 10 \cdot \mathbf{150} + 33 \\ \mathbf{150} &= 4 \cdot \mathbf{33} + 18 \\ \mathbf{33} &= 1 \cdot \mathbf{18} + 15 \\ \mathbf{18} &= 1 \cdot \mathbf{15} + 3 \\ \mathbf{15} &= 5 \cdot \mathbf{3} + 0. \end{aligned}$$

The last non-zero remainder is 3, and so $\gcd(1533, 150) = 3$. As the GCD is not affected by changing the signs of our initial numbers, we also conclude that

$$\gcd(1533, -150) = \gcd(-1533, 150) = \gcd(-1533, -150) = 3.$$

Exercise 6.2.11

In Example 6.2.4 we computed the GCD of 23814 and 8232 by factoring each number as a product of prime numbers. Re-compute the GCD by applying the Euclidean Algorithm.

Example 6.2.12 Suppose that x is a natural number *greater than* 1. What is the GCD of $x^7 - 1$ and $x^5 - 1$?

Solution
We can go ahead and apply the Euclidean Algorithm, but note that our quotients and remainders must be expressed in terms of the unknown number x. In order to do that, we use *polynomial division* with remainder:

$$\begin{aligned}
\mathbf{x^7 - 1} &= x^2 \cdot (\mathbf{x^5 - 1}) + (x^2 - 1)\\
\mathbf{x^5 - 1} &= (x^3 + x) \cdot (\mathbf{x^2 - 1}) + (x - 1)\\
\mathbf{x^2 - 1} &= (x + 1) \cdot (\mathbf{x - 1}) + 0.
\end{aligned}$$

Therefore, $\gcd(x^7 - 1, x^5 - 1) = x - 1$.

A fundamental and extremely useful property of the GCD, is that it can always be expressed as a *sum of integer multiples* of the two given numbers. This is the content of the following theorem.

Theorem 6.2.13 (Bézout's Identity) *Let a and b be two integers, not both zero. Then there exist $m, n \in \mathbb{Z}$ such that*

$$a \cdot m + b \cdot n = \gcd(a, b).$$

One of the proofs of Bézout's Identity is based on the *back-substitutions* strategy, demonstrated below. This strategy provides us with a procedure for finding appropriate values for m and n. We discuss a few examples before presenting the proof of the theorem.

Example 6.2.14 We have previously found that $\gcd(154, 35) = 7$. We rewrite the steps produced by the Euclidean Algorithm, keeping, in each step, the remainder on the right-hand-side, and moving the quotient term to the left. We also omit the last equality in which the remainder is zero:

$$\begin{aligned}
\mathbf{154} - 4 \cdot \mathbf{35} &= 14\\
\mathbf{35} - 2 \cdot \mathbf{14} &= 7.
\end{aligned}$$

Now, using the first equality, we replace the 14, in the second equation, with $154 - 4 \cdot 35$, and rearrange:

$$35 - 2 \cdot (154 - 4 \cdot 35) = 7 \Rightarrow 35 - 2 \cdot 154 + 8 \cdot 35 = 7 \Rightarrow 154 \cdot (-2) + 35 \cdot 9 = 7.$$

We have expressed the GCD of $a = 154$ and $b = 35$ as a sum of multiples of 154 and 35:

$$\underbrace{154}_{a} \cdot \underbrace{(-2)}_{m} + \underbrace{35}_{b} \cdot \underbrace{9}_{n} = \underbrace{7}_{\gcd(154,35)}.$$

Example 6.2.15 We also computed the GCD of $a = 1533$ and $b = 150$ to be 3. Rewriting the steps obtained by the Euclidean Algorithm (and omitting the last step), we obtain the following:

$$\begin{aligned} \mathbf{1533} - 10 \cdot \mathbf{150} &= 33 \\ \mathbf{150} - 4 \cdot \mathbf{33} &= 18 \\ \mathbf{33} - 1 \cdot \mathbf{18} &= 15 \\ \mathbf{18} - 1 \cdot \mathbf{15} &= 3. \end{aligned}$$

The back-substitution procedure leads to the following:

$$\begin{aligned} 18 - 1 \cdot (33 - 1 \cdot 18) = 3 \quad &\Rightarrow \quad 2 \cdot 18 - 33 = 3 \\ 2 \cdot (150 - 4 \cdot 33) - 33 = 3 \quad &\Rightarrow \quad 2 \cdot 150 - 9 \cdot 33 = 3 \\ 2 \cdot 150 - 9 \cdot (1533 - 10 \cdot 150) = 3 \quad &\Rightarrow \quad 1533 \cdot (-9) + 150 \cdot 92 = 3. \end{aligned}$$

And we obtained $1533 \cdot (-9) + 150 \cdot 92 = 3$, as needed.

Proof of Theorem 6.2.13 The back-substitution idea can be turned into an elegant proof using strong induction.

We assume, for now, that a and b are natural numbers, and perform induction *on the sum* $a + b$. The other cases, where a or b are zero or negative, are left as an exercise.

As the smallest sum of two natural numbers is 2, this is going to be our base case. If $a + b = 2$, then $a = b = 1$, and $\gcd(a, b) = \gcd(1, 1) = 1$. We can then pick $m = 1$ and $n = 0$ to satisfy the identity:

$$\underbrace{1}_{a} \cdot \underbrace{1}_{m} + \underbrace{1}_{b} \cdot \underbrace{0}_{n} = \underbrace{1}_{\gcd(a,b)} .$$

Now assume that the theorem is true for all $a, b \in \mathbb{N}$ with $2 \leq a + b \leq k$ (where $k \in \mathbb{N}$), and suppose that $a, b \in \mathbb{N}$ with $a + b = k + 1$. We prove the $(k + 1)$st case by considering the following three possibilities.

- If $a = b$, then $\gcd(a, b) = a$, and we can write $a \cdot 1 + b \cdot 0 = \gcd(a, b)$. That is, the theorem holds true with $m = 1$ and $n = 0$.
- If $a < b$, then a and $b - a$ are natural numbers, whose sum is $a + (b - a) = b < a + b$, and thus covered by the induction hypothesis. That is, we have

  $$a \cdot m + (b - a) \cdot n = \gcd(a, b - a).$$

  Remember though, that $\gcd(a, b - a) = \gcd(a, b)$ by Claim 6.2.8, and so, after rearranging the equation, we can write

  $$a \cdot (m - n) + b \cdot n = \gcd(a, b),$$

  which proves the $(k + 1)$st case.

- The last case, where $a > b$, is done similarly, by applying the induction hypothesis on the numbers $a - b$ and b.

We have proved the $(k+1)$st case, which completes the proof of the theorem. □

Remark. We could have used the Division Algorithm (Theorem 6.1.4) in the proof, so that it aligns with Examples 6.2.14 and 6.2.15, but we have chosen to present a simpler proof instead.

Exercise 6.2.16
Complete the proof for Bézout's Identity for when $a \leq 0$ or $b \leq 0$. You may need to handle several cases. Try to avoid induction, and instead rely on the case where both a and b are positive, which we have already proved.

We end this section with an application of Bézout's Identity.

Example 6.2.17 Prove that there are *integers* x, y, that solve the equation $1533x + 150y = 27$.

Solution
We computed the GCD of 1533 and 150 to be 3 (see Example 6.2.10 on page 172). According to Bézout's Identity, there are $m, n \in \mathbb{Z}$, for which $1533m + 150n = 3$ (we even found such m and n, but the actual values are not needed here). The numbers $x = 9m$ and $y = 9n$ solve the required equation, as

$$1533x + 150y = 1533 \cdot 9m + 150 \cdot 9n = 9 \cdot (1533m + 150n) = 9 \cdot 3 = 27.$$

6.3 The Fundamental Theorem of Arithmetic

In this section we present and discuss a central and well-known theorem about prime numbers. We begin with the following claim.

Claim 6.3.1 (Euclid's Lemma) *Let $p > 1$ be a prime number, and $a, b \in \mathbb{Z}$. If $p|ab$, then $p|a$ or $p|b$.*

Remark. In mathematics, a lemma is a proposition (or a "helping theorem") used as a stepping stone to prove a larger result, of possibly more interest.

Proof To prove the claim, we show that $p \nmid a$ implies $p|b$ (can you explain why this is sufficient?).

As p is a prime number, its only positive divisors are 1 and p. Since $p \nmid a$, the greatest common divisor of a and p must be 1, i.e., $\gcd(a, p) = 1$.

In other words, a and p are *relatively prime*. By Bézout's Identity, there are integers m, n for which

$$m \cdot a + n \cdot p = 1.$$

Multiplying both sides by b gives

$$m \cdot ab + n \cdot pb = b.$$

We know that $p|ab$, and so $p|(m \cdot ab)$. Also, $p|(n \cdot pb)$, and so p divides the whole left-hand side, $m \cdot ab + n \cdot pb$. Consequently, $p|b$, as needed. □

Example 6.3.2 The product of 30 and 259 is 7770, and hence divisible by 7. As 7 is prime, we are guaranteed, by Euclid's Lemma, that 7 divides either 30 or 259. Indeed, 259 is divisible by 7, as $259 = 7 \cdot 37$.

Remarks.

- Euclid's Lemma can be easily extended to products of more than two integers:

  If p divides a product of n integers, it divides at least one of the factors.

  The proof of this statement, by induction, is left as an exercise (see Problem 6.23).
- The requirement that p is a prime number is essential. If we take, for instance, $a = 4$ and $b = 9$, then their product, $ab = 36$, is divisible by 6. However, neither 4 nor 9 is divisible by 6.

Previously, we mentioned without proof, that the number $\sqrt{7}$ is irrational. That is, it cannot be expressed as a quotient of two integers. Now that we have Euclid's Lemma at our disposal, we are ready to prove this fact. The strategy we use can be applied to other radical numbers, such as $\sqrt{2}$, $\sqrt[5]{9}$, etc.

Claim 6.3.3 *The number $\sqrt{7}$ is irrational. That is, $\sqrt{7} \notin \mathbb{Q}$.*

Proof We prove the claim by contradiction. Assume that $\sqrt{7}$ is a rational number. Then $\sqrt{7} = \frac{a}{b}$ for some non-zero integers a, b. As $\sqrt{7} > 0$, we may assume that $a, b > 0$. Moreover, we assume that $\gcd(a, b) = 1$. If $\gcd(a, b) > 1$, we can simply divide a and b by $\gcd(a, b)$ to make the numerator and denominator relatively prime. See Problem 6.12.

When $\gcd(a, b) = 1$, we say that $\frac{a}{b}$ is in *lowest terms*, or *completely reduced*. From the equality $\sqrt{7} = \frac{a}{b}$, we get

$$7 = \frac{a^2}{b^2} \qquad \Rightarrow \qquad 7b^2 = a^2,$$

and hence $7|a^2$ (or $7|a \cdot a$). By Euclid's Lemma, $7|a$, and hence $a = 7n$ for some integer n. Replacing a by $7n$ gives

$$7b^2 = (7n)^2 \quad \Rightarrow \quad 7b^2 = 49n^2 \quad \Rightarrow \quad b^2 = 7n^2,$$

from which we conclude that $7|b^2$. Again, by Euclid's Lemma, we see that $7|b$, leading to a contradiction. The fraction $\frac{a}{b}$ was assumed to be in lowest terms, which is inconsistent with our conclusion, that both a and b are divisible by 7.

Consequently, our initial assumption must be false, and thus $\sqrt{7} \notin \mathbb{Q}$. □

We have seen, in Chapter 4, that every natural number, greater than 1, can be expressed as a product of prime numbers (see Theorem 4.5.1). In other words, prime numbers are, in some sense, the *building blocks* of all the natural numbers. If a product of prime numbers includes repetitions, we can use exponents to shorten the product, for instance:

$$\begin{aligned} 3969 &= 3 \cdot 3 \cdot 3 \cdot 3 \cdot 7 \cdot 7 = 3^4 \cdot 7^2 \\ 169400 &= 2 \cdot 2 \cdot 2 \cdot 5 \cdot 5 \cdot 7 \cdot 11 \cdot 11 = 2^3 \cdot 5^2 \cdot 7 \cdot 11^2. \end{aligned}$$

The Fundamental Theorem of Arithmetic, stated below, extends Theorem 4.5.1, adding that factoring a number as a product of primes can be done, essentially, in one way only. We can certainly reorder the exponents, and write 3969 as $7^2 \cdot 3^4$ instead of $3^4 \cdot 7^2$, but that is the only change we can make. The prime numbers appearing in the product, and the corresponding exponents, are determined uniquely.

Theorem 6.3.4 (The Fundamental Theorem of Arithmetic) *Every natural number $n \geq 2$ either is a prime, or can be expressed as a product of powers of distinct primes, in a unique way (except for reordering of the factors).*

Proof We have already proved the existence part of the theorem (Theorem 4.5.1). We thus proceed by proving *the uniqueness part*, using strong induction.

For the base case, $n = 2$, the theorem holds true, as 2 is a prime number. Assume that the uniqueness part of the theorem is true for $n = 2, 3, 4, \ldots, k$, and consider the number $k + 1$. We already know that $k + 1$ is a prime number, or can be expressed as a product of prime numbers. Our task is to prove that, if $k + 1$ is composite, then its factorization is unique.

We use a similar strategy to the one we used in the proof of the Division Algorithm (Theorem 6.1.4). Suppose that we can factor $k + 1$, as a product of distinct powers of primes, in two ways. That is

$$k + 1 = p_1^{a_1} \cdot p_2^{a_2} \cdots p_\ell^{a_\ell} = q_1^{b_1} \cdot q_2^{b_2} \cdots q_m^{b_m}, \qquad (*)$$

where $p_1, \ldots, p_\ell, q_1, \ldots, q_m$ are prime numbers, and $a_1, \ldots, a_\ell, b_1, \ldots, b_m$ are natural numbers.

Clearly, $p_1 | k+1$, as p_1 appears in the product $p_1^{a_1} \cdot p_2^{a_2} \cdots p_\ell^{a_\ell}$. Therefore, p_1 must divide the other representation of $k+1$:

$$p_1 | q_1^{b_1} \cdot q_2^{b_2} \cdots q_m^{b_m}.$$

Now, as p_1 is prime, we conclude, from Euclid's Lemma, that it must divide one of the numbers $q_1, \ldots, q_m$. Assume, for simplicity, that $p_1 | q_1$ (if p_1 divides one of the other qs, we can re-assign indices, so that $p_1 | q_1$).

Remember that q_1 is also a prime number, and so its only positive divisors are 1 and q_1. As $p_1 \neq 1$, we conclude that $p_1 = q_1$. We now divide the equalities $(*)$ by p_1 (or, equivalently, by q_1), and obtain

$$\frac{k+1}{p_1} = p_1^{a_1-1} \cdot p_2^{a_2} \cdots p_\ell^{a_\ell} = q_1^{b_1-1} \cdot q_2^{b_2} \cdots q_m^{b_m}.$$

Finally, we can use the induction hypothesis to complete the proof. As $\frac{k+1}{p_1}$ is smaller than $k+1$, it is covered by our hypothesis, and thus satisfies the uniqueness part of the theorem. This means that the ps, the qs, and the corresponding exponents must be the same. More explicitly, we have the following.

- The number of factors is the same, that is, $\ell = m$.
- The prime factors themselves have to be the same, though perhaps in some other arrangement. Thus, possibly after some re-indexing, we have $p_1 = q_1$, $p_2 = q_2, \ldots, p_\ell = q_\ell$.
- The exponents on the factors have to be the same, i.e., $a_1 - 1 = b_1 - 1$, $a_2 = b_2, \ldots, a_\ell = b_\ell$.

We conclude, from the above observations, that the two initial factorizations for $k+1$ (in $(*)$) were identical, as needed. We have thus proved the uniqueness part for $k+1$, which concludes the proof of the theorem. □

Remark. The Fundamental Theorem of Arithmetic implies that two integers, if not both zero, are relatively prime if and only if there is no prime number that divides them both. This explains the term *relatively prime* in Definition 6.2.1.

As a quick application of the Fundamental Theorem of Arithmetic, we prove that a certain logarithm is irrational.

Example 6.3.5 The number $\log_{48}(72)$ is irrational.

Proof Assume, by contradiction, that $\log_{48}(72)$ is rational, that is,

$$\log_{48}(72) = \frac{m}{n}$$

for some $m, n \in \mathbb{Z}$, with $n \neq 0$. In fact, we may assume that $m, n \in \mathbb{N}$, as $\log_{48}(72) > 0$.

Recall that the equality $\log_a(b) = c$ is equivalent to $b = a^c$, and so

$$\log_{48}(72) = \frac{m}{n} \qquad \Rightarrow \qquad 72 = 48^{m/n} \qquad \Rightarrow \qquad 72^n = 48^m.$$

We proceed by factoring 48 and 72 as a product of powers of distinct prime numbers:

$$(2^3 \cdot 3^2)^n = (2^4 \cdot 3)^m \qquad \Rightarrow \qquad 2^{3n} \cdot 3^{2n} = 2^{4m} \cdot 3^m.$$

Now we apply the uniqueness part of Theorem 6.3.4. As the last equality represents a number as a product of powers of distinct primes, the exponents must match:

$$\begin{array}{l} 3n = 4m \\ 2n = m \end{array} \qquad \Rightarrow \qquad 3n = 4 \cdot 2n \qquad \Rightarrow \qquad n = 0.$$

This is a contradiction, as $n \in \mathbb{N}$, and so we conclude that $\log_{48}(72)$ is indeed an irrational number, as needed. □

6.4 Problems

6.1 Some tend to confuse the divisibility symbol, $|$, with symbols such as — or /, used for fractions and division. Choose the correct notation in each of the phrases below. Explain your choice briefly.

a. As $\boxed{2|58}\boxed{58|2}\boxed{2/58}\boxed{58/2}$, we conclude that 58 is an even number.

b. Since $\boxed{60|15}\boxed{15|60}\boxed{60/15}\boxed{15/60}$, 60 is divisible by 15.

c. As $\boxed{7|84}\boxed{84|7}\boxed{7/84}\boxed{84/7}$ is an integer, 84 is divisible by 7.

d. If $\boxed{b|a}\boxed{a|b}\boxed{b/a}\boxed{a/b}$ (where $b \neq 0$), then $\boxed{b|a}\boxed{a|b}$ $\boxed{b/a}\boxed{a/b}$ is an integer.

6.2 Which of the following are *true* statements? Explain.

a. $6|54$
b. $33|3$
c. $-1|1$
d. $4 \nmid 2$
e. $4 \nmid 8$
f. $-11 \nmid -111$

6.3 Which number is divisible by 7^5? Explain.

- $210^3 \cdot 98^2$
- $7^2 \cdot 17^9$
- $77^4 \cdot 5^7$
- $27^5 \cdot 35^4$

6.4 Let a, b, d be three integers, with $d \neq 0$. For each statement, decide whether it is *true* or *false*. Provide a short proof or a counterexample.

a. If $d|a$ and $d|b$, then $d|(a+b)$.
b. If $d|(a+b)$, then $d|a$ and $d|b$.
c. If $d|(a+b)$, then $d|a$ or $d|b$.
d. If $d|(a+b)$, then either d divides both a and b, or d does not divide a or b.

6.5 Let a, b, d be three integers, with $d \neq 0$. For each statement, decide whether it is *true* or *false*. Provide a short proof or a counterexample.

a. If $d|a$ and $d|b$, then $d|a \cdot b$.
b. If $d|a \cdot b$, then $d|a$ or $d|b$.
c. If $d|a$ or $d|b$, then $d|a \cdot b$.
d. **(Harder!)** If $d^2|a^2$, then $d|a$.

6.6 In Example 6.1.6, we proved that for every $n \in \mathbb{N}$, if $5 \nmid n$, then $5|n^4 - 1$. Could we have used induction to prove this claim?

6.7 There are ways to generalize the Division Algorithm (Theorem 6.1.4), so that it can be applied to all integers (both positive and negative). Here is one possible generalization.

If $a, b \in \mathbb{Z}$, with $b \neq 0$, then there is a unique pair of integers, q and r, with $0 \leq r < |b|$, such that $a = q \cdot b + r$.

Note that the remainder is still required to be non-negative, so for instance, if we divide -21 by -4, the quotient is 6 and the remainder is 3, as $-21 = 6 \cdot (-4) + 3$.

a. Find the quotient and the remainder, obtained when dividing a by b.

- $a = 27$ and $b = -8$.
- $a = -15$ and $b = 2$.
- $a = -36$ and $b = -9$.
- $a = 5$ and $b = -7$.
- $a = -4$ and $b = 9$.

b. Prove the generalized version of the Division Algorithm given above. (Hint: Instead of using induction, proceed by cases, and refer to Theorem 6.1.4.)

6.8 Let a, b, c be three integers, such that $a^2 + b^2 = c^2$. Show that if c is even, then both a and b are even.

6.9 Choose a three-digit number (for instance, 273). Form a six-digit number, by repeating the digits of the original number twice (i.e., 273273). Prove that the resulting number is not prime. In fact, prove that it has at least three distinct prime factors.

6.10 In Definition 6.2.1, why do we require that a and b are not both zero? Why can we allow one of them to be zero?

6.11 Use prime factorization to compute the GCD of the following pairs of numbers (without a calculator).

a. 210 and 405
b. $10^6 \cdot 6^2 \cdot 5^{11}$ and $6 \cdot 15 \cdot 3^7$
c. $18^2 \cdot 21^3$ and $6 \cdot 10^3 \cdot 5^3$
d. $300 \cdot 35 \cdot 7^5$ and $33^5 \cdot 3 \cdot 64$

6.12 Let $a, b \in \mathbb{N}$ and $d = \gcd(a, b)$. Prove that $\gcd\left(\frac{a}{d}, \frac{b}{d}\right) = 1$.

6.13 a. Let n be a natural number. What is the GCD of $7n + 1$ and $8n + 1$?
b. Let $a, b \in \mathbb{Z}$, not both zero. Prove that $\gcd(a+b, a+2b) = \gcd(a, b)$.

6.14 Let n be a natural number.

a. What are all the possible values of $\gcd(n + 1, 2 - n)$? Explain.
b. What are all the possible values of $\gcd(7^n, 7^n + 4)$? Explain.

6.15 Find, with proof, all the integers a that satisfy the equation $\gcd(a, 10) = a$.

6.16 Suppose that a and b are two relatively prime natural numbers. Show that $a + b$ and $a \cdot b$ are relatively prime.

6.17 Use the Euclidean Algorithm to compute the GCD of the following pairs of numbers (without a calculator). Also, express the GCD as a *sum of multiples* of the two numbers.

a. 1872 and 300
b. 15477 and 15477154
c. 270028 and 27
d. 35530 and 355
e. 325299 and 325
f. 24 and $54 + 24^7$

6.18 Let a, b be two positive integers. We define the *least common multiple* (LCM) of a and b, denoted as $\mathrm{lcm}(a, b)$, to be the smallest positive integer, that is divisible by both a and b. For instance, $\mathrm{lcm}(12, 18) = 36$.

a. Find $\mathrm{lcm}(6, 15)$, $\mathrm{lcm}(4, 22)$ and $\mathrm{lcm}(9, 5)$ (do not use any computing device).
b. Let $a \in \mathbb{N}$. What are $\mathrm{lcm}(1, a)$, $\mathrm{lcm}(a, a)$ and $\mathrm{lcm}(a, a + 1)$? Explain.
c. Prove that if $a, b \in \mathbb{N}$, then $a \cdot b = \gcd(a, b) \cdot \mathrm{lcm}(a, b)$.
d. When would the LCM of two natural numbers be their product? Explain.

6.19 Complete the proof of Bézout's Identity in the cases where $a \leq 0$ or $b \leq 0$.

6.20 a. Let a, b, c be integers. Prove that the equation $ax + by = c$ has an integer solution (i.e., there are $x, y \in \mathbb{Z}$ solving the equation) if and only if $\gcd(a, b) | c$.
(Hint: One implication is quite simple. For the other, use Bézout's Identity.)

b. For which of the following values of m will the equation $12x+my=30$ have integer solutions?

- $m=12$
- $m=16$
- $m=18$
- $m=24$

6.21 Let a,b,c be integers. Prove that if a pair of integers (x_0,y_0) solves the equation $ax+by=c$, then the pair $(x_0+k\cdot b, y_0-k\cdot a)$ is also an integer solution of $ax+by=c$ (where k is an arbitrary integer). This shows that if $ax+by=c$ has an integer solution, then it has *infinitely many* solutions.

6.22 Let a,b,d be non-zero integers. Show that if $d|a$ and $d|b$, then $d|\gcd(a,b)$.

6.23 Prove, by induction, the following generalization of Euclid's Lemma (Claim 6.3.1):

If a prime p divides a product of n integers,
it must divide one of the factors.

6.24 Find three integers a,b,c, such that their product, abc, is divisible by 33, but none of them is divisible by 33. Does that contradict Euclid's Lemma (Claim 6.3.1)? Explain.

6.25 Let a,b,c be three integers. Prove that if a and c are relatively prime, and $c|ab$, then $c|b$. (Hint: Read carefully the proof of Euclid's Lemma.)

6.26 If j and k are natural numbers, and $37j=12k$, prove that $j+k$ is divisible by 7.

6.27 Prove that $\sqrt[3]{20}$ is an irrational number.

6.28 Prove that for every *odd* integer k, the number $\sqrt{2k}$ is irrational.

6.29 a. Prove that $\sqrt{11}$ is an irrational number.
b. Consider the function $f\colon \mathbb{N}\times\mathbb{N}\to\mathbb{R}$, $f(a,b)=a+b\cdot\sqrt{11}$. Is f *injective*? Is it *surjective*? Justify your arguments.

6.30 If $n,m\in\mathbb{N}$ such that $6^{2m+2}\cdot 3^n=4^n\cdot 9^{m+3}$, what are n and m? Explain.

6.31 Is the following statement *true* or *false*? Provide a proof or a counter-example:

"If $2^a\cdot 4^b=2^c\cdot 4^d$, then $a=c$ and $b=d$."

6.32 Consider the function $f\colon\mathbb{N}\times\mathbb{N}\to\mathbb{N}$ given by $f(n,m)=5^n\cdot 7^m$. Is f *injective*? Is f *surjective*? Justify your answer.

6.33 Consider the function $f\colon\mathbb{N}\times\mathbb{N}\to\mathbb{N}$ given by $f(a,b)=12^a\cdot 18^b$. Is f *injective*? Is f *surjective*? Justify your answer.

6.34 Prove that if p and q are two distinct (positive) prime numbers, then $\log_p(q)$ is irrational.

6.35 a. Show that $\log_{216}(36)$ is a *rational* number.

b. Obviously, if we try to prove that $\log_{216}(36)$ is irrational, by applying the strategy from Example 6.3.5, we will fail. Where exactly does the argument break? Explain.

6.5 Solutions to Exercises

Solution to Exercise 6.1.3

Since $a|b$, we have $b = a \cdot m$ for some $m \in \mathbb{Z}$. Similarly, as $b|a$, we have $a = b \cdot k$ for some $k \in \mathbb{Z}$. It follows that $a = a \cdot m \cdot k$, and since $a \neq 0$ we get $m \cdot k = 1$. We know that m and k are integers, and hence both are either 1 or -1. It follows that either $a = b$ or $a = -b$.

Solution to Exercise 6.2.6

If $n > 0$, then n is the largest integer which divides both n and 0, and so $\gcd(n, 0) = n$. If $n < 0$, then $\gcd(n, 0) = -n$. In conclusion, we have $\gcd(n, 0) = |n|$.

Solution to Exercise 6.2.11

We apply the Euclidean Algorithm on $a = 23814$ and $b = 8232$:

$$\begin{aligned}
\mathbf{23814} &= 2 \cdot \mathbf{8232} + 7350 \\
\mathbf{8232} &= 1 \cdot \mathbf{7350} + 882 \\
\mathbf{7350} &= 8 \cdot \mathbf{882} + 294 \\
\mathbf{882} &= 3 \cdot \mathbf{294} + 0,
\end{aligned}$$

and so $\gcd(23814, 8232) = 294$.

Solution to Exercise 6.2.16

If $a = 0$, then $b \neq 0$ and $\gcd(a, b) = |b|$. In this case, we can take $m = 0$ and $n = \pm 1$, according to the sign of b:

$$a \cdot 0 + b \cdot n = |b|.$$

A similar approach can be used when $b = 0$.

If both a and b are negative integers, then $\gcd(a, b) = \gcd(|a|, |b|)$. We apply Bézout's Identity on $|a|$ and $|b|$, which are natural numbers, to obtain $m, n \in \mathbb{Z}$ such that

$$|a| \cdot m + |b| \cdot n = \gcd(|a|, |b|) = \gcd(a, b).$$

We can adjust the signs of m and n, so that we have $a \cdot m + b \cdot n = \gcd(a, b)$, as needed.

7 Relations

In this chapter we discuss *relations*, a central notion in mathematics. As we will shortly see, we have already encountered many mathematical relations without using this terminology. We begin by formally defining what a relation is, and then introduce a special type of relation, *equivalence relations*, and the associated notion of *equivalence classes*. In Section 7.4, we study an important and useful equivalence relation: *congruence modulo n*.

7.1 The Definition of a Relation

The word *relation* (or *relationship*) is used in everyday language to describe a certain connection between two or more people or objects. For instance, some people are Canadian citizens, and some are not. Some are U.S. citizens, and some are not. "Being a citizen" is a relation between, or a way of associating, countries and people. Given a person x and a country y, x either is or is not a citizen of country y. In other words, given a pair (x, y) of a person and a country, the pair either satisfies the relation of "the first being a citizen of the second," or it does not.

Marriage is another example that comes to mind when thinking about relations between people. If two people are put in front of us, then either they are married, or they are not. Namely, every given pair of people either satisfies or does not satisfy the relation of "being married."

The relation "being the capital" can be applied to cities and countries. For instance, Ottawa is the capital of Canada, Paris is the capital of France, and Tel Aviv is *not* the capital of Israel. In other words, the pairs (Ottawa, Canada) and (Paris, France) satisfy the relation of "being the capital of," while (Tel Aviv, Israel) does not.

As we can see, a relation is, informally, a condition, connection, or property, that is satisfied by some pairs of objects or people, drawn from certain collections such as countries, age-group, cities, etc.

In this course, we focus on *mathematical relations*, and we have seen plenty of them already.

- Given two real numbers, x and y, either $x < y$ or $x \not< y$. The condition of "being less than" is an example of a relation in mathematics.

- Let X be a set, and $P(X)$ its power set. "Being an element of a set" is a relation between X and $P(X)$. Given a set $A \in P(X)$ and an object $a \in X$, either $a \in A$ or $a \notin A$.
- In Chapter 6 we encountered the notion of divisibility, which is another example of a relation. Given two integers a and b, with $b \neq 0$, either a is or is not divisible by b.

To be able to work with relations in mathematics, and prove statements about them, we need a formal definition. It is perfectly fine to think of relations as *conditions* or *properties* satisfied by certain objects, but in a formal definition, we must avoid notions such as "condition" or "connection."

To formalize the idea of a relation in mathematics, let us take a closer look at the example of cities, countries, and the relation of "being the capital." The two relevant sets in this example are cities and countries, so we denote:

$$A = \{\text{all cities}\} \qquad \text{and} \qquad B = \{\text{all countries}\}.$$

Whenever we form a pair of elements (x, y), where x is a city and y is a country, that pair either satisfies or does not satisfy the relation. In other words, a relation *separates* the collection of all pairs in $A \times B$ into two categories – pairs that satisfy our relation, and pairs that do not. This observation leads to a formal and elegant definition of a relation in mathematics. Instead of referring to a "condition" or a "property," we can simply *declare* which pairs satisfy the relevant condition, and define the relation as the collection of these pairs.

Definition 7.1.1 Let A and B be two sets. A *relation* R between A and B is a subset of $A \times B$. That is, a set R so that $R \subseteq A \times B$.

Remarks.

- In mathematics, we often refer to a subset of $A \times B$ as a *binary* relation, as it involves only two sets, A and B. More generally, we can define a relation on any number of sets (say, $A_1, A_2, \ldots, A_n$), as a *subset* $R \subseteq A_1 \times A_2 \times \cdots \times A_n$. However, in this book, we focus on binary relations only.
- When the sets A and B are equal, we simply say that *R is a relation on A*.

Example 7.1.2 Let $A = \{1, 2, 3, 4, 5\}$ and $B = \{6, 7, 8, 9, 10\}$. We define a relation R between A and B as follows:

$$R = \{(1,6), (1,7), (1,8), (2,6), (2,7), (3,6)\}.$$

Note that R is a subset of $A \times B$. The pair $(2, 6)$ satisfies the relation, while $(4, 8)$ does not. We can describe the relation R in a "condition-type" fashion, as

$$R = \{(x, y) : x + y \leq 9\},$$

and think of R as the relation of "having a sum less than or equal to 9." Nevertheless, simply listing the pairs is a perfectly valid definition.

Example 7.1.3 The *divisibility relation*, on $\mathbb{N}$, can be described as a set of pairs:

$$\{(a, b) : a \text{ is divisible by } b\}.$$

This is a subset of $\mathbb{N} \times \mathbb{N}$. The pair $(36, 4)$, for instance, satisfies the relation (namely, $(36, 4) \in R$), as 36 is divisible by 4. However, $(53, 8) \notin R$, as 53 is *not* divisible by 8.

Example 7.1.4 Let A be the set of all the words in the English language. Define the relation R between A and $\mathbb{N}$ as

$$R = \{(X, n) : \text{ the number of letters in } X \text{ is } n\} \subseteq A \times \mathbb{N}.$$

The pair (wisdom, 6) satisfies the relation, while (computer, 9) does not:

$$(\text{wisdom}, 6) \in R \qquad \text{and} \qquad (\text{computer}, 9) \notin R.$$

Symbols for Commonly Used Relations

The pair-notation is useful in certain cases. However, there are many relations in mathematics for which we have special symbols, and we tend to prefer them over pairs. The following are some examples.

- The symbol $|$ is used to denote *divisibility*, and we usually write $3|15$ and $6\not|14$, rather than $(15, 3) \in R$ and $(14, 6) \notin R$.
- The symbols $\in$ and $\subseteq$ are used to denote membership in a set, and inclusion of sets.
- Symbols such as $<$, $\leq$ and $>$ are used for *order relations* ("greater than," "less than or equal to," etc.).

The Definition of a Function

In Chapter 2 we defined a function to be "a rule" assigning elements in one set to elements in another set (see Definition 2.2.1). This informal definition can be made formal using relations.

If A and B are sets, then a pair (a, b), with $a \in A$ and $b \in B$, can be thought of as "assigning the element b to the element a." Thus, a function can be seen as a relation between A and B. We need, however, to be careful. In a function, *every* element in the domain must be paired with *exactly one* element in the codomain. This leads to the following formal definition of function.

Definition 7.1.5 Let A and B be sets, and $f \subseteq A \times B$ a relation. We say that f is a *function* from A to B if the following conditions hold.

- For every $a \in A$, there exists a pair $(x, y) \in f$ such that $x = a$.
- Whenever two pairs (a, b) and (a, c) are in f, it follows that $b = c$.

In other words, f is a function if for every $a \in A$ there is exactly one $b \in B$ such that $(a, b) \in f$. The notation $b = f(a)$ is commonly used in lieu of $(a, b) \in f$.

7.2 Equivalence Relations

In mathematics, we often encounter objects that represent the same abstract entity. For instance, the two simple fractions $\frac{4}{6}$ and $\frac{8}{12}$, though they look different, represent the same abstract number *two-thirds*. In fact, we even say that $\frac{2}{3}$ and $\frac{8}{12}$ are equal, and write $\frac{2}{3} = \frac{8}{12}$.

The triangles ABC and DEF in Figure 7.1 are congruent, and so can be thought of as being two copies of the same abstract triangle. In some sense, these two triangles are equal (or equivalent). The notion of an equivalence relation formalizes this idea. Here is the definition.

Definition 7.2.1 An *equivalence relation* R on a set S is a relation (that is, $R \subseteq S \times S$), such that the following hold.

1. For every $x \in S$, $(x, x) \in R$ (*reflexivity*).
2. For every $x, y \in S$, if $(x, y) \in R$, then $(y, x) \in R$ (*symmetry*).
3. For every $x, y, z \in S$, if $(x, y) \in R$ and $(y, z) \in R$, then $(x, z) \in R$ (*transitivity*).

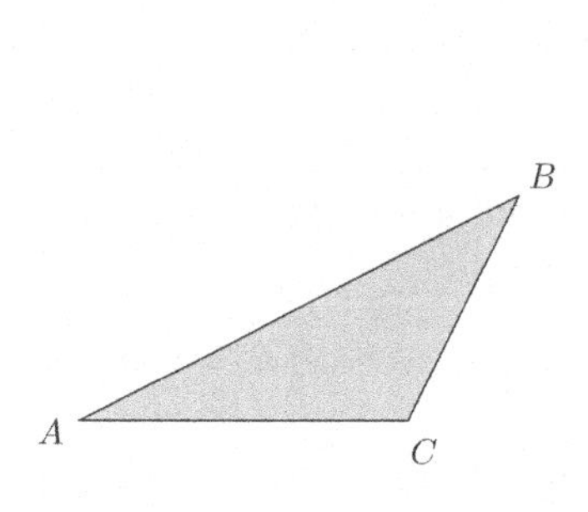

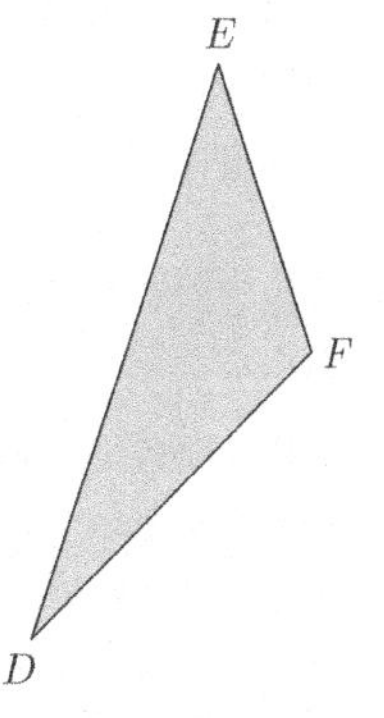

Figure 7.1 Congruent triangles.

In words, a relation that is *reflexive*, *symmetric* and *transitive*, is an equivalence relation. At the moment, it might be hard to see how these three properties capture the idea of equivalence. We will clarify this point later. For now, let us look at a few examples.

Example 7.2.2 The *order relation* $<$ ("less than") on the set of *real numbers* $\mathbb{R}$ is ...

- *not reflexive*, as for instance, $2 \not< 2$, and so $(2,2)$ is not part of the relation,
- *not symmetric*, since $5 < 7$ but $7 \not< 5$, that is, $(5,7)$ is in the relation, while $(7,5)$ is not,
- *transitive*, as if $x < y$ and $y < z$, then $x < z$, namely, if (x,y) and (y,z) satisfy the relation, then so does (x,z).

Therefore, the relation $<$ is *not* an equivalence relation. Note how all three requirements in Definition 7.2.1 involve the universal quantifier "for all." For this reason, we could use specific numbers to disprove reflexivity and symmetry, while we had to use variables to prove transitivity.

Example 7.2.3 The *inclusion relation* $\subseteq$ on a collection of sets is ...

- *reflexive*, as for every set B, we have $B \subseteq B$ (see remark on page 27),
- *not symmetric*, as we can find an example of a set that is properly contained in another, for instance, $\{1,2,3\} \subseteq \{1,2,3,4,5\}$, but $\{1,2,3,4,5\} \not\subseteq \{1,2,3\}$,
- *transitive*, since if A, B, C are sets, with $A \subseteq B$ and $B \subseteq C$, then $A \subseteq C$.

This relation is not symmetric, and hence *not* an equivalence relation.

Example 7.2.4 Define the following relation R on the set of real numbers:

$$R = \{(x,y) : x^2 + y^2 = 1\} \subseteq \mathbb{R} \times \mathbb{R}.$$

A number x satisfies this relation with respect to another number y, if and only if $x^2 + y^2 = 1$. For instance $\left(\frac{3}{5}, \frac{4}{5}\right) \in R$, while $\left(\frac{1}{2}, \frac{1}{2}\right) \notin R$. Is R reflexive? Symmetric? Transitive? Let us check.

- We have just mentioned that $\left(\frac{1}{2}, \frac{1}{2}\right)$ is not part of the relation R, and so R is *not reflexive*. Note that the pair $\left(\frac{1}{\sqrt{2}}, \frac{1}{\sqrt{2}}\right)$ does satisfy the relation, but reflexivity requires that *every* pair of the form (x,x) be in R, which is not the case here.
- The relation R *is symmetric*. If $(x,y) \in R$, then $x^2 + y^2 = 1$, or equivalently, $y^2 + x^2 = 1$. This shows that the pair (y,x) is in R as well, which proves the symmetry of the relation.

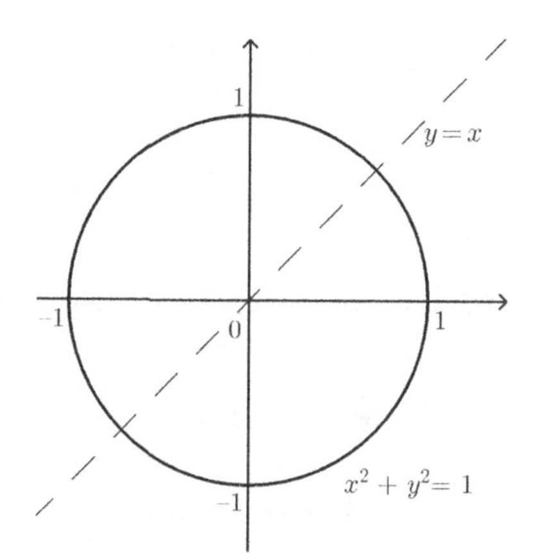

Figure 7.2 The unit circle.

- Transitivity, in many cases, is harder to detect. Here, we can experiment with a couple of pairs, and conclude that R is *not transitive*. Take, for instance, $x = -1$, $y = 0$ and $z = 1$. Then $(x, y) \in R$, as $(-1)^2 + 0^2 = 1$, and $(y, z) \in R$, as $0^2 + 1^2 = 1$. However, $(x, z) \notin R$, since $(-1)^2 + 1^2 \neq 1$, and therefore transitivity fails.

In this example, R is *not* an equivalence relation, as it is not reflexive or transitive.

A relation on a set of real numbers is a subset of the Cartesian product $\mathbb{R}^2 = \mathbb{R} \times \mathbb{R}$, and hence can be visualized as a curve or a region in the two-dimensional plane. Here, the relation R is nothing but the unit circle in $\mathbb{R}^2$. A reflexive relation on $\mathbb{R}$ must contain all the pairs of the form (x, x), and hence the line $y = x$. Our unit circle has only two points in common with the line $y = x$, which implies that R is *not reflexive* (see Figure 7.2). On the other hand, the unit circle is symmetric with respect to $y = x$ (or to switching x and y), and therefore represents a *symmetric relation*.

There is no simple way to detect transitivity from the graph of a relation on $\mathbb{R}$.

Example 7.2.5 Define a relation D on the set of *integers*, as follows:

$$D = \{(a, b) : a + b \text{ is an even number}\}.$$

We now prove that D is an equivalence relation, by showing that it is reflexive, symmetric and transitive.

- For every $a \in \mathbb{Z}$, we have $a + a = 2a$, which is an even number. Therefore, $(a, a) \in D$ and so D is reflexive.
- For every $a, b \in \mathbb{Z}$, if $a+b$ is an even number, then so is $b+a$. Consequently, $(a, b) \in D$ implies $(b, a) \in D$, which proves the symmetry of D.

Figure 7.3 Two line segments of the same length.

- Finally, we show that D is transitive. Let $a, b, c \in \mathbb{Z}$ such that $(a, b) \in D$ and $(b, c) \in D$ – that is, $a + b$ and $b + c$ are both even numbers. Our task is to prove that $(a, c) \in D$, namely, that $a + c$ is even, and we do so as follows. Write

$$a + c = (a + b) + (b + c) - 2b.$$

We know that $a + b$ and $b + c$ are even, and $2b$ is even as well. Sums and differences of even numbers are also even, and hence $a + c$ is an even number as needed.

Remarks.

1. On *every* set S, there is always an obvious equivalence relation, the *equality relation*! In other words, the pair (x, y) satisfies the relation if and only if $x = y$:

$$R = \{(x, x) : x \in S\}.$$

Equality is an equivalence relation, as it is reflexive, symmetric and transitive (check!). We often think of equivalence relations as a *generalization*, or *extension*, of the notion of equality.

2. We have mentioned in the beginning of the section equivalence relations that we have already encountered in previous studies. Congruence and similarity of triangles, and equivalence of fractions are examples of equivalence relations. They are all reflexive, symmetric and transitive.

   If two line segments, say AB and CD, have the same length, we often write $AB = CD$, even though strictly speaking, AB and CD are not equal as sets of points in the plane (see Figure 7.3). In fact, we are using the relation of "having the same length" to identify two line segments of the same length. This is another example of an equivalence relation.

3. For equivalence relations, we often use symbols such as $\equiv, \sim, \approx, \simeq$ and $\cong$ instead of the pair notation. These symbols resemble the equality symbol, and further emphasize the fact that elements satisfying an equivalence relation are often though of as being *equal* or *equivalent*. For instance, we may write $a \cong b$ or $a \equiv b$, instead of $(a, b) \in R$.

7.3 Equivalence Classes

We have already mentioned that given an equivalence relation on a set, pairs of elements satisfying the relation are often thought of as representing the same abstract idea. For instance, the fractions $\frac{4}{10}$ and $\frac{14}{35}$ represent the same number *two-fifths*. Congruent triangles represent copies of the same *abstract triangle*, and so on.

We can even find the notion of equivalence outside mathematics. For instance, young children, who learn about colors, are taught that a tomato's color is red, and that a strawberry is red as well. They are then able to identify objects as being red, as *they have the same color* as a tomato or a strawberry. For instance, a car or a shirt is red, because it has the same color as a tomato.

Children learn about colors without being given precise definitions. How would you define "the color red" without referring to physical objects? Proper definitions can be given using frequency or wavelengths of electromagnetic waves, but that is not how we first learn about colors. An understanding of what it means for two objects to have the same color, and knowing that a tomato is red, is enough in order to identify other red objects. If we define

$$S = \{\text{All objects, which have the same color as a tomato}\},$$

we may treat S as a set, representing the abstract notion of "the color red." That is, the notion "red" is defined as a property of a collection of objects, grouped together. Note that *having the same color* is an equivalence relation on a collection of objects.

Similarly, the abstract number *two-fifths* can be described as the set of all fractions $\frac{a}{b}$, in which $5a = 2b$.

Given an arbitrary equivalence relation on a set, we may group equivalent elements together, to form *equivalence classes*. Each equivalence class may be thought of as one abstract entity. Every member of the class is *a representative* of that class. Here is the definition.

Definition 7.3.1 Let R be an equivalence relation on a set S, and $x \in S$. The *equivalence class of* x is the set of all elements $y \in S$, which are equivalent to x:

$$\{y \in S\colon (x, y) \in R\}.$$

We denote the equivalence class of x by $[x]$.

Example 7.3.2 Define an equivalence relation $\sim$ on the set of real numbers, as follows:

$$x \sim y \qquad \text{if and only if} \qquad x \text{ and } y \text{ have } \textit{the same truncation}.$$

By *truncation* (or *integer part*), we mean the integer left after removing all digits to the right of the decimal point. Here are some examples.

- The truncation of 4.35 is 4, and the truncation of $\pi = 3.1415\ldots$ is 3.
- The truncation of -6.01 is -6, and the truncation of -9.99 is -9.
- The truncation of both 0.123 and -0.734 is 0.

Remark. Some real numbers have two decimal representations, for instance:

$$3 = 2.999\ldots, \quad -4 = -3.999\ldots, \quad 0.5 = 0.4999\ldots, \text{ etc.}$$

In this example, we use the convention of preferring the shorter decimal representation, rather that the one with $999\ldots$.

The numbers 5.11 and 5.9999 satisfy the relation, as they have the same truncation. We can then write $5.11 \sim 5.9999$. Similarly, we have $-0.5 \sim 0.5$. On the other hand, 4.1 and -4.1 do not have the same truncation, and hence do not satisfy the relation: $4.1 \not\sim -4.1$.

Having the same truncation is an equivalence relation (check!). What do equivalence classes look like for this relation?

For example, what is the equivalence class of the number 2.35? Well, by definition, we need to find all real numbers that have the same truncation as 2.35, which is 2. These are all the numbers between 2 and 3, including 2, but not including 3. In other words:

The equivalence class of 2.35 is the half-open and half-closed interval $[2, 3)$.

Similarly, the equivalence class of -6.5, is the set of all real numbers with truncation -6, and so:

The equivalence class of -6.5 is the interval $(-7, -6]$.

What is the equivalence class of 0? There are both positive and negative numbers, whose truncation is 0. In this case, the equivalence class is an open interval, containing all the numbers between -1 and 1:

The equivalence class of 0 is the open interval $(-1, 1)$.

By looking at a few more cases, we see that there are three types of equivalence classes.

- Classes with positive numbers: $[1, 2), [2, 3), [3, 4), [4, 5), \ldots$.
- Classes with negative numbers: $\ldots, (-4, -3], (-3, -2], (-2, -1]$.
- One class with both positive and negative numbers: $(-1, 1)$.

Figure 7.4 visualizes the equivalence classes on the number line.

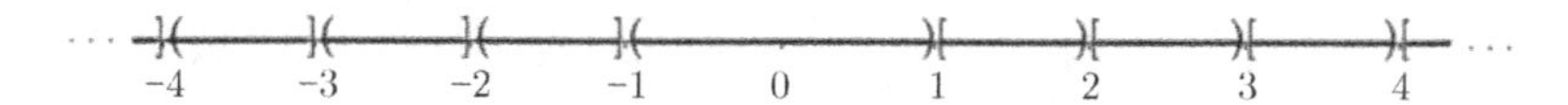

Figure 7.4 Equivalence classes for the relation of "having the same truncation."

Example 7.3.3 On the set of integers, we define the following equivalence relation:

$$a \equiv b \qquad \text{if and only if} \qquad a - b \text{ is divisible by 4.}$$

For instance, $15 \equiv 7$, as $15-7 = 8$ is divisible by 4, but $1 \not\equiv 8$, as $1-8 = -7$ is not divisible by 4. The relation $\equiv$ is indeed an equivalence relation (see Exercise 7.3.4 below).

Exercise 7.3.4
Let $a, b \in \mathbb{Z}$. Prove that $a - b$ is divisible by 4 if and only if a and b yield the same remainder when divided by 4. Use this observation to conclude that the above relation is indeed an equivalence relation.

What is [7]? Namely, what is the equivalence class of the number 7? By definition, we have

$$[7] = \{b \in \mathbb{Z} : 7 - b \text{ is divisible by } 4\}.$$

Which integers b satisfy the condition that $7 - b$ is divisible by 4? There are many, of course. If $b = 3$, then $7-3 = 4$ is divisible by 4. Also, if $b = 7$ or $b = -1$, then $7 - b$ is divisible by 4. Looking at a few more examples, we easily conclude that

$$[7] = \{\dots, -13, -9, -5, -1, 3, 7, 11, 15, 19, \dots\}.$$

These are precisely the numbers which yield a remainder of 3 when divided by 4. Note that the equivalence class of 3 or -9 or any number which is equivalent to 7 will be identical to [7]. That is,

$$[7] = [3] = [-9].$$

What about the equivalence class of 4? Well, the number $4 - b$, where $b \in \mathbb{Z}$, is divisible by 4 if and only if b itself is divisible by 4, and so the equivalence class of 4 is simply the set of all multiples of 4:

$$[4] = \{\dots, -12, -8, -4, 0, 4, 8, 12, 16, 20, \dots\}.$$

Having covered all multiples of 4, and all numbers leaving a remainder of 3 when divided by 4, we realize that there are two more equivalence classes that have not been mentioned. These are the ones containing integers with remainder 1, and with remainder 2:

$$[1] = \{\dots, -11, -7, -3, 1, 5, 9, 13, 17, \dots\}$$
$$[2] = \{\dots, -10, -6, -2, 2, 6, 10, 14, 18, \dots\}.$$

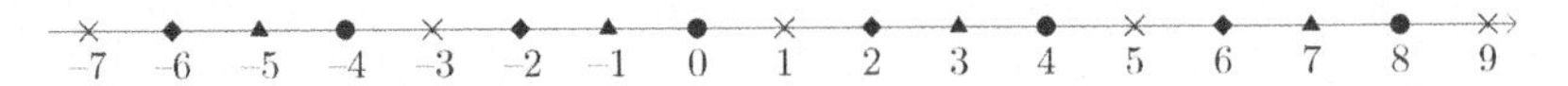

Figure 7.5 Equivalence classes for the relation "$a - b$ is divisible by 4."

Overall, there are *four* equivalence classes for this relation. To visualize them, we use four different symbols (•, x, ♦ and ▲) to mark integers belonging to the same equivalence class (Figure 7.5).

Looking at the two examples above, we make two important observations.

- The equivalence classes *cover* our original set. In the first example, the equivalence classes are intervals, covering all the real numbers. In the second example, the four equivalence classes (or more precisely, their union) contain all the integer numbers.
- In both examples, the equivalence classes *do not overlap*. Namely, the intersection of any two different classes is empty.

These two observations are completely general, and apply to all equivalence relations.

Theorem 7.3.5 *Let $\sim$ be an equivalence relation on a set S. Then every $x \in S$ belongs to some equivalence class, and every two different equivalence classes are disjoint.*

Proof Given an element $x \in S$, we have $x \sim x$, as an equivalence relation is reflexive, and so $x \in [x]$. This shows that each element in x belongs to some equivalence class (namely, its own class).

The second part of the theorem claims that every two different equivalence classes are disjoint. We prove the contrapositive. That is, we show that two equivalence classes which are *not disjoint*, must be *equal to each other*. Let $[x]$ and $[y]$ be two non-disjoint equivalence classes: $[x] \cap [y] \neq \emptyset$. As their intersection is not empty, there is an element $z \in [x] \cap [y]$.

From the fact that $z \in [x]$ and $z \in [y]$, we conclude that $x \sim z$ and $y \sim z$. Symmetry of the relation implies that $z \sim y$, and transitivity implies that $x \sim y$ (see Figure 7.6).

We will now prove that $[x] = [y]$ by mutual inclusion (after all, $[x]$ and $[y]$ are *sets*). If $u \in [y]$, then $y \sim u$. But we also know that $x \sim y$. From transitivity of the relation, we conclude that $x \sim u$, and hence $u \in [x]$ (can you construct a diagram representing the argument in the last three sentences?). We have thus proved the inclusion $[x] \supseteq [y]$. The inclusion $[x] \subseteq [y]$ is proved in a similar way, from which we conclude that $[x] = [y]$, as needed. □

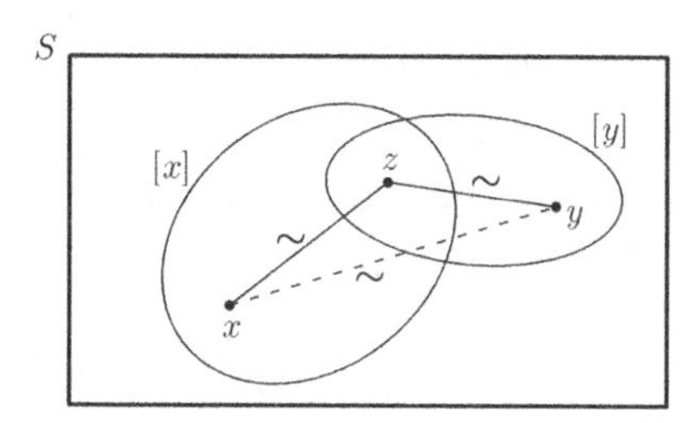

Figure 7.6 If $z \in [x] \cap [y]$, then $x \sim y$.

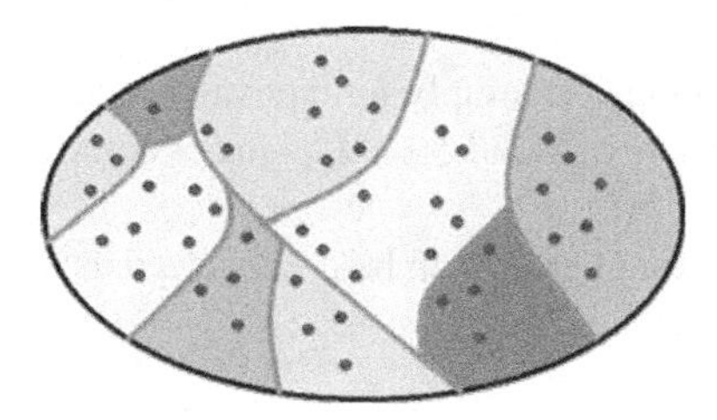

Figure 7.7 An equivalence relation induces a partition on the underlying set.

Note how the three defining properties of equivalence relations – reflexivity, symmetry and transitivity – are used in the proof. Without them, the theorem will not be valid.

Another way to phrase the theorem, is to say that an equivalence relation induces a *partition* of S into equivalence classes, i.e., it splits S into non-overlapping regions (see Figure 7.7). Elements in the same class (or region) represent the same abstract entity.

Remark. Given an equivalence relation $\sim$ on a set S, we can form a set whose elements are the corresponding equivalence classes. This set is often called the *quotient set* (or the *quotient space*), and is denoted as $S/\sim$. That is,

$$S/\sim = \{[x] : x \in S\}.$$

Note that an element of $S/\sim$ is an equivalence class, which is itself a subset of S.

Example 7.3.6 In Example 7.3.3 we discussed the following relation on $\mathbb{Z}$:

$$a \equiv b \qquad \text{if and only if} \qquad a - b \text{ is divisible by 4.}$$

This relation partitions the integers into *four* equivalence classes, and so the quotient set $\mathbb{Z}/\equiv$ has four elements:

$$\mathbb{Z}/\equiv \; = \{[7], [4], [1], [2]\}.$$

We could, of course, write instead

$$\mathbb{Z}/\equiv \; = \{[0], [1], [2], [3]\},$$

since 0 and 4 represent the same class, as do 3 and 7 (that is, $[0] = [4]$ and $[3] = [7]$).

Example 7.3.7 The rational numbers can be defined as equivalence classes of pairs of integers. On the set $S = \mathbb{Z} \times (\mathbb{Z} \setminus \{0\})$, we define the following equivalence relation:

$$(a, b) \approx (c, d) \qquad \text{if and only if} \qquad ad = bc.$$

The condition $ad = bc$ is our way to write $\frac{a}{b} = \frac{c}{d}$ without referring to division or fractions.

Now, a rational number is simply an equivalence class under this relation. The notation $\frac{a}{b}$, for $a, b \in \mathbb{Z}$ and $b \neq 0$, is just a common way to refer to the equivalence class $[(a, b)]$.

The set of rational numbers $\mathbb{Q}$ can be now viewed as the quotient set $S/\approx$.

Exercise 7.3.8
Prove that indeed, as argued in Example 7.3.7, the relation

$$(a, b) \approx (c, d) \qquad \text{if and only if} \qquad ad = bc$$

is an equivalence relation on $S = \mathbb{Z} \times (\mathbb{Z} \setminus \{0\})$.

7.4 Congruence Modulo n

In this section, we focus on one fundamental and widely used equivalence relation: congruence modulo n. It is frequently used within mathematics in number theory and discrete mathematics, and has numerous applications in related fields, such as computer science and cryptography. In fact, the relation in Example 7.3.3 on page 193 is a special case of congruence. Here is the definition.

Definition 7.4.1 Let $n \in \mathbb{N}$. The relation, on the set of *integers*, defined by:

$$a \equiv b \pmod{n} \qquad \text{if and only if} \qquad a - b \text{ is divisible by } n,$$

is called *congruence modulo n*.

If $a, b \in \mathbb{Z}$ satisfy this relation, we say that they are *congruent modulo n*.

Example 7.4.2

- $7 \equiv 10 \pmod 3$, since 3 divides $7-10$. Recall that the statement "3 divides $7-10$" can be written as $3 | 7-10$.

- $35 \equiv 15 \pmod 5$, since $5|35-15$. We say that 35 is *congruent to* 15 *modulo* 5.
- $4 \equiv 4 \pmod 7$, since 7 divides $4-4 = 0$.
- The number 3 is *not* congruent to 8 modulo 2, as $3-8$ is *not* divisible by 2. We denote this by writing $3 \not\equiv 8 \pmod 2$.

One can prove that if integers a and b are congruent modulo n, then they produce the same remainder when divided by n. Exercise 7.3.4 is really a special case of this observation for $n = 4$.

We now proceed by showing that congruence modulo n is an equivalence relation.

Theorem 7.4.3 *For every $n \in \mathbb{N}$, congruence modulo n is an equivalence relation on $\mathbb{Z}$.*

Proof To prove the theorem, we need to show that congruence modulo n is a reflexive, symmetric and transitive relation on integers.

- *Reflexivity* For every $k \in \mathbb{Z}$, $k \equiv k \pmod n$, as $k - k = 0$ is divisible by every positive integer. Therefore, this relation is reflexive.
- *Symmetry* If $a \equiv b \pmod n$, then $a - b$ is divisible by n. But then $b - a = -(a-b)$ is also divisible by n, which implies that $b \equiv a \pmod n$, as needed.
- *Transitivity* Suppose that $a \equiv b \pmod n$ and $b \equiv c \pmod n$. By definition, we get that *both* $a - b$ and $b - c$ are divisible by n. Consequently, their sum, $(a - b) + (b - c) = a - c$ is also divisible by n, which shows that $a \equiv c \pmod n$, as needed. □

As well as being an equivalence relation, congruence has other useful properties. The following claim shows that congruence behaves well with respect to addition and multiplication of integers.

Claim 7.4.4 *Let $n \in \mathbb{N}$ and $a, b, r, s \in \mathbb{Z}$. If $a \equiv r \pmod n$ and $b \equiv s \pmod n$, then $a + b \equiv r + s \pmod n$ and $a \cdot b \equiv r \cdot s \pmod n$.*

Proof We are given that $a \equiv r \pmod n$ and $b \equiv s \pmod n$, which means that both $a - r$ and $b - s$ are divisible by n:

$$a - r = n \cdot m \qquad \text{and} \qquad b - s = n \cdot k \qquad \text{for some } m, k \in \mathbb{Z}.$$

By adding these two equalities, we get

$$(a - r) + (b - s) = n \cdot m + n \cdot k \qquad \Rightarrow \qquad (a + b) - (r + s) = n \cdot (m + k).$$

We see that $(a+b)-(r+s)$ is divisible by n, and hence $a+b \equiv r+s \pmod{n}$, which proves the first part of the claim.

To prove the multiplication part, we rewrite the equalities $a - r = n \cdot m$ and $b - s = n \cdot k$ as

$$a = r + n \cdot m \qquad \text{and} \qquad b = s + n \cdot k,$$

and then multiply them:

$$a \cdot b = (r+n \cdot m) \cdot (s+n \cdot k) \qquad \Rightarrow \qquad a \cdot b = r \cdot s + r \cdot n \cdot k + n \cdot m \cdot s + n^2 \cdot m \cdot k.$$

Rearranging the last equality, we get

$$a \cdot b - r \cdot s = n \cdot (r \cdot k + m \cdot s + n \cdot m \cdot k),$$

which shows that $a \cdot b - r \cdot s$ is divisible by n, and hence $a \cdot b \equiv r \cdot s \pmod{n}$, as needed. □

By applying the multiplication part of the claim repeatedly, with $a = b$, we obtain the following conclusion.

Conclusion Let $a, r \in \mathbb{Z}$ and $n \in \mathbb{N}$. If $a \equiv r \pmod{n}$ and $k \in \mathbb{N}$, then $a^k \equiv r^k \pmod{n}$.

Exercise 7.4.5
Use mathematical induction to provide a proof of the above conclusion.

Here are a few examples, demonstrating the use of this equivalence relation, and Claim 7.4.4.

Example 7.4.6 What is the unit (i.e., the rightmost) digit of the number 7^{20}?

Solution
Given a natural number, the unit (or rightmost) digit equals the remainder obtained when dividing the number by 10 (assuming the number is written in the usual base 10 form). For instance, when the number 14523 is divided by 10, the quotient is 1452 and the remainder is 3, since

$$14523 = 1452 \cdot 10 + 3.$$

To find the unit digit of 7^{20}, we therefore use the *congruence modulo* 10 equivalence relation, as follows:

$$\begin{aligned} & 7 \equiv -3 \pmod{10} && \text{(as } 7-(-3) \text{ is divisible by 10)} \\ \Rightarrow\ & 7^4 \equiv (-3)^4 \equiv 81 \equiv 1 \pmod{10} && \text{(by the Conclusion)} \\ \Rightarrow\ & (7^4)^5 \equiv 1^5 \pmod{10} && \text{(by the Conclusion as well)} \\ \Rightarrow\ & 7^{20} \equiv 1 \pmod{10}. && \end{aligned}$$

Therefore, the remainder obtained when 7^{20} is divided by 10 is 1, and hence 1 is the unit digit of 7^{20}.

Remark. A chain of equivalences of the form

$$A \equiv B \equiv C \equiv \cdots \equiv D \pmod{n}$$

indicates that every pair of terms from the chain are congruent modulo n.

Example 7.4.7 What is the remainder obtained when the product $7697 \cdot 9154$ is divided by 5?

Solution

An integer is divisible by 5 if and only if it ends with either 5 or 0. Therefore, we have:

$$7697 \equiv 2 \pmod{5} \qquad \text{(as } 7697-2 \text{ is divisible by 5)}$$
$$9154 \equiv 4 \pmod{5} \qquad \text{(as } 9154-4 \text{ is divisible by 5).}$$

By Claim 7.4.4, we get:

$$7697 \cdot 9154 \equiv 2 \cdot 4 \equiv 8 \equiv 3 \pmod{5}.$$

Consequently, the required remainder is 3.

Note that the remainder must be either $0, 1, 2, 3$ or 4.

Example 7.4.8 Let $a \in \mathbb{Z}$. Show that if $a^2 + 5$ is *not* divisible by 7, then $a-3$ is also *not* divisible by 7.

Solution

We apply Claim 7.4.4 repeatedly, to prove the contrapositive. Namely, we show that if $a-3$ is divisible by 7, then so is a^2+5 (try to justify each of the following steps):

$$\begin{aligned} & a-3 \equiv \pmod{7} \\ \Rightarrow \quad & a \equiv 3 \pmod{7} \\ \Rightarrow \quad & a^2 \equiv 9 \equiv 2 \pmod{7} \\ \Rightarrow \quad & a^2 + 5 \equiv 7 \equiv 0 \pmod{7}. \end{aligned}$$

From which it follows that $a^2 + 5$ is divisible by 7, as needed.

7.5 Problems

7.1 Define, on the set of integers, the following equivalence relation:

$$k \sim \ell \qquad \text{if and only if} \qquad |k| = |\ell|.$$

a. Prove that the above relation is indeed an equivalence relation.
b. Describe the equivalence classes for this relation.

7.2 For each of the following relations R on *the set of real numbers*, decide whether it is reflexive, symmetric and/or transitive. Justify your arguments. Is the relation an equivalence relation? Explain.

a. $(x, y) \in R$ if and only if $|x - y| \leq 3$.
b. $(x, y) \in R$ if and only if $x \cdot y > 0$.
c. $(x, y) \in R$ if and only if $x^2 - y = y^2 - x$.
d. $(x, y) \in R$ if and only if $(x - y)(x^2 + y^2 - 1) = 0$.
e. $(x, y) \in R$ if and only if $|x + y| = |x| + |y|$.

7.3 Let us define the following relation on the set $\mathbb{Z} \setminus \{0\}$:

$$a \sim b \quad \text{if and only if} \quad a|2b \text{ or } b|2a.$$

Is this relation reflexive? Is it symmetric? Is it transitive? Justify your answers.

7.4 a. Define a relation R on $\mathbb{R} \setminus \{0\}$ as follows:

$$(x, y) \in R \text{ if and only if } x + \frac{1}{y} = y + \frac{1}{x}.$$

Is this relation reflexive? symmetric? transitive? Is it an equivalence relation? Explain.

b. Generalize the observation from Part a. Let $f : A \to B$ be an arbitrary function. Prove that the relation

$$x \sim y \quad \text{if and only if} \quad f(x) = f(y),$$

on the set A, is an equivalence relation.

7.5 Define a relation R on $\mathbb{Z}$ as follows: $(m, n) \in R$ if and only if $m + n$ is odd. Is the relation reflexive? symmetric? transitive? Is it an equivalence relation? Explain.

7.6 Define a relation R on $\mathbb{Z}$ as follows: $(m, n) \in R$ if and only if $m + n$ is divisible by 3. Is the relation reflexive? symmetric? transitive? Is it an equivalence relation? Explain.

7.7 Define a relation R on the set $\{2, 3, 4, \ldots\}$, as follows:

$(x, y) \in R$ if and only if x and y have a common factor greater than 1.

Is this relation reflexive? symmetric? transitive? Is this an equivalence relation? Justify your arguments.

7.8 Is the following relation on $\mathbb{Z}$ reflexive? symmetric? transitive? Is it an equivalence relation?

$$a \simeq b \quad \text{if and only if} \quad 3a + 5b \text{ is divisible by } 8.$$

Prove your arguments.

7.9 Let $\simeq$ be a relation on $\mathbb{Z}$ defined as follows:

$$a \simeq b \quad \text{if and only if} \quad 2a + 3b \text{ is divisible by } 5.$$

a. Show that $\simeq$ is an equivalence relation.
b. What is the equivalence class of 0 ?

7.10 Define a relation on $\mathbb{Q} \setminus \{0\}$ as follows:

$$x \sim y \qquad \Leftrightarrow \qquad \frac{x}{y} = 2^k \text{ for some } k \in \mathbb{Z}.$$

Prove that this is an equivalence relation.

7.11 Consider the following equivalence relation on $\mathbb{R} \setminus \{0\}$:

$$a \sim b \quad \text{if and only if} \quad \frac{a}{b} \in \mathbb{Q}.$$

In the following list of equivalence classes, find two classes which are *equal*. Explain your choice briefly.

- $[\sqrt{3}]$
- $[1]$
- $[\sqrt{12}]$
- $[\sqrt{6}]$

7.12 Consider the following equivalence relation on $\mathbb{Z}$:

$$n \approx m \qquad \text{if and only if} \qquad \cos(n) \cdot \cos(m) > 0.$$

a. Prove that this is an equivalence relation.
b. How many equivalence classes are there for this relation? Explain.

7.13 Consider the following relation on $\mathbb{R}$:

$$x \sim y \qquad \text{if and only if} \qquad x^2 - y^2 \in \mathbb{Z}.$$

a. Prove that this is an equivalence relation.
b. Let C be the equivalence class of 0 and I the closed interval $[5, 6]$. How many elements are in $C \cap I$? Explain.

7.14 On the set $\mathbb{N} \times \mathbb{N}$, define the following relation:

$$(a, b) \sim (c, d) \quad \text{if and only if} \quad a + d = b + c.$$

a. Show that this is an equivalence relation.
b. Describe the *equivalence class* of $(1, 1)$.

7.15 Consider the equivalence relation *congruence mod* 5, on the set of integers.

a. Describe the equivalence class of 33.
b. How many different equivalence classes are there for this relation?
c. Which $a \in \mathbb{Z}$ satisfy the condition $[a] = [17]$? Explain.

7.16 For each statement, decide whether it is *true* or *false*. Justify your answer briefly.

a. For every $n \in \mathbb{N}$ and $k \in \mathbb{Z}$, we have $k \equiv k \pmod{n}$.
b. For every $n \in \mathbb{N}$ and $k \in \mathbb{Z}$, we have $k \equiv n \pmod{n}$.
c. For every $a, b \in \mathbb{Z}$, we have $a \equiv b \pmod{1}$.
d. For every $m \in \mathbb{Z}$, we have $(m+3)^2 \equiv m^2 + 3^2 \pmod{6}$.
e. For every $m \in \mathbb{Z}$, we have $(m+3)^2 \equiv m^2 + 3^2 \pmod{9}$.
f. For every $a, b \in \mathbb{Z}$ and $n \in \mathbb{N}$, if $a^2 \equiv b^2 \pmod{n}$, then $a \equiv b \pmod{n}$.

7.17 The following statement is *false*:

"If a relation is both symmetric and transitive, then it is also reflexive."

Find the mistake in the following false proof.

Suppose that R is a relation on a set A, which is both symmetric and transitive, and let $x \in A$. Choose an element $y \in A$, for which $(x, y) \in R$. By symmetry, we also have $(y, x) \in R$, and since the relation R is transitive, we conclude from $(x, y) \in R$ and $(y, x) \in R$, that $(x, x) \in R$. Therefore, the relation R is also reflexive.

7.18 For each statement, decide whether it is *true* or *false*. Justify your answers.

a. For every two integers a and b, if $a \equiv b \pmod{3}$ and $a \equiv b \pmod{5}$, then $a \equiv b \pmod{15}$.
b. For every two integers a and b, if $a \equiv b \pmod{4}$ and $a \equiv b \pmod{6}$, then $a \equiv b \pmod{24}$.

7.19 Let a and b be two integers, and p a prime number. Show that if $ab \equiv 0 \pmod{p}$, then $a \equiv 0 \pmod{p}$ or $b \equiv 0 \pmod{p}$.
(Hint: This is a restatement of a result from Chapter 6.)

7.20 Let X, Y be two non-empty sets, $\sim$ an equivalence relation on X, and $f: X \to Y$ a function. Suppose that for every $a, b \in X$, we have $f(a) = f(b)$ whenever $a \sim b$. Show that there exists a function $\overline{f}: X/\sim \to Y$ for which $f(a) = \overline{f}([a])$ for every $a \in X$.

In other words, a function f on X which maps equivalent elements to the same image induces a function $\overline{f}$ on the quotient space $X/\sim$.

Can you relate this observation to Example 2.5.3 from Chapter 2?

7.21 a. Find the unit (rightmost) digit of 29^{8000}.
b. Find the remainder of 5^{2467} when divided by 3.

7.22 For each statement, decide whether it is *true* or *false*. Justify your answers.

a. For every $a, b \in \mathbb{Z}$, we have $(a+b)^4 \equiv a^4 + b^4 \pmod 4$.
b. For every $a, b \in \mathbb{Z}$, we have $(a+b)^5 \equiv a^5 + b^5 \pmod 5$.

7.23 Let $a, b \in \mathbb{Z}$ and $n \in \mathbb{N}$. For each statement decide whether it is necessarily true, or whether it can be false.

a. If $a \equiv b \pmod n$, then $\gcd(a, n) = \gcd(b, n)$.
b. If $\gcd(a, n) = \gcd(b, n)$, then $a \equiv b \pmod n$.

7.6 Solutions to Exercises

Solution to Exercise 7.3.4
Using the Division Algorithm (Theorem 6.1.4), we can write $a = 4q_1 + r_1$ and $b = 4q_2 + r_2$, with $q_1, q_2 \in \mathbb{Z}$ and $r_1, r_2 \in \{0, 1, 2, 3\}$. Consequently, we have

$$a - b = 4(q_1 - q_2) + r_1 - r_2.$$

If $r_1 = r_2$, then $a - b = 4(q_1 - q_2)$, which is clearly divisible by 4.

Conversely, if $a - b$ is divisible by 4, then $r_1 - r_2$ must be divisible by 4. However, as $-3 \leq r_1 - r_2 \leq 3$, we conclude that $r_1 = r_2$.

Now we argue that $\equiv$ is an equivalence relation. If $a, b \in \mathbb{Z}$ are integers, and $a = b$, then clearly they yield the same remainder when divided by 4. This proves reflexivity. Also, "having the same remainder" is certainly a symmetric relation. Finally, if $a, b, c \in \mathbb{Z}$, and we know that both pairs a, b and b, c produce equal remainders when divided by 4, then so does the pair a, c.

Solution to Exercise 7.3.8
We show that the relation $\approx$ is reflexive, symmetric and transitive, and hence an equivalence relation.

- If $(a, b) \in S$, then $(a, b) \approx (a, b)$ as $ab = ba$.
- If $(a, b) \approx (c, d)$ for $(a, b), (c, d) \in S$, then

$$ad = bc \qquad \Rightarrow \qquad cb = da \qquad \Rightarrow \qquad (c, d) \approx (a, b).$$

- Finally, if $(a, b), (c, d), (e, f) \in S$, with $(a, b) \approx (c, d)$ and $(c, d) \approx (e, f)$, then

$$ad = bc \qquad \text{and} \qquad cf = de.$$

We multiply the first equation by f

$$adf = bcf$$

and then use the second equation to replace cf by de:

$$adf = bde.$$

As $d \neq 0$, we can cancel it from both sides, which gives $af = be$. This shows that $(a, b) \approx (e, f)$, which proves transitivity of the relation.

Solution to Exercise 7.4.5

Suppose that $a \equiv r \pmod{n}$. We use induction on k to prove the conclusion. If $k = 1$, there is nothing to do, as we already know that $a^1 \equiv r^1 \pmod{n}$. Assume that $a^k \equiv r^k \pmod{n}$ for some k, and apply Claim 7.4.4:

$$a^k \cdot a \equiv r^k \cdot r \pmod{n} \qquad \Rightarrow \qquad a^{k+1} \equiv r^{k+1} \pmod{n}.$$

We have established the conclusion for $k + 1$, as needed, and so our proof is complete.

PART II
Additional Topics

8 Elementary Combinatorics

Combinatorics is an area of mathematics that focuses on counting arguments. It is often used to solve problems involving finite or countable sets in statistics, computer science, logic and other areas of mathematics. Combinatorics is viewed as part of discrete mathematics, a broad area in mathematics concerned with the study of finite or countable structures such as logical and algebraic structures, graphs and more. In this chapter, we discuss a few common tools for counting elements in a finite set. The terminology, notation, and proof techniques discussed in earlier chapters, in the context of sets and functions, will prove to be important and useful in addressing counting problems.

8.1 Counting Arguments: Selections, Arrangements and Permutations

Many problems in mathematics, science and everyday life involve counting the number of elements in a finite set. Quite often, the focus is on counting the number of possible outcomes of a certain process, involving selections or arrangements of objects according to given rules. In this section we describe common types of counting tasks that arise in such problems, and develop some strategies for approaching them.

We begin by looking at four elementary examples of counting problems.

Example 8.1.1 In how many different ways can a quiz be answered, if it contains ten multiple choice questions with four choices each?

Example 8.1.2 A club has 20 members. The offices of president, vice president, secretary and treasurer are to be filled, and no member may serve in more than one office. How many slates of candidates are possible?

Example 8.1.3 If a club has 20 members, how many four-member committees are possible?

Example 8.1.4 A child is allowed to choose three candies out of eight given types of candies. How many different choices are possible?

Let us take a closer look at the type of counting that needs to be done in each case.

Example 8.1.1

This example involves a quiz with ten questions, each of which is a multiple-choice question with four choices. If we label the questions as Q1, Q2, ..., Q10, and the choices as A, B, C and D, then a solved quiz may be described as a table like this one:

Q1	Q2	Q3	Q4	Q5	Q6	Q7	Q8	Q9	Q10
A	B	B	A	B	C	D	A	C	D

The table shows the choice picked for each question. In other words, a possible way of answering the quiz can be described as a *10-letter sequence* (often referred to as a *10-tuple*) involving only A, B, C and D. For instance, the above solution may be written as

ABBABCDACD.

Each letter represents a choice selected for one of the questions of the quiz. Note that *the order of the letters matters*. For instance, the sequence ABCCCCCCCC represents a different way of answering the quiz than BACCCCCCCC. In order to solve Example 8.1.1 we thus need to count the number of all such sequences. Can you now figure out the answer?

Exercise 8.1.5
Answer the question posed in Example 8.1.1.

In your solution to Example 8.1.1 you may have used a version of the following principle.

The Rule of Product. *Let k and $n_1, n_2, \ldots, n_k$ be natural numbers. Suppose a process involves k steps, and that there are n_i ways of performing the ith step. Then the number of different ways of executing the whole process is $n_1 \cdot n_2 \cdots n_k$.*

This rule is quite intuitive, and so we will not discuss the proof here. It can be proved from a more elementary rule called the *Rule of Sum*, mentioned later in this chapter (see Proposition 8.4.1 in Section 8.4).

Note that the Rule of Product can be used *even when the outcome of one step influences the possible outcomes of the next step.*

Exercise 8.1.6
Suppose that $k \in \mathbb{N}$ and $A_1, A_2, \ldots, A_k$ are finite sets. Explain how the Rule of Product implies that

$$|A_1 \times A_2 \times \cdots \times A_k| = |A_1| \cdot |A_2| \cdots |A_k|.$$

Example 8.1.2

Here, we have a club with 20 members, which we denote as M1, M2, ..., M20. We need to choose a member for each of the four offices, and so a slate of candidates can be described as a table with four columns.

President	Vice President	Secretary	Treasurer
M3	M13	M5	M18

Again, a slate can be viewed as a sequence of four elements (or a 4-tuple) from the set of club members, but here the situation is a bit different. As members cannot serve in more than one office, we should not count slates in which a member appears more than once (for instance, the sequence M2, M5, M9, M5 should not be counted). This means that once we choose a member to serve as president, we can choose a vice president from the remaining 19 members. And once the vice president is chosen, we have 18 members left to choose from for secretary. The Rule of Product can be applied here, keeping in mind that the number of choices decreases in each step.

Note that, as in Example 8.1.1, *the order of club members in a slate matters*, as each entry corresponds to a specific office.

Exercise 8.1.7
Answer the question posed in Example 8.1.2.

Definition 8.1.8 Let S be a finite set, and $k \in \mathbb{N}$. An *arrangement* (or *variation*) of k objects from S is a sequence formed by k distinct elements from S. Such a sequence is also called *a k-permutation of S*.

In Example 8.1.2 we counted the number of arrangements of 4 objects from the set $S = \{M1, M2, \ldots, M20\}$. We can generalize the strategy used in the example to obtain a general formula.

Theorem 8.1.9 *Let S be a non-empty set with n elements and $1 \leq k \leq n$. The number of arrangements of k objects from S is equal to*

$$n \cdot (n-1) \cdot (n-2) \cdots (n-k+1),$$

or, equivalently, to

$$\frac{n \cdot (n-1) \cdots (n-k+1) \cdot [(n-k) \cdot (n-k-1) \cdots 2 \cdot 1]}{(n-k) \cdot (n-k-1) \cdots 2 \cdot 1} = \frac{n!}{(n-k)!}.$$

Proof The proof is obtained from the Rule of Product. As the elements in an arrangement are distinct, we have n ways of choosing the first element, $n-1$ ways to choose the second, and so on. For the last element, there are $n-k+1$ ways. Multiplying these quantities yields

$$n \cdot (n-1) \cdot (n-2) \cdots (n-k+1).$$

If we now multiply and divide this product by

$$(n-k) \cdot (n-k-1) \cdots 2 \cdot 1$$

we obtain the equivalent expression in terms of factorials. □

Remark. An arrangement of all the elements of a finite set S is called a *permutation*. If S has k elements, then there are $k!$ permutations that can be formed from the elements of S. This follows directly from Theorem 8.1.9 when $n = k$.

Exercise 8.1.10
A mother is taking her four children to the park. At lunch, she asks them to sit next to each other on a bench. In how many ways can the four children sit?

Example 8.1.3

Moving on, we now look at Example 8.1.3, which appears to be quite similar to Example 8.1.2. The difference is that now we are required to count the number of four-member committees from a club that has 20 members. There is no mentioning of specific offices that need to be filled. How would that affect the solution?

A possible slate of members can be again described as a four-element sequence, say M5, M7, M3, M14. But now, the order in which we list the members is irrelevant. Here, we are required to simply count four-member committees, and so sequences such as M7, M3, M14, M5 and M14, M7, M3, M5 will describe the same committee.

In fact, if we pick four club members, we can list them in a sequence in $4! = 24$ ways. After all, this is the number of permutations formed from a set with four elements. Consequently, we should divide our answer to Example 8.1.2 by $4!$ to obtain the answer to Example 8.1.3. Specifically, the number of four-member committees that can be formed at a club with 20 members is

$$\frac{20 \cdot 19 \cdot 18 \cdot 17}{4!} = \frac{116280}{24} = 4845.$$

In other words, we had to count the number of four-element subsets (rather than sequences) that can be formed from a set with 20 elements.

Definition 8.1.11 Let S be a finite set of size n, and $0 \leq k \leq n$. A *selection* (or *combination*) of k elements from S is a subset of S of size k. The number of such selections is denoted as $\binom{n}{k}$ and reads *n choose k*.

Theorem 8.1.12 *Let $n, k \in \mathbb{Z}$ with $0 \leq k \leq n$. The number of selections of k elements from a set of size n is given by*

$$\binom{n}{k} = \frac{n!}{(n-k)! \cdot k!}.$$

Proof First, we check the case $k = 0$. The number of subsets of S with 0 elements is 1 (only the empty set), and so $\binom{n}{0} = 1$, which aligns with the right-hand side of the identity, as

$$\frac{n!}{(n-0)! \cdot 0!} = \frac{n!}{n! \cdot 1} = 1.$$

Now suppose $1 \leq k \leq n$. By Theorem 8.1.9, the number of arrangements of k elements from a set of n elements is

$$\frac{n!}{(n-k)!}.$$

In a selection, the order does not matter. The number of ways to list k elements in a sequence is $k!$ (see Remark on page 210). Therefore, there are $k!$ arrangements that correspond to a single given selection. Consequently, to count the number of selections, we divide the above formula by $k!$, and get

$$\frac{n!}{(n-k)!} \cdot \frac{1}{k!} = \frac{n!}{(n-k)! \cdot k!},$$

as needed. □

Exercise 8.1.13

On a shelf, there are ten books in English and ten books in French. In how many ways can we choose three books in English and four books in French?

Example 8.1.4

Finally, we address the last example from the beginning of the section. Here, a child can choose three candies out of eight given types of candies. We need to figure out how many different choices are possible. Denote the types of candies by T1, T2, ..., T8.

A choice of three candies is a choice of three types. For instance, T2, T5, T8 describes one possible choice. Note that the order does not matter here. The sequence T5, T8, T2 represents the same choice. Moreover, repetitions are allowed, as the child can choose two candies of the same type, or perhaps all the three candies to have the same type. Therefore, we are required to count the number of possible *selections with repetitions.*

Here the situation is a little tricky. A choice of three candies can be represented using eight bins, each corresponding to a type of candy. For instance, the following describes a choice of one candy of type T2 and two candies of type T5.

T1	T2	T3	T4	T5	T6	T7	T8
	1			2			

Alternatively, we can represent the above choice as follows:

$$0 \mid 1 \mid 0 \mid 0 \mid 2 \mid 0 \mid 0 \mid 0$$

Here, we removed the first row, and used vertical bars to separate the columns. We can still see clearly the choice being represented: one candy of type T2 and two candies of type T5. Additionally, we can simply replace the numbers 1 and 2 by dots representing single candies, and obtain the following diagram:

$$\mid \bullet \mid\mid\mid \bullet \bullet \mid\mid\mid$$

The last diagram determines uniquely the choice made by the child. A choice of three candies of types T1, T5 and T8 will look as follows:

$$\bullet \mid\mid\mid\mid \bullet \mid\mid\mid \bullet$$

And a choice where all the three candies are of type T2 corresponds to the following diagram:

$$\mid \bullet \bullet \bullet \mid\mid\mid\mid\mid\mid$$

Note that all the diagrams have 7 bars separating the 8 candy types, and three dots representing the candies. Altogether there are $7+3=10$ symbols that are needed to set a 10-element sequence. Once we choose where to put the dots, the location for the bars is determined. Therefore, the total number of choices is given by selecting three objects (the candies) out of a set with 10 elements (the 10 available slots) and so there are $\binom{10}{3}=120$ possible choices. A similar strategy can be used to determine a general formula.

This method, of representing a selection with repetitions using dots and bars, is often referred to as the *sticks and stones* method. Other common names are *stars and bars*, *balls and bars* and *dots and dividers*.

Theorem 8.1.14 *Let n, k be two non-negative integers. The number of ways to select k elements from a set with n elements, where repetitions are allowed and the order does not matter is* $\binom{n+k-1}{k}$.

Exercise 8.1.15

Prove Theorem 8.1.14 by using the strategy applied in answering Example 8.1.4.

Example 8.1.16 How many non-negative integer solutions does the equation $x+y+z+w=30$ have?

Solution

A solution is obtained by starting with 30 "units" and assigning each of them to either x, y, z or w. This is the same as selecting 30 elements from a set of four elements, with repetitions. By Theorem 8.1.14, the number of such selections is $\binom{4+30-1}{30} = 5456$.

Example 8.1.17 Obtain a general formula for the number of non-negative integer solutions to the equation $x_1 + x_2 + \cdots + x_n = k$, where $k, n \in \mathbb{N}$.

Solution

Again, the number of solutions is the same as the number of selections, with repetitions, of k elements from a set with n elements (the order does not matter here). Thus, the number of solutions is simply $\binom{n+k-1}{k}$.

We end the section with a summary of the formulas we developed for counting the number of possible ways to make k choices from a set of n elements (see Table 8.1). In each scenario, we indicate the appropriate formula, depending on whether we are allowed repetitions, and whether the order in which elements are chosen matters.

Example 8.1.18 In the English alphabet there are 26 letters. How many four-letter words are there, in which at least two of the letters are identical?

In this context, a word is *any finite ordered list* of letters from the alphabet (even if it is meaningless).

Table 8.1 Choosing k elements from a set with n elements

	With repetitions	**Without repetitions**
Order matters	n^k	$\frac{n!}{(n-k)!}$
Order does not matter	$\binom{n+k-1}{k}$	$\binom{n}{k}$

Solution

We offer two different solutions to this example.

First solution

Let us count the number of words as follows.

- The number of words in which all letters are the same is 26.
- If exactly three of the letters are identical, we obtain

$$26 \cdot 25 \cdot 4 = 2600$$

  words. This is because there are 26 ways to choose the repeated letter, 25 ways to choose the other letter, and four options of where to place the non-repeated letter in the word.
- Suppose that exactly two of the four letters are identical (for instance, ABBJ, CLCO, PIOP). How many such words do we have? There are 26 ways to choose the repeated letter, and $\binom{25}{2}$ to choose the other two. After choosing the letters, there are $\binom{4}{2} = 6$ ways to place the repeated letters in the four-letter word. There are then two ways to place the remaining letters in the vacant places. Overall, we get

$$26 \cdot \binom{25}{2} \cdot 6 \cdot 2 = 26 \cdot \frac{25 \cdot 24}{2} \cdot 12 = 93600.$$

- If the word has two pairs of identical letters (such as GHHG or MMPP), we have $\binom{26}{2}$ ways to choose two distinct letters from the alphabet, and $\binom{4}{2}$ ways to decide where to place one pair of identical letters. Overall, we have

$$\binom{26}{2} \cdot 6 = 1950$$

  such words.

Overall, the answer to the question is

$$26 + 2600 + 93600 + 1950 = 98176.$$

Second solution

Here is a shorter solution, using a different strategy. Denote by U the set of all four-letter words (without restrictions), and by A the set of all four-letter words in which all letters are distinct. Then, using the formulas developed, we have

$$|U| = 26^4 = 456976 \qquad \text{and} \qquad |A| = 26 \cdot 25 \cdot 24 \cdot 23 = 358800.$$

The set of words in which two or more letters are identical is the complement $U \setminus A$. Therefore, we have

$$|U \setminus A| = |U| - |A| = 456976 - 358800 = 98176,$$

which is the same answer obtained using the first method.

8.2 The Binomial Theorem and Pascal's Triangle

In the previous section we introduced the notation $\binom{n}{k}$ for integers satisfying $0 \leq k \leq n$, which represents the number of possible selections of k elements, without repetitions, from a set with n elements. We now see how these quantities show up naturally in an important algebraic identity.

Suppose we expand the expression $(x+y)^5$:

$$(x+y)^5 = (x+y)(x+y)(x+y)(x+y)(x+y).$$

Once we multiply out the terms, we obtain a long sum of products involving x and y. For instance, x^5, x^2y^3 and xy^4 are some of the terms that show up in the expansion. Every summand is a product of five factors, each being either x or y. For example, the term x^2y^3 is obtained when we use the x in the first and second occurrence of $x+y$, and the y from remaining terms.

However, the term x^2y^3 will appear more than once. If we pick y from the second, third and fifth factors of $(x+y)^5$, and x from the other factors, we get again $xyyxy = x^2y^3$. In fact, any choice of y from three of the five $(x+y)$ factors, and x from the others, will result in x^2y^3. Consequently, the number of terms of the form x^2y^3 is the same as the number of ways of selecting three out of five possible $(x+y)$ factors. We thus conclude that the term x^2y^3 appears $\binom{5}{3} = 10$ times in the expansion of $(x+y)^5$. Similarly, the term x^4y appears $\binom{5}{1} = 5$ times in the expansion. In general, a term of the form x^iy^j, where i, j are non-negative integers satisfying $i+j=5$, appears $\binom{5}{j}$ times in the expansion of $(x+y)^5$, and so we get:

$$\begin{aligned}(x+y)^5 &= \binom{5}{0}x^5 + \binom{5}{1}x^4y + \binom{5}{2}x^3y^2 + \binom{5}{3}x^2y^3 + \binom{5}{4}xy^4 + \binom{5}{5}y^5 \\ &= x^5 + 5x^4y + 10x^3y^2 + 10x^2y^3 + 5xy^4 + y^4.\end{aligned}$$

A similar argument leads to the general formula for $(x+y)^n$.

Theorem 8.2.1 (The Binomial Theorem.) *Let $n \in \mathbb{N}$ and $x, y \in \mathbb{R}$. Then*

$$(x+y)^n = \binom{n}{0}x^n + \binom{n}{1}x^{n-1}y + \binom{n}{2}x^{n-2}y^2 + \cdots + \binom{n}{n-1}xy^{n-1} + \binom{n}{n}y^n.$$

Equivalently, we can write

$$(x+y)^n = \sum_{j=0}^{n} \binom{n}{j} x^{n-j}y^j.$$

Proof The expression $(x+y)^n$ is a product of n factors. The expansion will be a sum of terms, each having the form $x^{n-j}y^j$ (where $0 \leq j \leq n$). The term

$x^{n-j}y^j$ comes from choosing y from j of the $x+y$ factors, and x from the other factors. Thus, there are $\binom{n}{j}$ occurrences of $x^{n-j}y^j$, which justifies the formula

$$(x+y)^n = \sum_{j=0}^{n} \binom{n}{j} x^{n-j} y^j.$$

□

Remarks.

1. The Binomial Theorem remains valid in other contexts, such as when x and y are complex numbers, or simply treated as variables.
2. The proof presented uses a combinatorial argument. That is, it involves counting. There are other ways to prove the theorem. In Problem 8.8 you are required to use mathematical induction to prove the theorem.
3. Due to Theorem 8.2.1, the quantities $\binom{n}{j}$ are often called the *binomial coefficients*.

Exercise 8.2.2
The term x^3y^k appears in the expansion of $(x+y)^8$. Find the value of k, and the coefficient of that term in the expansion.

The binomial coefficients have many properties and satisfy various relations. We present some of the commonly used ones in the following proposition.

Proposition 8.2.3 *Let $n \in \mathbb{N}$. Then the following hold.*

1. $\binom{n}{0} + \binom{n}{1} + \binom{n}{2} + \cdots + \binom{n}{n-1} + \binom{n}{n} = 2^n$.
2. *For every $0 \le k \le n$, $\binom{n}{k} = \binom{n}{n-k}$ (symmetry).*
3. *For every $1 \le k \le n$, $\binom{n+1}{k} = \binom{n}{k-1} + \binom{n}{k}$ (Pascal's Identity).*
4. *For every $1 \le k \le n$, $k\binom{n}{k} = n\binom{n-1}{k-1}$ (the chairperson identity).*

Proof 1. One quick way to obtain this identity is by setting $x = y = 1$ in the Binomial Theorem. Then the left-hand side becomes 2^n, and the right-hand side becomes

$$\sum_{j=0}^{n} \binom{n}{j},$$

as needed.

Alternatively, we offer the following combinatorial argument. We have already proved (see Claim 4.1.12 on page 105) that the number of subsets of a set with n elements is 2^n. On the other hand, each term in the expression

$$\binom{n}{0}+\binom{n}{1}+\binom{n}{2}+\cdots+\binom{n}{n-1}+\binom{n}{n}$$

represents the number of subsets with a prescribed number of elements: $\binom{n}{1}$ is the number of subsets with one element, $\binom{n}{2}$ is the number of subsets with two elements, and so on. Clearly, the sum gives the total number of subsets, and hence is equal to 2^n.

2. This identity follows immediately from the algebraic expression for the binomial coefficients (Theorem 8.1.12):

$$\binom{n}{n-k}=\frac{n!}{[n-(n-k)]!\cdot(n-k)!}=\frac{n!}{k!\cdot(n-k)!}=\binom{n}{k}.$$

Again, we can also use a combinatorial argument to justify the identity. Suppose S is a set with n elements. Then $\binom{n}{k}$ represents the number of selections of k elements from S, and $\binom{n}{n-k}$ can be viewed as representing the number of ways of omitting k elements from S. As a subset of k elements is determined uniquely by deciding which elements to omit, the two binomial coefficients are equal.

3. We prove Pascal's Identity using a combinatorial argument. Suppose S is a set with $n+1$ elements, which we denote by $x_1, x_2, \ldots, x_n$ and y:

$$S=\{x_1, x_2, \ldots, x_n, y\}.$$

Every subset of S with k elements either does or does not include the element y. Subsets with k elements that do not include y are, in fact, subsets of $\{x_1, \ldots, x_n\}$ and so there are $\binom{n}{k}$ of those. Subsets that do include y can be constructed by first choosing $k-1$ elements from $\{x_1, \ldots, x_n\}$ and then adding y as the kth element. Consequently, there are $\binom{n}{k-1}$ such subsets. As every subset of S either includes or does not include y, we obtain Pascal's Identity:

$$\underbrace{\binom{n+1}{k}}_{\text{subsets of } S}=\underbrace{\binom{n}{k-1}}_{\substack{\text{subsets of } S\\ \text{that include } y}}+\underbrace{\binom{n}{k}}_{\substack{\text{subsets of } S \text{ that}\\ \text{do not include } y.}}$$

4. To prove the last identity, and also justify its name, we explain how both sides represent the number of ways to form a k-person committee with a designated chair from a group of n people. There are $\binom{n}{k}$ ways to select a k-person committee, and then k ways of selecting a chair from the chosen committee. Combining these two steps (and using the Rule of Product), we get the left-hand side $k\binom{n}{k}$. Alternatively, we can first choose a person from the group to be the committee's chair, and then select the remaining $k-1$ members. This

approach leads to the right-hand side of the identity, $n\binom{n-1}{k-1}$, as there are n ways to choose a chair, and then $\binom{n-1}{k-1}$ to complete the committee. □

Exercise 8.2.4
Use Theorem 8.1.12 to provide an algebraic proof of Pascal's Identity and the chairperson identity.

Pascal's Triangle

The binomial coefficients can be listed in a structure called *Pascal's Triangle*. Starting with $n = 0$, each line of the triangle contains the binomial coefficients $\binom{n}{k}$ for $0 \leq k \leq n$. Here are the first few lines of this triangle.

$$
\begin{array}{lccccccccccccc}
n=0 & & & & & & & 1 & & & & & & \\
n=1 & & & & & & 1 & & 1 & & & & & \\
n=2 & & & & & 1 & & 2 & & 1 & & & & \\
n=3 & & & & 1 & & 3 & & 3 & & 1 & & & \\
n=4 & & & 1 & & 4 & & 6 & & 4 & & 1 & & \\
n=5 & & 1 & & 5 & & 10 & & 10 & & 5 & & 1 & \\
n=6 & 1 & & 6 & & 15 & & 20 & & 15 & & 6 & & 1 \\
\vdots & & \vdots & & & & \vdots & & \vdots & & & & \vdots &
\end{array}
$$

Pascal's Identity from Proposition 8.2.3 indicates how each entry in the triangle (other than the 1s) is the sum of the two entries immediately above it. For instance, in the row corresponding to $n = 4$, the numbers $4, 6, 4$ are obtained as sums of entries from the previous row:

$$4 = 1 + 3, \quad 6 = 3 + 3, \quad 4 = 3 + 1.$$

Also note how the symmetry property $\binom{n}{k} = \binom{n}{n-k}$ can be seen in the triangle as symmetry about the vertical axis passing through the vertex of the triangle.

Exercise 8.2.5
Use Pascal's Identity to calculate the entries in row $n = 7$ from the previous row.

In our next example, we use mathematical induction to prove another identity involving binomial coefficients.

Example 8.2.6 Let n, j be two integers satisfying $0 \leq j \leq n$. Prove that

$$\sum_{i=j}^{n} \binom{i}{j} = \binom{n+1}{j+1}.$$

Solution

We perform an induction on n. As n is a non-negative integer, our base case is $n = 0$. In that case, j must be zero, and we get:

$$\sum_{i=0}^{0} \binom{i}{0} = \binom{0}{0} = 1 = \binom{1}{1}.$$

This verifies the base case.

Now suppose that the given identity holds true for some non-negative integer $n = k$ and for all $0 \leq j \leq k$. We prove the identity for $n = k + 1$. Using the induction hypothesis, we get for $0 \leq j \leq k$

$$\sum_{i=j}^{k+1} \binom{i}{j} = \left[\sum_{i=j}^{k} \binom{i}{j}\right] + \binom{k+1}{j} = \binom{k+1}{j+1} + \binom{k+1}{j}$$

and using Pascal's Identity

$$= \binom{k+2}{j+1},$$

as needed. The case $j = k + 1$ can be checked directly. This completes the proof of the identity.

8.3 The Pigeonhole Principle

The Pigeonhole Principle is a relatively intuitive principle, stating that if more than n objects are placed into n boxes, then at least one box must contain two objects or more. We often think of pigeons as our objects who are being placed into pigeonholes, that is, our boxes.

Theorem 8.3.1 (The Pigeonhole Principle) *Let $n, m \in \mathbb{N}$ with $m > n$. If m objects are placed into n boxes, then at least one box contains two objects or more.*

Although the principle is quite intuitive, we present a proof by contradiction.

Proof Denote by $B_1, B_2, \ldots, B_n$ our n boxes (formally speaking, these are just sets). Suppose, by contradiction, that every box contains less than two elements. Then the total number of objects satisfies

$$|B_1| + |B_2| + \cdots + |B_n| \leq 1 + 1 + \cdots + 1 = n,$$

which is impossible, as we are told that the number of objects, m, is greater than n. Therefore, $|B_i| \geq 2$ for at least one $1 \leq i \leq n$. □

The Pigeonhole Principle may seem, at first, too elementary to be of any interest or use in advanced applications. Surprisingly, applying this principle cleverly can lead to far-reaching and often unexpected consequences. Here are a few examples.

Example 8.3.2 In every set of eight integers, there are two numbers whose difference is divisible by 7.

Solution

Suppose $A = \{k_1, k_2, k_3, k_4, k_5, k_6, k_7, k_8\}$ is a set of eight integers. According to the Division Algorithm (Theorem 6.1.4), each of these numbers can be divided by 7 with remainder. The possible remainders in this case are $0, 1, 2, 3, 4, 5, 6$. As there are eight numbers (our pigeons) and seven possible remainders (our pigeonholes), we conclude, from the Pigeonhole Principle, that one remainder must be obtained at least twice. That is, there exist two numbers in A, say k_i and k_j, with $i \neq j$, such that

$$k_i = 7q + r \qquad \text{and} \qquad k_j = 7q' + r$$

(where $q, q', r \in \mathbb{Z}$ and $0 \leq r \leq 6$). Consequently, we get $k_i - k_j = 7q - 7q'$, which is divisible by 7, as needed.

Example 8.3.3 Let $n \in \mathbb{N}$, and S a set of $2n + 2$ real numbers in the interval $[0, n]$. Prove that there exists a pair of numbers in S at a distance of less than $\frac{1}{2}$ from each other.

Solution

Divide the interval $[0, n]$ into $2n + 1$ closed sub-intervals of equal width. Since S has $2n + 2$ numbers, two of them must fall into the same sub-interval, by the Pigeonhole Principle. The length of each sub-interval is $\frac{n}{2n+1} < \frac{1}{2}$, and so the distance between the two numbers in the same sub-interval must be less than $\frac{1}{2}$, as needed.

The Pigeonhole Principle can be generalized in several ways. Here is one possible generalization.

Theorem 8.3.4 (The Generalized Pigeonhole Principle) *Let $n, m, k \in \mathbb{N}$ with $m > kn$. If m objects are placed into n boxes, then at least one box contains $k + 1$ objects or more.*

Exercise 8.3.5

Prove the Generalized Pigeonhole Principle.

Example 8.3.6 Let $n \in \mathbb{N}$ and S a subset of $\{1, 2, 3, \ldots, 3n\}$ having $2n + 1$ elements. Prove that S contains three consecutive numbers.

Solution

Consider the following partition of the set $\{1, 2, 3, \ldots, 3n\}$ into n subsets (these are our boxes):

$$\{1, 2, 3\}, \{4, 5, 6\}, \ldots, \{3n - 2, 3n - 1, 3n\}.$$

As S has $2n+1$ elements, at least one of the subsets must contain three elements from S (or else the total number of elements in S would be at most $2n$, which is a contradiction). Therefore, S must contain three consecutive numbers.

8.4 The Inclusion-Exclusion Principle

In this section we present another counting principle, that generalizes the following proposition.

Proposition 8.4.1 (The Rule of Sum) *If A and B are two disjoint finite sets, then the number of elements in $A \cup B$ is the sum of the number of elements in A and in B. In other words:*

$$|A \cup B| = |A| + |B| \qquad \text{if } A \cap B = \varnothing.$$

Proof Suppose that the sets A and B have m and n elements, respectively, where $m, n \in \mathbb{N} \cup \{0\}$.

If $n = 0$, then $B = \varnothing$, and we have

$$|A \cup B| = |A \cup \varnothing| = |A| + |B|.$$

Similarly, the statement holds true if $m = 0$.

Now assume that m and n are both non-zero. Then, we have bijections

$$f \colon \{1, 2, \ldots, m\} \to A \qquad \text{and} \qquad g \colon \{1, 2, \ldots, n\} \to B.$$

We can now define a function $h \colon \{1, 2, \ldots, m + n\} \to A \cup B$ as follows:

$$h(i) = \begin{cases} f(i) & \text{if } 1 \leq i \leq m \\ g(i - m) & \text{if } m + 1 \leq i \leq m + n. \end{cases}$$

This function is a bijection. It is surjective as

$$\begin{aligned} h(\{1, 2, \ldots, m + n\}) &= h(\{1, 2, \ldots, m\}) \cup h(\{m + 1, m + 2, \ldots, m + n\}) \\ &= f(\{1, 2, \ldots, m\}) \cup g(\{1, 2, \ldots, n\}) = A \cup B. \end{aligned}$$

Moreover, h is injective. Suppose that $h(i) = h(j)$ for some $1 \leq i, j \leq m + n$. We look at the following cases.

- If $1 \leq i, j \leq m$, then $h(i) = h(j)$ implies $f(i) = f(j)$, and since f is injective we get $i = j$.

- Similarly, if $m+1 \leq i,j \leq m+n$ then we get $g(i-m) = g(j-m)$ which gives us again $i = j$.
- Finally, we cannot have $i \leq m$ and $j > m$ (or vice versa), as this implies that $h(i) \in A$ and $h(j) \in B$. This is impossible, as $h(i) = h(j)$ and $A \cap B = \varnothing$.

We see that $h(i) = h(j)$ necessarily implies $i = j$, which proves the injectivity of h.

We have constructed a bijection between $\{1, 2, \ldots, m+n\}$ and $A \cup B$, from which it follows that

$$|A \cup B| = m + n = |A| + |B|,$$

as needed. □

Example 8.4.2 How many elements are in the following set?

$$C = \{-100, -99, -98, \ldots, 99, 100\} \setminus \{-20, -19, \ldots, 9, 10\}$$

Solution

By the definition of difference of sets, the set C is equal to

$$\{-100, -99, \ldots, -21\} \cup \{11, 12, \ldots, 100\}.$$

The set $A = \{-100, -99, \ldots, -21\}$ has 80 elements, and $B = \{11, 12, \ldots, 100\}$ has 90 elements. As $A \cap B = \varnothing$, the Rule of Sum implies that $|C| = |A| + |B| = 80 + 90 = 170$.

The Rule of Sum, however, cannot be applied directly with two sets which are not necessarily disjoint. Here is an example.

Example 8.4.3 How many integers between 1 and 101 are divisible by 5 or 7 (or both)?

Solution

Let us denote by A and B the integers between 1 and 101 which are divisible by 5 and by 7, respectively. That is,

$$A = \{5, 10, 15, 20, \ldots, 100\} \qquad \text{and} \qquad B = \{7, 14, 21, 28, \ldots, 98\}.$$

The set A has 20 elements, and the set B has 14 elements. Can we conclude that $A \cup B$ has $20 + 14 = 34$ elements? Not quite! The Rule of Sum does not apply here, as $A \cap B \neq \varnothing$. There are numbers which are divisible by both 5 and 7. Specifically, these are the multiples of 35. Therefore, we have $A \cap B = \{35, 70\}$. This means that when we added $|A|$ and $|B|$, the numbers 35 and 70, which belong to both sets, were counted twice. Consequently, $|A \cup B| = 34 - 2 = 32$.

In general, the following formula can be used to calculate the number of elements in a union of two finite sets:

$$|A \cup B| = |A| + |B| - |A \cap B|.$$

Exercise 8.4.4
Use the Rule of Sum to prove the above formula.

With three sets, things get a little bit more complicated. Suppose that A, B and C are finite sets, and that we wish to find a formula for $|A \cup B \cup C|$. We can start by looking at $|A| + |B| + |C|$. In this sum, elements in more than one of the sets are counted multiple times, so we might wish to adjust this formula and consider

$$|A| + |B| + |C| - |A \cap B| - |B \cap C| - |A \cap C|.$$

This will almost fix the problem, as elements present in exactly two sets are now counted only once. However, elements in the intersection $A \cap B \cap C$ are counted and subtracted three times in the above formula, and so overall they are not counted. We thus must add the term $|A \cap B \cap C|$ to obtain the correct formula.

Proposition 8.4.5 *Let A, B and C be finite sets. Then*

$$|A \cup B \cup C| = |A| + |B| + |C| - |A \cap B| - |B \cap C| - |A \cap C| + |A \cap B \cap C|.$$

Proof We prove this identity by making sure that every element in $A \cup B \cup C$ is counted exactly once in the formula on the right-hand side. An alternative proof is discussed in the exercise below.

Suppose $x \in A \cup B \cup C$. If x is in exactly one of the sets A, B or C, it will be counted once in either $|A|$, in $|B|$ or in $|C|$, but not in any of the other terms.

If x appears in exactly two of the three sets, it is counted twice in $|A| + |B| + |C|$, but is then removed once, as it is present in either $A \cap B$, $B \cap C$ or $A \cap C$. If x appears in all three sets, it is counted three times in $|A| + |B| + |C|$, then subtracted three times through the terms $-|A \cap B| - |B \cap C| - |A \cap C|$, and finally added once in $|A \cap B \cap C|$. Overall, x is counted once on the right-hand side. □

Exercise 8.4.6
Prove Proposition 8.4.5 by writing

$$A \cup B \cup C = A \cup (B \setminus (A \cap B)) \cup (C \setminus (C \cap (A \cup B)))$$

and then use the Rule of Sum and the above result for two sets.

Exercise 8.4.7

Can you derive a formula for the union of four finite sets? Can you justify it through an arument similar to the proof of Proposition 8.4.5?

The Inclusion-Exclusion Principle, stated below, generalizes the above discussion to an arbitrary (but finite) collection of finite sets. That is, given finite sets $A_1, A_2, \ldots, A_n$, the principle provides a formula for $|\bigcup_{i=1}^{n} A_i|$ in terms of the sizes of the A_i and their intersections.

Note that for $n = 2, 3, 4$ we already have formulas for the size of the union, and we may write them as follows.

For $n = 2$:

$$\left|\bigcup_{i=1}^{2} A_i\right| = \left(\sum_{i=1}^{2} |A_i|\right) - |A_1 \cap A_2|.$$

For $n = 3$:

$$\left|\bigcup_{i=1}^{3} A_i\right| = \left(\sum_{1 \le i \le 3} |A_i|\right) - \left(\sum_{1 \le i < j \le 3} |A_i \cap A_j|\right) + |A_1 \cap A_2 \cap A_3|.$$

And for $n = 4$:

$$\begin{aligned} \left|\bigcup_{i=1}^{4} A_i\right| &= \left(\sum_{1 \le i \le 4} |A_i|\right) - \left(\sum_{1 \le i < j \le 4} |A_i \cap A_j|\right) \\ &+ \left(\sum_{1 \le i < j < k \le 4} |A_i \cap A_j \cap A_k|\right) - |A_1 \cap A_2 \cap A_3 \cap A_4|. \end{aligned}$$

At this point, the pattern should be clear. We start by adding up the number of elements in each set A_i, then subtract the number of elements in each intersection $A_i \cap A_j$, then add the number of elements in each intersection $A_i \cap A_j \cap A_k$, and so on. To be able to write the general formula in a clear way, we introduce the following notation.

- Denote by S_1 the sum $\sum_{i=1}^{n} |A_i|$.
- Now, denote by S_2 the sum of the number of elements in all possible intersections of two different sets:

$$S_2 = \sum_{1 \le i < j \le n} |A_i \cap A_j|.$$

- Next, denote by S_3 the sum of the number of elements in all possible intersections of three different sets:

$$S_3 = \sum_{1 \le i < j < k \le n} |A_i \cap A_j \cap A_k|.$$

- And so on. Note that $S_n = |A_1 \cap A_2 \cap \cdots \cap A_n|$.

We are now ready to state and prove the Inclusion-Exclusion Principle.

Theorem 8.4.8 (The Inclusion-Exclusion Principle) *Let $A_1, A_2, \ldots, A_n$ be finite sets. Then, using the quantities S_i defined above, we have*

$$\left|\bigcup_{i=1}^{n} A_i\right| = S_1 - S_2 + S_3 - S_4 + \cdots + (-1)^{n-1} S_n.$$

Or, equivalently,

$$\left|\bigcup_{i=1}^{n} A_i\right| = \sum_{j=1}^{n} (-1)^{j-1} S_j.$$

We offer two proofs for the theorem. The first uses a counting argument and is somewhat informal. The second proof uses induction. It is more formal, but also more technical and harder to follow.

Proof 1 (using counting) Let $x \in \bigcup_{i=1}^{n} A_i$, and suppose that x is present in exactly k of the sets $A_1, \ldots, A_n$. On the left-hand side, the element x is counted once. What about the right-hand side?

- In S_1, the element x is counted k times (or $\binom{k}{1}$ times).
- In S_2, x is counted whenever it is present in an intersection $A_i \cap A_j$. There are precisely $\binom{k}{2}$ such intersections, so x is counted $\binom{k}{2}$ times.
- Similarly, in S_3, x is counted $\binom{k}{3}$ times, and so on.

Overall, the number of times x is counted on the right-hand side is

$$\binom{k}{1} - \binom{k}{2} + \binom{k}{3} - \binom{k}{4} + \cdots + (-1)^{k-1}\binom{k}{k}$$

(note that x is not counted at all in S_i if $i > k$). The Binomial Theorem can be used to show that this sum is equal to 1 (see Exercise 8.4.9 below), and so x is counted once on the right-hand side, as needed. □

And here is the second proof.

Proof 2 (by induction) The theorem is valid when $n = 1$, and we have already established the case $n = 2$ (see Exercise 8.4.4). Suppose that the statement is true for $n = k$, where $k \geq 2$, and consider the case where $n = k+1$.

Suppose that $A_1, A_2, \ldots, A_{k+1}$ are finite sets. Using the case of two sets and De Morgan's Laws, we get

$$\left|\bigcup_{i=1}^{k+1} A_i\right| = \left|\left(\bigcup_{i=1}^{k} A_i\right) \cup A_{k+1}\right| = \left|\bigcup_{i=1}^{k} A_i\right| + |A_{k+1}| - \left|\left(\bigcup_{i=1}^{k} A_i\right) \cap A_{k+1}\right|$$

$$= \left|\bigcup_{i=1}^{k} A_i\right| + |A_{k+1}| - \left|\bigcup_{i=1}^{k} (A_i \cap A_{k+1})\right|.$$

We now use the following notation.

- Let S'_j denote, for $1 \leq j \leq k$, the sum of the number of elements in all possible intersections of j sets from $A_1, \ldots, A_k$. That is, we define:

$$S'_1 = \sum_{i=1}^{k} |A_i|$$

$$S'_2 = \sum_{1 \leq i < j \leq k} |A_i \cap A_j|$$

$$\vdots$$

$$S'_k = |A_1 \cap \cdots \cap A_k|.$$

- Similarly, let S''_j denote the analogous sum for the sets $A_1 \cap A_{k+1}, \ldots, A_k \cap A_{k+1}$.
- Finally, let S_j denote the analogous sum for the sets $A_1, \ldots, A_{k+1}$.

Using the induction hypothesis for $n = k$, we get

$$= \left[\sum_{j=1}^{k} (-1)^{j-1} S'_j\right] + |A_{k+1}| - \left[\sum_{\ell=1}^{k} (-1)^{\ell-1} S''_\ell\right].$$

Now we take out the term for $j = 1$ from the first sum, adjust the indices in the second sum, and combine the two sigmas:

$$= S'_1 + \left[\sum_{j=2}^{k} (-1)^{j-1} S'_j\right] + |A_{k+1}| + \left[\sum_{\ell=2}^{k+1} (-1)^{\ell-1} S''_{\ell-1}\right]$$

$$= (S'_1 + |A_{k+1}|) + \left[\sum_{j=2}^{k} (-1)^{j-1} \cdot (S'_j + S''_{j-1})\right] + (-1)^k \cdot S''_k.$$

Finally, observe that

$S_1 = S'_1 + |A_{k+1}|$,
$S_{k+1} = S''_k$, and
$S_j = S'_j + S''_{j-1}$ for all $2 \leq j \leq k$.

The last equality is valid as each S_j' counts intersections not involving A_{k+1}, while S_{j-1}'' counts intersections that do involve A_{k+1}. We conclude that

$$= \quad S_1 + \left[\sum_{j=2}^{k}(-1)^{j-1}S_j\right] + (-1)^k S_{k+1} = \sum_{j=1}^{k+1}(-1)^{j-1}S_j,$$

which proves the $n = k+1$ case, and thus completes the proof of the theorem. □

Exercise 8.4.9
Let $k \in \mathbb{N}$. Prove the following identity used in the proof of Theorem 8.4.8:

$$\binom{k}{1} - \binom{k}{2} + \binom{k}{3} - \binom{k}{4} + \cdots + (-1)^{k-1}\binom{k}{k} = 1.$$

(Hint: Use the Binomial Theorem with $x = 1$ and $y = -1$.)

Example 8.4.10 Let $n \in \mathbb{N}$. In how many ways can we color n *distinct* objects with *four* colors, such that *all the four colors are used*?

Solution
Let us label the colors by $1, 2, 3$ and 4, and denote by A_i the set of all possible colorings in which the ith color is *not* used. Then we have the following.

- For every $i = 1, 2, 3, 4$, $|A_i| = 3^n$, as each object may be given one of three colors (note that the order here matters, as the objects are distinct, and repetitions are allowed, since different objects can be given the same color).
- For every $i \neq j$ in $\{1, 2, 3, 4\}$ we have $|A_i \cap A_j| = 2^n$, as this is the number of ways to color n distinct objects with 2 colors.
- Finally, for every $i < j < k$ in $\{1, 2, 3, 4\}$ we have $|A_i \cap A_j \cap A_k| = 1$.

Using the Inclusion-Exclusion Principle, we get (see Exercise 8.4.7):

$$|A_1 \cup A_2 \cup A_3 \cup A_4| = 4 \cdot 3^n - 6 \cdot 2^n + 4 \cdot 1 - 0 = 4 \cdot 3^n - 6 \cdot 2^n + 4.$$

This is the number of possible colorings that do not use all four colors. As the total number of ways to color n distinct objects with four colors is 4^n, the answer to the given question is

$$4^n - (4 \cdot 3^n - 6 \cdot 2^n + 4) = 4^n - 4 \cdot 3^n + 6 \cdot 2^n - 4.$$

Example 8.4.11 Let $X = \{1, 2, 3, 4, 5\}$ and $Y = \{1, 2, 3\}$. How many surjective functions are there from X to Y?

Solution

Here, we first try to figure out the number of non-surjective functions from X to Y. For $i = 1, 2, 3$ denote

$$B_i = \{f : X \to Y \text{ such that } f(x) \neq i \text{ for all } x \in X\}.$$

The set $B_1 \cup B_2 \cup B_3$ contains all non-surjective functions from X to Y. Using the Inclusion-Exclusion Principle:

$$\begin{aligned} |B_1 \cup B_2 \cup B_3| &= \sum_{i=1}^{3} |B_i| - \sum_{1 \leq i < j \leq 3} |B_i \cap B_j| + |B_1 \cap B_2 \cap B_3| \\ &= 3 \cdot 2^5 - 3 \cdot 1 + 0 = 3 \cdot 32 - 3 = 93. \end{aligned}$$

This is the number of functions which are not surjective. As there are 3^5 functions from X to Y (with no restrictions), the number of surjections is $3^5 - 93 = 150$.

8.5 Problems

8.1 Nancy just gave birth to triplets: Charlie, Josh and Danny. Her sister bought five pyjamas for the babies in different colors: blue, purple, yellow, white and grey. In how many different ways can Nancy dress her boys with these five pyjamas?

8.2 Three women and five men are divided into two groups of four people each. In how many ways can this be done, if each group must contain at least one women?

8.3 Pens are available for purchase in four different colors. In how many ways can we purchase 10 pens?

8.4 We have defined a permutation of a set S to be an arrangement of all elements of that set. An equivalent definition is given below.

Definition 8.5.1 A *permutation* of a set S (which may be finite or infinite) is a bijection from S to itself.

a. If $n \in \mathbb{N}$, how many permutations are there on the set $S = \{1, 2, \ldots, n\}$.
b. Let $A = \{1, 2, 3, 4, 5, 6\}$. How many permutations $f : A \to A$ are there for which $f(2) = 2$ and $f(5) = 5$?
c. Let $B = \{1, 2, 3, 4, 5\}$. How many permutations $g : B \to B$ are there for which $g(3) \neq 3$?

8.5 A *derangement* of a set S is a permutation $f\colon S \to S$ in which $f(x) \neq x$ for all $x \in S$.

a. List all the derangements of the set $S = \{a, b, c, d\}$.

b. For each $n \in \mathbb{N}$, denote by d_n the number of all derangements of the set $B_n = \{1, 2, 3, \dots, n\}$. Use the Inclusion-Exclusion Principle to prove that

$$d_n = n! \cdot \left(\frac{1}{0!} - \frac{1}{1!} + \frac{1}{2!} - \frac{1}{3!} + \frac{1}{4!} - \dots + \frac{(-1)^n}{n!}\right) = n! \cdot \sum_{k=0}^{n} \frac{(-1)^k}{k!}.$$

(Hint: For each $k \in \{1, 2, \dots, n\}$, let A_k be the set of all permutations $f\colon B_n \to B_n$ satisfying $f(k) = k$.)

c. (This part requires familiarity with infinite series.) Use a result from calculus, related to the infinite series $\sum_{k=0}^{\infty} \frac{x^k}{k!}$, to calculate the limit $\lim_{n\to\infty} \frac{d_n}{n!}$. The result you obtained approximates the probability that a randomly chosen permutation of B_n, when n is large, is a derangement.

8.6 Find the number of functions from $A = \{1, 2, \dots, 5\}$ to $B = \{11, 12, 13, \dots, 20\}$. How many of these functions are injective?

8.7 Let $k, n \in \mathbb{N}$. How many *positive* integer solutions are there for the equation $x_1 + x_2 + \dots + x_n = k$?
Hint: For each $1 \leq i \leq n$, define $x'_i = x_i - 1$, and rewrite the equation in terms of the x'_i.

8.8 Prove the Binomial Theorem (Theorem 8.2.1) by induction on n.

8.9 Prove the following identities involving the binomial coefficients. For each identity, provide two proofs, an algebraic and a combinatorial proof.

a. For every integer $0 \leq j \leq k \leq n$, $\binom{n}{k}\binom{k}{j} = \binom{n}{j}\binom{n-j}{k-j}$.

b. For non-negative integers k, n, m such that $0 \leq k \leq n + m$, we have

$$\binom{n+m}{k} = \sum_{j=0}^{k} \binom{n}{j}\binom{m}{k-j}.$$

Here we use the convention that $\binom{i}{l} = 0$ if $l > i$.

c. For every $n \in \mathbb{N}$, $n^2 = 2\binom{n}{2} + n$.

8.10 Prove the following identity (for integers $0 \leq k \leq n$):

$$\sum_{k=0}^{n} (-1)^k \binom{n}{k} 2^{n-k} = 1.$$

8.11 Prove that if n is an odd positive integer, then $\binom{n}{(n-1)/2} = \binom{n}{(n+1)/2}$.

8.12 Generalize Example 8.3.2 and show that in every set of $n+1$ integers (where $n \in \mathbb{N}$), there are two numbers whose difference is divisible by n.

8.13 Let A be a subset of $n+1$ numbers from $\{1, 2, \ldots, 2n\}$, where $n \in \mathbb{N}$. Show that there are two numbers in A which are relatively prime.

8.14 On a field 400 yards long, ten people each mark off football fields of length 100 yards. Prove that some point belongs to at least four of the fields. Represent the fields using closed intervals.

8.15 Prove that every set of five points in a square of area 1 has two points separated by distance at most $\frac{1}{\sqrt{2}}$.

8.16 Find the number of functions from $X = \{0, 2, 4, 6, 8, 10\}$ to $\{1, 3, 5, 7, 9\}$. How many of these functions are surjective?

8.17 How many ways are there to place 10 (distinct) people in four distinct rooms? How many ways are there to place 10 (distinct people) in four distinct rooms so that every room receives at least one person?
(Hint: For the second part, use the Inclusion-Exclusion Principle.)

8.18 The *Erdős–Szekeres Theorem* states the following.
Let $n \in \mathbb{N}$. Every sequence of n^2+1 distinct real numbers has a monotonic subsequence (either increasing or decreasing) consisting of $n+1$ elements. For instance, here is a sequence of $3^2+1 = 10$ real numbers:

$$7,\ 4,\ 1,\ 8,\ 5,\ 2,\ 6.5,\ 9,\ 6,\ 3.$$

The sequence $7, 6.5, 6, 3$ is a decreasing subsequence, while the sequence $1, 2, 6.5, 9$ is an increasing subsequence. If we remove the element 6.5, there is no monotonic subsequence of length 4. This demonstrates the theorem for $n = 3$.
Prove the theorem by following the steps below.

a. Suppose that $x_1, x_2, \ldots, x_{n^2+1}$ is a sequence of n^2+1 distinct real numbers.
b. For each $1 \leq i \leq n^2+1$, denote by a_i and b_i the following:
   - a_i is the longest increasing subsequence ending with x_i,
   - b_i is the longest decreasing subsequence ending with x_i.
c. Assume, by contradiction, that $1 \leq a_i \leq n$ and $1 \leq b_i \leq n$ for all i.
d. Use the Pigeonhole Principle to conclude that there exist two pairs (a_i, b_i) and (a_j, b_j), with $i \neq j$, which are equal.
e. From the fact that $a_i = a_j$ and $b_i = b_j$ obtain a contradiction.

8.6 Solutions to Exercises

Solution to Exercise 8.1.5
Denote by S the set containing the letters A, B, C, D. A 10-element sequence containing letters from S is simply an element of the Cartesian product

$$S^{10} = S \times S \times S \times \cdots \times S.$$

As S contains four elements, the number of 10-tuples in S^{10} is $4^{10} = 1048576$. Hence, the quiz can be answered in 1,048,576 ways.

Solution to Exercise 8.1.6
Picking an element from the Cartesian product $A_1 \times A_2 \times \cdots \times A_k$, which is a k-tuple, can be thought of as a k-steps process. We first choose an element from A_1, then an element from A_2, etc. In step i, we choose an element from A_i, and hence have $|A_i|$ ways of doing so. According to the Rule of Product, the number of ways of choosing an element from the Cartesian product is $|A_1| \cdot |A_2| \ldots |A_k|$.

Solution to Exercise 8.1.7
Our process has four steps. We have 20 ways of choosing the president, 19 ways for the vice president, 18 for the secretary, and 17 for the treasurer. By the Rule of Product, there are $20 \cdot 19 \cdot 18 \cdot 17 = 116280$ possible slates of candidates.

Solution to Exercise 8.1.10
Counting the number of ways the children can sit is the same as counting the number of permutations that can be formed from a set with four elements. Therefore, the answer is $4! = 24$.

Solution to Exercise 8.1.13
There are $\binom{10}{3} = 120$ ways to choose three books in English, and $\binom{10}{4} = 210$ ways to choose four books in French. According to the Rule of Product, the total number of possible choices is $120 \cdot 210 = 25200$.

Solution to Exercise 8.1.15
A selection of k elements from a set of n elements corresponds to creating a *bars and dots* diagram. The bars ($n - 1$ of them) separate the n possible choices, and dots represent the elements chosen. Overall, we need to place k dots in sequence of $(n - 1) + k$ available slots, and there are $\binom{n+k-1}{k}$ ways to do so.

Solution to Exercise 8.2.2
Every term in the expansion of $(x + y)^8$ has the form $x^i y^j$ where $i + j = 8$. Therefore, we must have $3 + k = 8$, implying that $k = 5$. From the Binomial Theorem we conclude that the coefficient of x^3y^5 is

$$\binom{8}{5} = \frac{8!}{5!3!} = 56.$$

Solution to Exercise 8.2.4

Here is an algebraic proof of Pascal's Identity:

$$\binom{n}{k-1} + \binom{n}{k} = \frac{n!}{(n-k+1)!(k-1)!} + \frac{n!}{(n-k)!k!}$$

$$= \frac{n!}{(n-k+1)!(k-1)!} \cdot \frac{k}{k} + \frac{n!}{(n-k)!k!} \cdot \frac{n-k+1}{n-k+1}$$

$$= \frac{n!}{(n-k+1)!k!} \cdot (k+n-k+1) = \frac{(n+1)!}{(n+1-k)!k!} = \binom{n+1}{k}.$$

To prove the chairperson identity, we make use of the fact that $n! = n \cdot (n-1)!$ and $k! = k \cdot (k-1)!$,

$$k\binom{n}{k} \;=\; k \cdot \frac{n!}{(n-k)!k!} = n \cdot \frac{(n-1)!}{(n-k)!(k-1)!} = n\binom{n-1}{k-1}.$$

Solution to Exercise 8.2.5

The row for $n = 7$ is obtained by adding pairs of consecutive terms in row $n = 6$ of Pascal's Triangle:

$1 \quad 1+6=7 \quad 6+15=21 \quad 15+20=35 \quad 20+15=35 \quad 15+6=21 \quad 6+1=7 \quad 1$

Solution to Exercise 8.3.5

Denote by $B_1, B_2, \ldots, B_n$ our n boxes. Suppose, by contradiction, that every box contains less than $k+1$ elements. Then the total number of objects satisfies

$$|B_1| + |B_2| + \cdots + |B_n| \leq k + k + \cdots + k = kn,$$

which is impossible, as we are told that the number of objects, m, is greater than kn. Therefore, $|B_i| \geq k+1$ for at least one $1 \leq i \leq n$.

Solution to Exercise 8.4.4

First, note that if C and D are finite sets, and $D \subseteq C$, then $(C \setminus D) \cup D = C$, and the Rule of Sum implies that

$$|C \setminus D| + |D| = |C| \qquad \Rightarrow \qquad |C \setminus D| = |C| - |D|.$$

Now, we can write $A \cup B = (A \setminus (A \cap B)) \cup B$, and conclude that

$$|A \cup B| = |A \setminus (A \cap B)| + |B| = |A| - |A \cap B| + |B|,$$

as needed.

Solution to Exercise 8.4.6

The sets A, $B \setminus (A \cap B)$ and $C \setminus (C \cap (A \cup B))$ are disjoint, so the Rule of Sum, applied repeatedly, gives:

$$|A \cup B \cup C| = |A| + |B \setminus (A \cap B)| + |C \setminus (C \cap (A \cup B))|.$$

The Rule of Sum also implies that if D and E are finite sets, and $D \subseteq E$, then $|E \setminus D| = |E| - |D|$. We thus conclude that

$$|A \cup B \cup C| = |A| + |B| - |A \cap B| + |C| - |C \cap (A \cup B)|.$$

Finally, we have $C \cap (A \cup B) = (C \cap A) \cup (C \cap B)$, so using a previous result for the union of two sets, we get

$$|C \cap (A \cup B)| = |A \cap C| + |B \cap C| - |A \cap B \cap C|.$$

Substituting the right-hand side in the formula for $|A \cup B \cup C|$, we get

$$|A \cup B \cup C| = |A| + |B| + |C| - |A \cap B| - |A \cap C| - |B \cap C| + |A \cap B \cap C|,$$

as needed.

Solution to Exercise 8.4.7

If A, B, C, D are four finite sets, then

$$\begin{aligned} |A \cup B \cup C \cup D| = |A| + |B| + |C| + |D| & \\ & - |A \cap B| - |A \cap C| - |A \cap D| - |B \cap C| - |B \cap D| - |C \cap D| \\ & + |A \cap B \cap C| + |A \cap B \cap D| + |A \cap C \cap D| + |B \cap C \cap D| \\ & - |A \cap B \cap C \cap D|. \end{aligned}$$

This formula can be proved in a similar way to the proof of Proposition 8.4.5. We set an arbitrary $x \in A \cup B \cup C \cup D$, and argue that this element is counted once on the right-hand side. We consider four different cases, in which x is present in exactly one, two, three or all four sets.

Solution to Exercise 8.4.9

Setting $x = 1$ and $y = -1$ in the Binomial Theorem (and replacing n with k) gives

$$0 = \sum_{j=0}^{k} \binom{k}{j} (-1)^j$$

or

$$\binom{k}{0} - \binom{k}{1} + \binom{k}{2} - \binom{k}{3} + \cdots + (-1)^k \binom{k}{k} = 0.$$

Replacing $\binom{k}{0}$ with 1, and moving all the other terms to the right-hand side results in

$$1 = \binom{k}{1} - \binom{k}{2} + \binom{k}{3} - \binom{k}{4} + \cdots + (-1)^{k-1} \binom{k}{k},$$

as needed.

9 Preview of Real Analysis – Limits and Continuity

Real analysis is a branch of mathematics focusing on the study of real numbers and related objects. Sets of real numbers, sequences, functions and series of real numbers are at the core of the subject. The notions of limits and convergence are central in analysis and are used to investigate such objects. Learning real analysis means, in part, deepening our understanding and study of the theoretical foundations of calculus topics. For these reasons, many view real analysis as *a rigorous version of calculus*.

In this chapter, we look at how limits of sequences and function can be formally defined. The precise definitions may require some effort to grasp, but this is absolutely essential for advanced studies in mathematics and related fields. Formal definitions of limits not only allow us to prove various statements (such as the Extreme and the Intermediate Value Theorems, often proved in a real analysis course), but also allow us to investigate more complicated functions and sequences.

Our experience with proof writing and logical statements will be invaluable for our discussion. We also highlight the use of limits to defining continuity and differentiability of functions.

9.1 The Limit of a Sequence

Sequences have been mentioned several times in previous chapters. Informally speaking, a sequence, in the context of this chapter, is an infinite ordered list of real numbers. More formally, a sequence can be defined as a function.

Definition 9.1.1 An *infinite sequence of real numbers* (or, in short, a *sequence*) is a function from $\mathbb{N}$ to $\mathbb{R}$.

Elements of a sequence are often denoted by a lowercase letter and a subscript. For instance, $(a_n)_{n=1}^{\infty}$, or simply (a_n), refer to a sequence whose elements are $a_1, a_2, a_3, \ldots$.

Roughly speaking, if the elements of a sequence (a_n) are getting closer and closer to a fixed number L, as n grows larger and larger, we say that the limit of (a_n) is L, and write $\lim a_n = L$.

Quite often, if the sequence is simple enough, we can easily *guess* its limit.

Example 9.1.2 Suppose that (a_n) is the following sequence:

$$3.9, 3.99, 3.999, 3.9999, \ldots.$$

That is, $a_1 = 3.9$, $a_2 = 3.99$, etc. We can also describe this sequence using a formula, by observing that

$$\begin{aligned} 3.9 &= 4 - 0.1 \\ 3.99 &= 4 - 0.01 \\ 3.999 &= 4 - 0.001 \end{aligned}$$

and, in general, $a_n = 4 - 0.1^n$ for every $n \in \mathbb{N}$.

Intuitively, it is quite clear that as n becomes larger and larger, a_n gets closer and closer to 4, and so 4 is the limit of (a_n). We write $\lim a_n = 4$.

Example 9.1.3 Here is another example of a sequence we call (b_n):

$$\frac{1}{11}, \frac{2}{12}, \frac{3}{13}, \frac{4}{14}, \ldots, \frac{52}{62}, \frac{53}{63}, \ldots.$$

In this sequence, every element is obtained by increasing the numerator and denominator of the previous element by 1. Here, it is harder to "see" the limit of the sequence. If we write the first 20 elements in decimal notation, we get

$$0.091, 0.167, 0.231, 0.286, 0.333, 0.375, 0.412, 0.444, 0.474, 0.500,$$
$$0.524, 0.545, 0.565, 0.583, 0.600, 0.615, 0.630, 0.643, 0.655, 0.667, \ldots.$$

Looking at these numbers, it is not even clear whether this sequence will approach a certain number as n grows. We can, however, write down a formula for this sequence, and manipulate it algebraically, as you may have done in a previous calculus course. For every $n \in \mathbb{N}$, we have

$$b_n = \frac{n}{n+10} = \frac{1}{1 + \frac{10}{n}}.$$

Now, we can argue informally that the limit of (b_n) must be 1, as the term $\frac{10}{n}$ gets closer and closer to 0 as n grows.

In order to investigate more complicated sequences, and prove general statements about them, we need a formal definition of what a limit of a sequence is. Let us take a closer look at Example 9.1.3. By saying that the limit of this sequence is 1, we mean that the elements of the sequence are getting closer and closer to 1, as n becomes larger and larger.

How close will b_n be to 1? As close as we want. That is, we can achieve *any degree of closeness* of b_n to 1, by making sure n is sufficiently large.

Example 9.1.4 For the sequence (b_n), given by $b_n = \frac{n}{n+10}$, how large should n be, in order for b_n to be less than 0.1 units away from 1?

Solution

If the distance of b_n from 1 is less than 0.1, we can write $|b_n - 1| < 0.1$, which is equivalent to

$$\left|\frac{n}{n+10} - 1\right| < 0.1 \qquad \Leftrightarrow \qquad \frac{10}{n+10} < 0.1 \qquad \Leftrightarrow \qquad 90 < n.$$

So as long as $n \geq 91$, we have $|b_n - 1| < 0.1$.

Example 9.1.5 And how large should n be, so that $|b_n - 1| < 0.03$?

Solution

Again, we solve the inequality $|b_n - 1| < 0.03$ and get

$$\frac{10}{n+10} < 0.03 \qquad \Leftrightarrow \qquad 323\tfrac{1}{3} < n.$$

That is, if $n \geq 324$, then $|b_n - 1| < 0.03$.

The examples above suggest that we may be able to achieve *any degree of closeness* of b_n to 1, which justifies our intuition that $\lim b_n = 1$.

Exercise 9.1.6

How large should n be, if we want the distance from b_n to 1 to be less than $\frac{1}{5000}$?

The formal definition of limits of sequences is based on the idea that *any degree of closeness* to a proposed limit should be achievable by making n sufficiently large.

Definition 9.1.7 Let (a_n) be a sequence and L a real number. We say that *the limit of* (a_n) *is* L, or that (a_n) *converges* to L, if for every positive number ε, there exists a natural number N, such that $|a_n - L| < \varepsilon$ whenever $n \geq N$. In that case, we write $\lim a_n = L$. If (a_n) does not have a limit we say that it *diverges*.

In this definition, we think of ε as representing the desired degree of closeness to L, and of N as representing the location in the sequence from which this degree of closeness is achieved. Other common notations for the limit of a sequence are

$$\lim_{n\to\infty} a_n = L, \qquad a_n \xrightarrow[n\to\infty]{} L \qquad \text{and} \qquad a_n \to L.$$

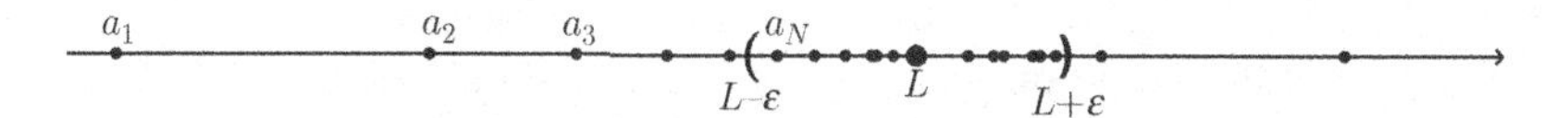

Figure 9.1 Convergence of a sequence.

We can also interpret this definition geometrically (see Figure 9.1). The condition $|a_n - L| < \varepsilon$ is equivalent to $L-\varepsilon < a_n < L+\varepsilon$, or to $a_n \in (L-\varepsilon, L+\varepsilon)$. This means that for every $\varepsilon > 0$, the sequence (a_n) *eventually* enters the open interval $(L-\varepsilon, L+\varepsilon)$, and stays there!

Example 9.1.8 We now prove formally that $\lim b_n = 1$, where $b_n = \frac{n}{n+10}$.

Solution

Suppose that ε is an arbitrary positive number. The inequality $|b_n - 1| < \varepsilon$ is equivalent to

$$\left|\frac{n}{n+10} - 1\right| < \varepsilon \quad \Leftrightarrow \quad \frac{10}{n+10} < \varepsilon \quad \Leftrightarrow \quad \frac{10}{\varepsilon} - 10 < n.$$

Let us pick a natural number N, satisfying $N > \frac{10}{\varepsilon} - 10$. And now, indeed, for every natural number $n \geq N$, we have

$$n \geq N > \frac{10}{\varepsilon} - 10,$$

and so

$$|b_n - 1| < \varepsilon,$$

as needed.

Take a close look at what we have done here. Given an arbitrary positive number ε, we proved that one can pick an $N \in \mathbb{N}$, such that $|b_n - 1| < \varepsilon$ for all $n \geq N$. We thus conclude that $\lim b_n = 1$.

As we can see, to prove that $\lim a_n = L$, we must show that every $\varepsilon > 0$ can be matched with an index (or location) N from which the distance between a_n and L is less than ε.

Remark. In Example 9.1.8, we used the fact that the natural numbers are *unbounded*. More specifically, we used the following:

For every $x \in \mathbb{R}$, there exists a number $n \in \mathbb{N}$, for which $n > x$.

This property can be proved formally from certain axioms used to define (or characterize) the real numbers. In this chapter, we will rely on this fact without proving it.

Here is another example.

Example 9.1.9 Let $a_n = \sqrt{n+1} - \sqrt{n}$. Prove that $\lim a_n = 0$.

Before writing the proof, let us figure out how should N be chosen for a given $\varepsilon > 0$. The expression $|a_n - 0|$ can be manipulated using the trick of *multiplying by the conjugate*:

$$|a_n - 0| = \sqrt{n+1} - \sqrt{n} = \frac{1}{\sqrt{n+1} + \sqrt{n}}.$$

Note that

$$\sqrt{n+1} + \sqrt{n} > \sqrt{n} + \sqrt{n} = 2\sqrt{n}$$

and so

$$|a_n - 0| < \frac{1}{2\sqrt{n}}.$$

To make sure that $|a_n - 0| < \varepsilon$, it is enough to ensure that

$$\frac{1}{2\sqrt{n}} \leq \varepsilon \qquad \Leftrightarrow \qquad n \geq \frac{1}{4\varepsilon^2}.$$

Now we are ready to write our proof.

Proof Let $\varepsilon > 0$. Choose $N \in \mathbb{N}$ such that $N \geq \frac{1}{4\varepsilon^2}$. Then, for every $n \geq N$, we have

$$|a_n - 0| = \frac{1}{\sqrt{n+1} + \sqrt{n}} < \frac{1}{2\sqrt{n}} \leq \frac{1}{2\sqrt{N}} \leq \varepsilon,$$

as needed. We have thus proved that $\lim a_n = 0$. □

Can we have a sequence with more than one limit? In other words, will our definition allow for two different numbers to be limits for a given sequence? The answer is no, and we prove this below. We also show that convergent sequences must be bounded.

Proposition 9.1.10 *Let (a_n) be a sequence.*

1. *If $\lim a_n = L$ and $\lim a_n = L'$, then $L = L'$. That is, the limit of (a_n), when it exists, is unique.*
2. *If (a_n) converges, then it is bounded. Namely, there exists a real number M such that $|a_n| \leq M$ for all $n \in \mathbb{N}$.*

Proof 1. Suppose that both L and L' are limits of (a_n), and let $\varepsilon > 0$. As $\lim a_n = L$, there exists an $N \in \mathbb{N}$, such that $|a_n - L| < \varepsilon$ whenever $n \geq N$. Similarly, there exists an $N' \in \mathbb{N}$, such that $|a_n - L'| < \varepsilon$ whenever $n \geq N'$.

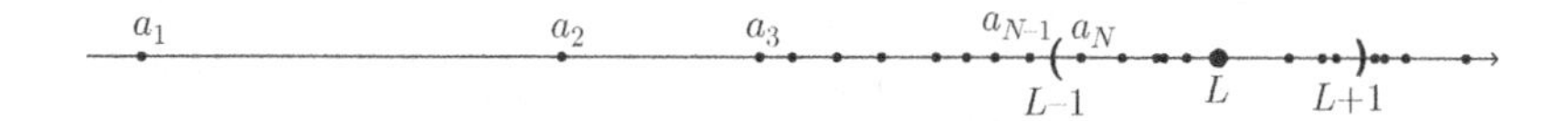

Figure 9.2 A convergent sequence is bounded.

Now take an $n \in \mathbb{N}$ that is larger than both N and N' (for instance, $n = N + N'$). Then, using the Triangle Inequality, we get

$$|L - L'| = |L - a_n + a_n - L'| \leq |a_n - L| + |a_n - L'| < \varepsilon + \varepsilon = 2\varepsilon.$$

Remember that ε was an arbitrary positive number, and so we conclude that $|L - L'| < 2\varepsilon$ for all $\varepsilon > 0$. This implies that $|L - L'| = 0$ (see Exercise 9.1.11 below), from which it follows that $L = L'$, as needed.

2. Suppose that (a_n) converges to some real number L. Let $\varepsilon = 1$. It follows, from Definition 9.1.7, that there exists an $N \in \mathbb{N}$, such that $|a_n - L| < 1$ whenever $n \geq N$. Using the reverse Triangle Inequality, we get:

$$|a_n| - |L| \leq |a_n - L| < 1 \qquad \Rightarrow \qquad |a_n| < |L| + 1.$$

In other words, we showed that $|L| + 1$ is a bound for all the terms of (a_n) starting from $n = N$. There are only finitely many remaining terms not accounted for: $a_1, a_2, \ldots, a_{N-1}$ (see Figure 9.2). We thus take $M = \max\{|L| + 1, |a_1|, |a_2|, \ldots, |a_{N-1}|\}$, which is indeed a bound for the whole sequence: $|a_n| \leq M$ for all $n \in \mathbb{N}$. □

Exercise 9.1.11
Show that if a real number a satisfies

$$0 \leq a < 2\varepsilon \ \text{ for all } \ \varepsilon > 0,$$

then $a = 0$.

Proposition 9.1.10 can be used to quickly construct examples of sequences that have no limit. For instance, the sequences (c_n) and (d_n) given by

$$c_n = 2n^2 + 1 \qquad \text{and} \qquad d_n = 1 - \sqrt{n}$$

are unbounded (i.e., not bounded), and so $\lim c_n$ and $\lim d_n$ do not exist. However, bounded sequences can also have no limit.

Example 9.1.12 Suppose $a_n = (-1)^{n+1}$ for every $n \in \mathbb{N}$. We show that $\lim a_n$ does not exist.

The sequence (a_n) is alternating between 1 and -1:

$$1, -1, 1, -1, 1, -1, \ldots.$$

To prove that the sequence has no limit, we use the following strategy and proof by contradiction. If $\lim a_n = L$, we should be able to make the distance between a_n and L less than $\frac{1}{2}$ for large enough n:

$$|a_n - L| < \frac{1}{2} \qquad \text{or} \qquad L - \frac{1}{2} < a_n < L + \frac{1}{2} \qquad \text{for large enough } n.$$

Note that the interval $\left(L - \frac{1}{2}, L + \frac{1}{2}\right)$ is of length 1, and since our sequence alternates between 1 and -1, there is no way both of these numbers are in this interval.

Proof Suppose that $\lim a_n = L$. Then, for $\varepsilon = \frac{1}{2}$, there is an $N \in \mathbb{N}$, such that $|a_n - L| < \frac{1}{2}$ whenever $n \geq N$. Pick two natural numbers, n_1 and n_2, satisfying

$$n_1, n_2 \geq N \qquad \text{and} \qquad n_1 \text{ is even} \qquad \text{and} \qquad n_2 \text{ is odd}.$$

Using the Triangle Inequality, we get

$$|a_{n_1} - a_{n_2}| \leq |a_{n_1} - L| + |a_{n_2} - L| < \frac{1}{2} + \frac{1}{2} = 1,$$

which contradicts the fact that

$$|a_{n_1} - a_{n_2}| = |(-1) - 1| = 2.$$

Therefore, $\lim a_n$ does not exist. □

We now state a theorem that justifies various algebraic manipulations we normally use to quickly calculate limits of sequences. In practice, we often use this theorem, and other tools, to find limits of sequences whenever possible. Working directly with Definition 9.1.7 requires more time and effort, and so we mainly use it to prove theorems about sequences and when there are no other tools we can apply.

Theorem 9.1.13 *Let (a_n) and (b_n) be two sequences, and $L, M \in \mathbb{R}$. If $\lim a_n = L$ and $\lim b_n = M$, then*

1. $\lim(a_n + b_n) = L + M$,
2. $\lim(k \cdot a_n) = k \cdot L$ *for every* $k \in \mathbb{R}$,
3. $\lim(a_n \cdot b_n) = L \cdot M$,
4. $\lim \frac{a_n}{b_n} = \frac{L}{M}$ *provided that* $M \neq 0$.

Proof We prove Parts 1 and 3. The others are left as an exercise (see Problems 9.6 and 9.7).

1. Let ε be an arbitrary positive number. To prove that $\lim(a_n + b_n) = L + M$, we need to show that the difference $|(a_n + b_n) - (L + M)|$ can be made smaller than ε for large enough n.

   Using the Triangle Inequality, we obtain

$$|(a_n + b_n) - (L + M)| = |a_n - L + b_n - M| \leq |a_n - L| + |b_n - M|.$$

As the limits of (a_n) and (b_n) are, respectively, L and M, the quantities $|a_n - L|$ and $|b_n - M|$ can be made as small as we want. We require that each is smaller than $\frac{\varepsilon}{2}$ and so their sum will be smaller than ε. We thus proceed as follows.

As $\lim a_n = L$ and $\lim b_n = M$, there are $N_1, N_2 \in \mathbb{N}$ such that

$$|a_n - L| < \frac{\varepsilon}{2} \text{ for } n \geq N_1 \qquad \text{and} \qquad |b_n - M| < \frac{\varepsilon}{2} \text{ for } n \geq N_2.$$

Now set $N = N_1 + N_2$. For every $n \geq N$ we have both $n \geq N_1$ and $n \geq N_2$, and so

$$|(a_n + b_n) - (L + M)| \leq |a_n - L| + |b_n - M| \leq \frac{\varepsilon}{2} + \frac{\varepsilon}{2} = \varepsilon,$$

as needed. This completes the proof of $\lim(a_n + b_n) = L + M$.

3. This part is a little trickier. We begin by using the Triangle Inequality again as follows:

$$\begin{aligned} |a_n \cdot b_n - L \cdot M| &= |a_n \cdot b_n - L \cdot b_n + L \cdot b_n - L \cdot M| \\ &= |(a_n - L) \cdot b_n + L \cdot (b_n - M)| \\ &\leq |a_n - L| \cdot |b_n| + |L| \cdot |b_n - M|. \end{aligned}$$

Proposition 9.1.10 implies that (b_n) is a bounded sequence. That is, there is a real number K such that $|b_n| \leq K$ for all $n \in \mathbb{N}$. We conclude that

$$|a_n \cdot b_n - L \cdot M| \leq |a_n - L| \cdot K + |L| \cdot |b_n - M|$$

for all $n \in \mathbb{N}$.

Now let $\varepsilon > 0$. As (a_n) and (b_n) converge to L and M, there exist $N_1, N_2 \in \mathbb{N}$ such that

$$|a_n - L| < \frac{\varepsilon}{2 \cdot (K + 1)} \quad \text{for } n \geq N_1,$$

and

$$|b_n - M| < \frac{\varepsilon}{2 \cdot (|L| + 1)} \quad \text{for } n \geq N_2$$

(we used $K + 1$ and $|L| + 1$ in the denominators to ensure that we are not dividing by zero).

Let $N = N_1 + N_2$. If $n \geq N$, we get

$$\begin{aligned} |a_n \cdot b_n - L \cdot M| &\leq |a_n - L| \cdot K + |L| \cdot |b_n - M| \\ &< \frac{\varepsilon}{2(K+1)} \cdot K + |L| \cdot \frac{\varepsilon}{2 \cdot (|L| + 1)} \\ &= \frac{\varepsilon}{2} \cdot \frac{K}{K+1} + \frac{\varepsilon}{2} \cdot \frac{|L|}{|L| + 1} < \frac{\varepsilon}{2} + \frac{\varepsilon}{2} = \varepsilon. \end{aligned}$$

This completes the proof of $\lim(a_n \cdot b_n) = L \cdot M$. □

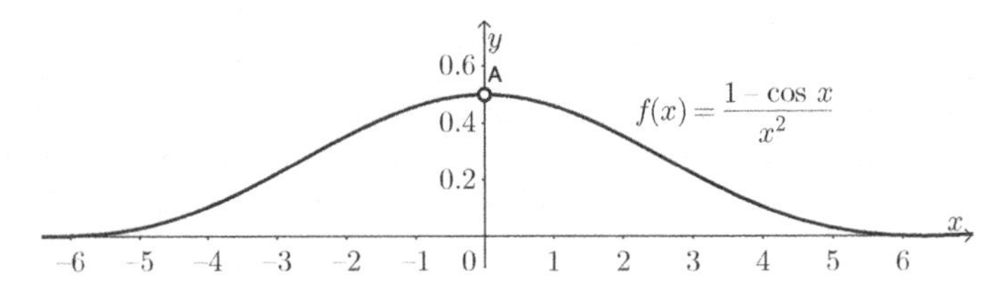

Figure 9.3 The function f is undefined at $x = 0$ but has a limit there.

Exercise 9.1.14
Use the fact that $\lim \frac{1}{n} = 0$, together with Theorem 9.1.13 to calculate the limit of $a_n = \frac{3n^2+5}{7n^2-3}$.

9.2 The Limit of a Function

The limit of a function is a fundamental notion in calculus and analysis, and is used to describe the behaviour of a function near a particular value. For instance, consider the function

$$f\colon \mathbb{R} \setminus \{0\} \to \mathbb{R}, \qquad f(x) = \frac{1 - \cos x}{x^2}.$$

Although undefined at $x = 0$, it seems that $f(x)$ is getting closer and closer to 0.5 as x approaches 0.

x	−0.4	−0.1	−0.03	0	0.02	0.05	0.2
$f(x)$	0.49337	0.49958	0.49996	undefined	0.49998	0.4999	0.49834

If we attempt to draw the graph of f (Figure 9.3), it looks like a curve with a *hole* at the point $(0, 0.5)$. We say that the limit of $f(x)$ is 0.5 as x approaches 0, and write

$$\lim_{x \to 0} f(x) = 0.5.$$

Informally, the notation $\lim_{x \to c} f(x) = L$, where c and L are real numbers, means:

"$f(x)$ can be made arbitrarily close to L, by making x close enough, but not equal, to c."

This imprecise description of limits was used by Newton and Leibniz in their development of calculus in the seventeenth and eighteenth centuries. Only in the middle of the nineteenth century, the German mathematician Karl Weierstrass introduced the formal definition of limits which we use today, known as the epsilon-delta definition. This definition avoids vague terms such as "arbitrarily close" and "close enough," and is similar in style to Definition 9.1.7 defining limits of sequences.

Let us look at a specific example. Take $f(x) = x^2$, and consider the limit at $x = 3$. Intuitively, it seems that

$$\lim_{x \to 3} f(x) = 9,$$

as when x is made closer and closer to 3, x^2 becomes closer and closer to 9. Alternatively, one may say:

"$f(x)$ can be made as close as we want to 9,
by making x close enough, but not equal, to 3."

Let us examine this statement more carefully. "Closeness" to the number 9 can be described using symmetric open intervals around 9. For instance, the interval $(8.9, 9.1)$ contains all the real numbers which are within 0.1 units of 9. Similarly, the interval $(2.5, 3.5)$ contains all numbers within 0.5 units of 3. If we remove the number 3, we obtain the set

$$(2.5, 3.5) \setminus \{3\} = (2.5, 3) \cup (3, 3.5)$$

which contains all numbers within 0.5 units of 3 and that are *different* from 3. The intervals $(8.9, 9.1)$ and $(2.5, 3.5)$ are called *neighbourhoods* of 9 and 3, respectively. The set $(2.5, 3.5) \setminus \{3\}$ is called a *punctured neighbourhood* of 3. More generally, we introduce the following definitions.

Definition 9.2.1 Let $c \in \mathbb{R}$. An interval of the form $(c - r, c + r)$, where $r > 0$, is called an *r-neighbourhood* of c and is denoted as $N_r(c)$. The number r is called the *radius* of the neighbourhood.

If we remove c from the interval, we obtain the set

$$(c - r, c + r) \setminus \{c\} = (c - r, c) \cup (c, c + r),$$

which we call a *punctured r-neighbourhood* of c, and denote it as $N_r^*(c)$.

Using neighbourhoods, we attempt to improve our informal description of $\lim_{x \to 3} f(x) = 9$:

"Given a neighbourhood of 9, $f(x)$ can be made to lie in that neighbourhood,
by choosing the x-values from a specific punctured neighbourhood of 3."

For example, $N_{0.5}(9) = (8.5, 9.5)$ is a neighbourhood of 9. Can we find a neighbourhood of 3 consisting of x-values mapped by f into $N_{0.5}(9)$? Figure 9.4 suggests that every neighbourhood of 3 contained in the interval $(\sqrt{8.5}, \sqrt{9.5})$ will do. For instance, the neighbourhood $N_{0.05}(3) = (2.95, 3.05)$ will work.

We summarize as follows.

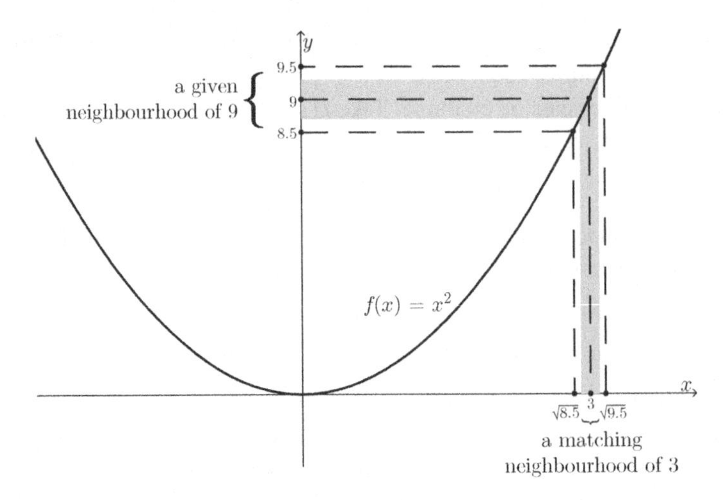

Figure 9.4 A neighbourhood of 9 is matched with a neighbourhood of 3.

The values of $f(x)$ lie in the open interval $(8.5, 9.5)$ if our xs are in the interval $(2.95, 3.05)$. In other words, $f(x)$ is within 0.5 units of 9 if we choose xs which are within 0.05 units of 3.

This process can be repeated for other neighbourhoods of 9. For instance, $N_{0.01}(9) = (8.99, 9.01)$ is a smaller neighbourhood of 9. Can we find a matching neighbourhood of 3? Yes we can.

Exercise 9.2.2
Find a neighbourhood of 3 such that each x-value in that neighbourhood is mapped by $f(x) = x^2$ into the interval $(8.99, 9.01)$.

We will soon show that *every* neighbourhood of 9, of any size, can be matched with a corresponding neighbourhood of 3. In other words, $f(x)$ can be made *as close as we want* to 9, by making sure x is close enough to 3. We can now present the formal definition of limits of functions using the language of neighbourhoods.

Definition 9.2.3 Let f be a real-valued function defined on some punctured neighbourhood of $c \in \mathbb{R}$. We say that the *limit of f as x approaches c is L*, and write $\lim_{x \to c} f(x) = L$, if for every neighbourhood $N_\varepsilon(L)$, there exists a punctured neighbourhood $N_\delta^*(c)$, such that $f(N_\delta^*(c)) \subseteq N_\varepsilon(L)$.

Figure 9.5 provides a geometric visualization of the definition. The neighbourhood $N_\varepsilon(L)$ can be thought of as a vertical open interval on the y-axis

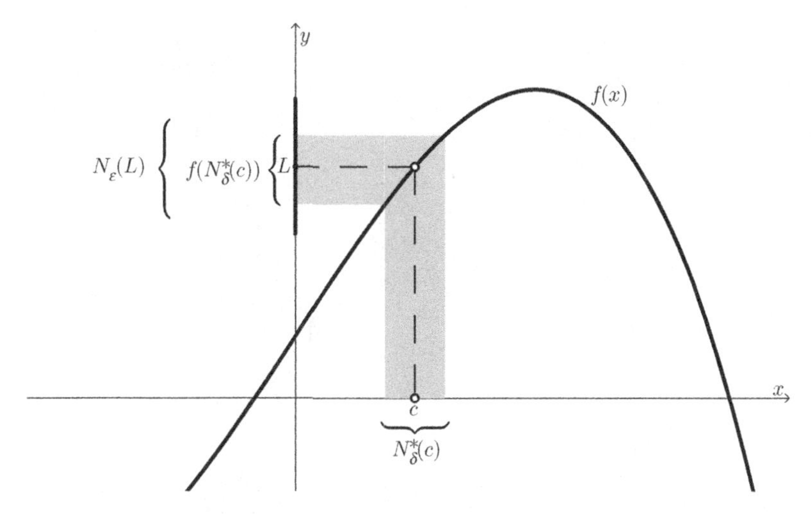

Figure 9.5 The definition of the limit of a function.

around L. If each such neighbourhood can be matched with a neighbourhood $N_\delta^*(c)$, on the x-axis, which is mapped by $f(x)$ into $N_\varepsilon(L)$, we say that the limit of $f(x)$ as x approaches c is L.

Note how our definition requires the existence of a *punctured* neighbourhood of c for a given neighbourhood of L. The reason for this is that the behaviour of $f(x)$ at the point c is irrelevant, and should not impact the existence or the value of the limit. Limits are supposed to measure behaviour of $f(x)$ *near* $x = c$ regardless of what happens at the point $x = c$ itself. As we have seen at the beginning of this section, $f(x)$ may even be undefined for some $x = c$ but still have a limit at that point.

We now prove formally, using our new definition, that $\lim_{x\to 3} x^2 = 9$.

Proof Suppose that $N_\varepsilon(9) = (9 - \varepsilon, 9 + \varepsilon)$ is an arbitrary neighbourhood of 9. We distinguish between two cases.

Case 1: If $9-\varepsilon \geq 0$ (that is, if $\varepsilon \leq 9$), then the interval $(\sqrt{9-\varepsilon}, \sqrt{9+\varepsilon})$ will be mapped by $f(x)$ into $N_\varepsilon(9)$ (Figure 9.6). This interval is not symmetric around 3 and thus cannot serve as a matching neighbourhood. However, we can choose a $\delta > 0$ such that

$$\delta \leq 3 - \sqrt{9-\varepsilon} \qquad \text{and} \qquad \delta \leq \sqrt{9+\varepsilon} - 3$$

so that the resulting neighbourhood $N_\delta^*(3) = (3 - \delta, 3 + \delta) \setminus \{3\}$ is mapped by $f(x)$ into $N_\varepsilon(9)$. To be more specific, we can define

$$\delta = \min\{3 - \sqrt{9-\varepsilon}, \sqrt{9+\varepsilon} - 3\}.$$

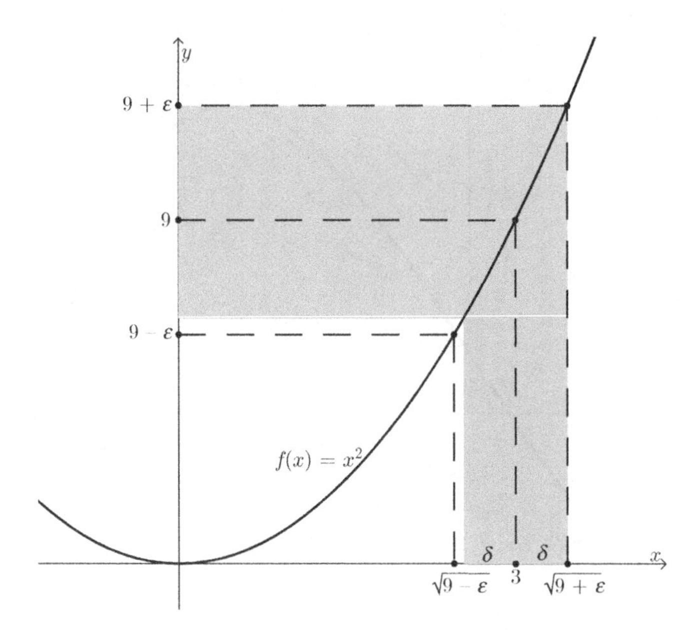

Figure 9.6 Case 1: $9 - \varepsilon \geq 0$.

That is, δ is the distance between 3 and the closer endpoint of $(\sqrt{9-\varepsilon}, \sqrt{9+\varepsilon})$. Let us re-confirm that our choice of δ works. Indeed, if $x \in N^*_\delta(3)$, then

$$3 - \delta < x < 3 + \delta.$$

But, from the definition of δ, we have

$$3 + \delta \leq 3 + \sqrt{9+\varepsilon} - 3 = \sqrt{9+\varepsilon}$$

and

$$3 - \delta \geq 3 - (3 - \sqrt{9-\varepsilon}) = \sqrt{9-\varepsilon},$$

from which it follows that

$$\sqrt{9-\varepsilon} < x < \sqrt{9+\varepsilon} \quad \Rightarrow \quad 9 - \varepsilon < x^2 < 9 + \varepsilon \quad \Rightarrow \quad f(x) \in N_\varepsilon(9).$$

We conclude that $f(N^*_\delta(3)) \subseteq N_\varepsilon(9)$, as needed.

Case 2: If $9 - \varepsilon < 0$ (Figure 9.7), we can simply choose

$$\delta = \sqrt{9+\varepsilon} - 3.$$

In this case, if $x \in N^*_\delta(3)$, we get

$$x < 3 + \delta = \sqrt{9+\varepsilon}$$

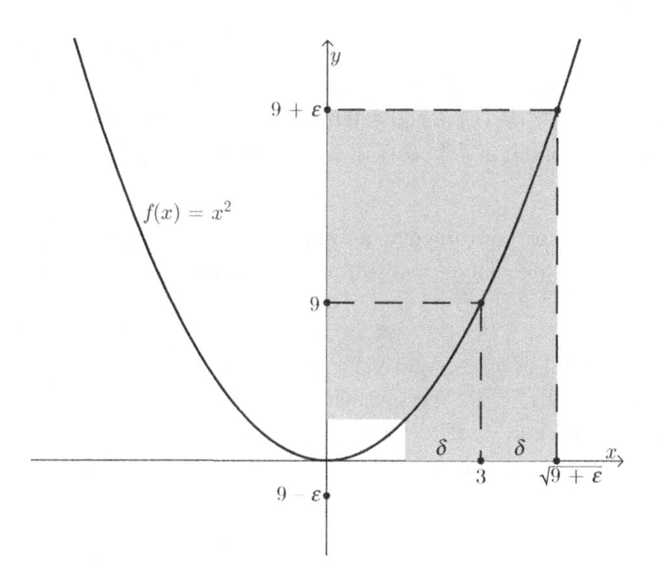

Figure 9.7 Case 2: $9 - \varepsilon < 0$.

and

$$x > 3 - \delta = 6 - \sqrt{9+\varepsilon} > -\sqrt{9+\varepsilon},$$

and so

$$-\sqrt{9+\varepsilon} < x < \sqrt{9+\varepsilon} \quad \Rightarrow \quad 9-\varepsilon < 0 \le x^2 < 9+\varepsilon,$$

which implies that $f(x) \in N_\varepsilon(9)$ as needed. □

Exercise 9.2.4
Show that for every $0 < \varepsilon < 9$ we have

$$\sqrt{9+\varepsilon} - 3 < 3 - \sqrt{9-\varepsilon}.$$

This implies that we could have simplified the definition of δ and just let $\delta = \sqrt{9+\varepsilon} - 3$, in both cases.

In Definition 9.2.3 we used the notion of neighbourhoods to define limits. We can, however, rephrase this definition to obtain the well-known *epsilon-delta definition*. To do that, we make the following observations.

- A neighbourhood of a real number L is determined solely by its radius, and so instead of writing "for every neighbourhood $N_\varepsilon(L)$" we can write "for every $\varepsilon > 0$."
- Similarly, we may replace the phrase "there exists a punctured neighbourhood $N^*_\delta(c)$" with "there exists a $\delta > 0$."

- The statement $x \in N_\delta^*(c)$ is equivalent to $0 < |x-c| < \delta$. The condition $0 < |x-c|$ ensures that $x \neq c$ as we are dealing with a *punctured* neighbourhood. Similarly, $f(x) \in N_\varepsilon(L)$ is equivalent to $|f(x) - L| < \varepsilon$.
- Finally, the condition $f(N_\delta^*(c)) \subseteq N_\varepsilon(L)$ can be written as "$|f(x) - L| < \varepsilon$ whenever $0 < |x - c| < \delta$."

Making all these adjustments in Definition 9.2.3 leads to the following equivalent and somewhat more algebraic description of limits.

Definition 9.2.5 Let f be a real-valued function defined on some punctured neighbourhood of $c \in \mathbb{R}$. We say that the *limit of f as x approaches c is L*, and write $\lim_{x \to c} f(x) = L$, if for every $\varepsilon > 0$, there exists a $\delta > 0$, such that $|f(x) - L| < \varepsilon$ whenever $0 < |x - c| < \delta$.

We demonstrate the use of Definition 9.2.5 in the following examples.

Example 9.2.6 Prove that $\lim_{x \to 4} (5x - 3) = 17$.

Solution

Our goal is to show that the quantity $|(5x - 3) - 17|$ can be made as small as we want, by making $|x - 4|$ small enough. Consequently, we are looking to relate the two quantities. In our case, we observe that

$$|(5x - 3) - 17| = |5x - 20| = 5|x - 4|,$$

and so if we want to make $|(5x - 3) - 17|$ less than ε, we need to make sure that $|x - 4|$ is less than $\frac{\varepsilon}{5}$. We are now ready to write down the proof.

Proof Let $\varepsilon > 0$. Choose $\delta = \frac{\varepsilon}{5}$. Then, whenever x is a real number satisfying $0 < |x - 4| < \delta$, we get

$$|(5x - 3) - 17| = |5x - 20| = 5|x - 4| < 5 \cdot \frac{\varepsilon}{5} = \varepsilon,$$

as needed. □

Example 9.2.7 Prove that $\lim_{x \to 0} \frac{6}{x^2+3} = 2$.

Solution

Again, in order to use the definition above, we need to consider an arbitrary $\varepsilon > 0$ (representing a neighbourhood of 2) and match it with a suitable $\delta > 0$ (representing a punctured neighbourhood of 0). We start with some rough work in order to discover a δ that would work for a given ε. The idea is to manipulate the expression $\left|\frac{6}{x^2+3}-2\right|$ and relate it somehow to the expression $|x-0|$:

$$\left|\frac{6}{x^2+3}-2\right| = \left|\frac{6-2x^2-6}{x^2+3}\right| = \frac{2x^2}{x^2+3} \leq 2x^2 = 2|x|^2.$$

In the proof, $|x-0| = |x|$ will be restricted by δ. That is, $|x| < \delta$, and hence $2|x|^2 < 2\delta^2$. On the other hand, our goal is to make $2|x|^2$ less than ε. We can achieve this goal by making sure that $2\delta^2 \leq \varepsilon$, or equivalently, $\delta \leq \sqrt{\frac{\varepsilon}{2}}$. We now proceed to presenting the proof, in which we set $\delta = \sqrt{\frac{\varepsilon}{2}}$.

Proof Let $\varepsilon > 0$, and set $\delta = \sqrt{\frac{\varepsilon}{2}}$. If x satisfies the condition $0 < |x-0| < \delta$, then

$$\left|\frac{6}{x^2+3}-2\right| = \frac{2x^2}{x^2+3} \leq 2|x|^2 < 2\delta^2 = 2\left(\sqrt{\frac{\varepsilon}{2}}\right)^2 = \varepsilon$$

as needed. Therefore, $\lim\limits_{x\to 0} \frac{6}{x^2+3} = 2$. □

As you can see, most of the effort was put into finding a matching δ for an arbitrary positive ε. Once such a δ is discovered, writing the proof becomes quite straightforward.

Limits of functions do not always exist. For instance, consider the following function

$$h\colon \mathbb{R} \to \mathbb{R}, \qquad h(x) = \begin{cases} x+1 & \text{if } x \geq 0 \\ x-1 & \text{if } x < 0. \end{cases}$$

By looking at the graph (Figure 9.8), we see a "jump" at $x = 0$, which implies that there is no single number the function is approaching as x gets closer and closer to zero. In order to prove that $\lim\limits_{x\to 0} h(x)$ does not exist, we first negate Definition 9.2.5:

> Suppose that $f(x)$ is a real-valued function, defined in a punctured neighboorhood of $x = c$, and L a real number. Then $\lim\limits_{x\to c} f(x) \neq L$ if there exists an $\varepsilon > 0$, such that for every $\delta > 0$, there is an x, satisfying
>
> $$0 < |x-c| < \delta \text{ and } |f(x) - L| \geq \varepsilon.$$

We now proceed to proving that the function $h(x)$, defined above, has no limit at $x = 0$.

Proof Let $L \in \mathbb{R}$. We show that $\lim\limits_{x\to 0} h(x) \neq L$ by using the negation stated above. Set $\varepsilon = 1$, and let $\delta > 0$ be an arbitrary positive number. Define $x_1 =$

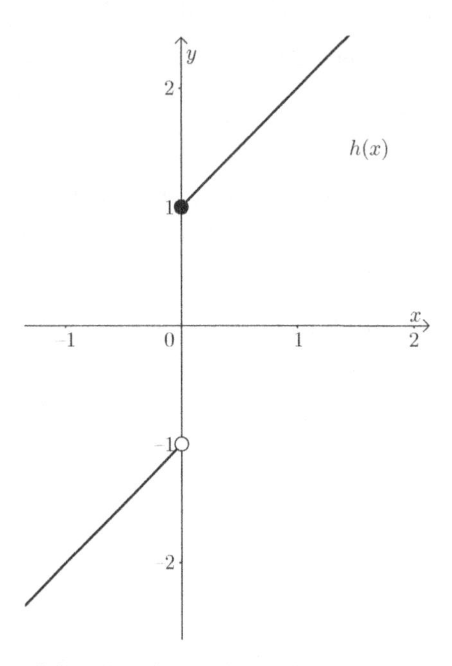

Figure 9.8 A function with one-sided limits at $x = 0$.

$-\frac{\delta}{2}$ and $x_2 = \frac{\delta}{2}$. Both x satisfy the condition $0 < |x - 0| < \delta$, and we show that one of them also satisfies the inequality $|h(x) - L| \geq \varepsilon$.

Note that since $x_1 < 0$ and $x_2 > 0$, we have $h(x_1) \leq -1$ and $h(x_2) \geq 1$, from which it follows that $|h(x_1) - h(x_2)| \geq 2$. Using the Triangle Inequality, we get

$$2 \leq |h(x_1) - h(x_2)| = |h(x_1) - L + L - h(x_2)| \leq |h(x_1) - L| + |h(x_2) - L|.$$

As both $|h(x_1) - L|$ and $|h(x_2) - L|$ are non-negative, and their sum is greater than or equal to 2, at least one of the two quantities must be greater than or equal to 1.

We have found an x satisfying $0 < |x - 0| < \delta$ and $|h(x) - L| \geq 1 = \varepsilon$, and thus proved that $\lim_{x \to 0} h(x) \neq L$. □

One-Sided Limits

Let us take another careful look at the function $h(x)$ above. We just proved that $\lim_{x \to 0} h(x)$ does not exist, meaning there is no single number that $h(x)$ approaches as x gets closer and closer to zero. However, it seems that if we approach $x = 0$ from either the negative or the positive side, the values of $h(x)$ will get closer to a single number: -1 if we come from the left, and 1 if we come from the right. This observation leads to the following definition of one-sided limits.

Definition 9.2.8

- Let f be a real-valued function, defined on an interval of the form $(c, c+r)$ for some $c \in \mathbb{R}$ and $r > 0$. We say that the *right-hand limit* of f is $L \in \mathbb{R}$, and write $\lim_{x \to c^+} f(x) = L$, if for every $\varepsilon > 0$ there is a $\delta > 0$, such that $|f(x) - L| < \varepsilon$ whenever $c < x < c + \delta$.
- Let f be a real-valued function, defined on an interval of the form $(c - r, c)$ for some $c \in \mathbb{R}$ and $r > 0$. We say that the *left-hand limit* of f is $L \in \mathbb{R}$, and write $\lim_{x \to c^-} f(x) = L$, if for every $\varepsilon > 0$ there is a $\delta > 0$, such that $|f(x) - L| < \varepsilon$ whenever $c - \delta < x < c$.

Note how similar these definitions are to Definition 9.2.5. The main difference is that the inequality $0 < |x - c| < \delta$ is replaced by $c < x < c + \delta$ or $c - \delta < x < c$, which force x to be either on the right or on the left side of c.

Using the notion of one-sided limits, we can now assert that

$$\lim_{x \to 0^+} h(x) = 1 \qquad \text{and} \qquad \lim_{c \to 0^-} h(x) = -1.$$

Exercise 9.2.9
For the function $h(x)$ defined above, prove that $\lim_{x \to 0^+} h(x) = 1$ and $\lim_{c \to 0^-} h(x) = -1$.

The next theorem states, informally, that limits of functions behave nicely under addition, subtraction, multiplication and division. It is the analogue of Theorem 9.1.13 for functions.

Theorem 9.2.10 *Let f and g be two real-valued functions, defined on some punctured neighbourhood of a real number c. If $\lim_{x \to c} f(x) = L$ and $\lim_{x \to c} g(x) = M$, then the following hold.*

1. $\lim_{x \to c} [f(x) + g(x)] = L + M$.
2. $\lim_{x \to c} [k \cdot f(x)] = k \cdot L$ *for every* $k \in \mathbb{R}$.
3. $\lim_{x \to c} [f(x) \cdot g(x)] = L \cdot M$.
4. $\lim_{x \to c} \frac{f(x)}{g(x)} = \frac{L}{M}$ *provided that* $M \neq 0$.

This theorem can be proved directly using similar arguments to those used to prove Theorem 9.1.13. However, the above theorem follows directly from a result we discuss in the next section, so we defer the proof until then.

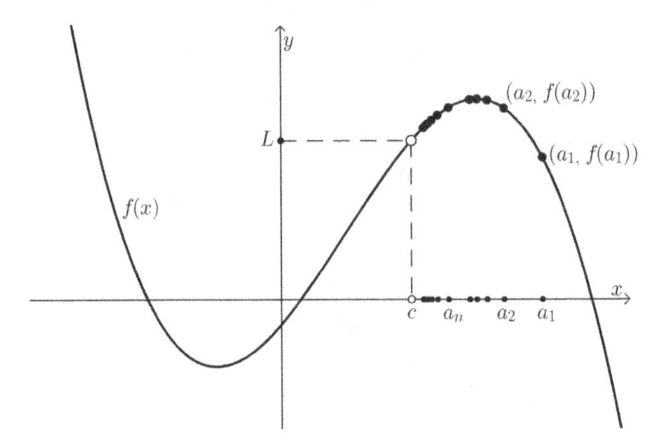

Figure 9.9 The sequence $(f(a_n))$ must "follow" the shape of the graph of f.

9.3 The Relation between Limits of Functions and Sequences

Limits of sequences and functions are intimately related. A sequence of points on the graph of a function must follow the shape of the graph, and consequently, may be forced to approach a certain height. The next theorem formalizes the connection between limits of sequences and functions.

Theorem 9.3.1 *Let f be a real-valued function, defined on some punctured neighbourhood $N_r^*(c)$ of c, and L a real number. Then the following are equivalent.*

1. $\lim\limits_{x \to c} f(x) = L$.
2. *For every sequence (a_n) in $N_r^*(c)$ that converges to c, we have* $\lim f(a_n) = L$.

Proof

- We first show that 1⇒2. Suppose that $\lim\limits_{x \to c} f(x) = L$, and that (a_n) is a sequence in $N_r^*(c)$ that converges to c (Figure 9.9). We need to show that the sequence $(f(a_n))$ converges to L.

  Let $\varepsilon > 0$. As $\lim\limits_{x \to c} f(x) = L$, there exists a $\delta > 0$, such that for all $x \in N_r^*(c)$:

$$0 < |x - c| < \delta \quad \Rightarrow \quad |f(x) - L| < \varepsilon.$$

On the other hand, as $\lim a_n = c$, there exists an $N \in \mathbb{N}$ such that for all $n \in \mathbb{N}$:

$$n \geq N \qquad \Rightarrow \qquad |a_n - c| < \delta.$$

Remember that the sequence (a_n) is in $N_r^*(c)$, and so $a_n \neq c$ for every n. Therefore, if n is a natural number satisfying $n \geq N$, then $0 < |a_n - c| < \delta$. But then, thinking of a_n as our x, we conclude that $|f(a_n) - L| < \varepsilon$.

We thus showed that $\lim f(a_n) = L$, as needed.

- To prove the other implication, 2⇒1, we use contrapositive. That is, we show that if $\lim_{x \to c} f(x) \neq L$, then there exists a sequence (a_n) in $N_r^*(c)$, converging to c, for which $\lim f(a_n) \neq L$.

  From the assumption that $\lim_{x \to c} f(x) \neq L$, we know that there exists an $\varepsilon_0 > 0$ that has no corresponding $\delta > 0$. In other words, for every $\delta > 0$, there is an $x \in N_r^*(c)$, such that

  $$0 < |x - c| < \delta \qquad \text{but} \qquad |f(x) - L| \geq \varepsilon_0.$$

  For every $n \in \mathbb{N}$, if $\delta = \frac{1}{n}$, there must exist an x-value $x = a_n \in N_r^*(c)$, such that

  $$0 < |a_n - c| < \delta = \frac{1}{n} \qquad \text{and} \qquad |f(a_n) - L| \geq \varepsilon_0.$$

  We complete the proof by observing the following. First, $\lim a_n = c$, as the distance between a_n and c can be made as small as we want by making n sufficiently large. Second, $\lim f(a_n) \neq L$, as the distance between $f(a_n)$ and L can never be made smaller than ε_0 for large enough values of n. Therefore, we have proved the contrapositive, which concludes the proof of the theorem. □

Theorem 9.3.1 can be used to show that certain limits of functions do not exist. Here is an example.

Example 9.3.2 Prove that $\lim_{x \to 0} \cos\left(\frac{1}{x}\right)$ does not exist.

Solution

Suppose, by contradiction, that $\lim_{x \to 0} \cos\left(\frac{1}{x}\right)$ exists and equals a real number L. Then, according to Theorem 9.3.1, whenever (a_n) is a sequence satisfying $a_n \to 0$ and $a_n \neq 0$ for all $n \in \mathbb{N}$, it follows that $\lim \cos\left(\frac{1}{a_n}\right) = L$.

For instance, the sequences $x_n = \frac{1}{2\pi n}$ and $y_n = \frac{1}{2\pi n + \pi}$ are both positive and converge to zero. This means that

$$\lim \cos\left(\frac{1}{x_n}\right) = \lim \cos\left(\frac{1}{y_n}\right) = L.$$

On the other hand,

$$\lim \cos\left(\frac{1}{x_n}\right) = \lim \cos(2\pi n) = 1$$

and

$$\lim \cos\left(\frac{1}{y_n}\right) = \lim \cos(2\pi n + \pi) = -1,$$

which gives a contradiction, as $1 \neq -1$.

We conclude that the limit $\lim_{x\to 0} \cos\left(\frac{1}{x}\right)$ does not exist.

The relation between limits of functions and limits of sequences can often be used to prove statements about limits of functions from analogous results on limits of sequences. For example, here is a proof of Part 3 of Theorem 9.2.10.

Proof of Theorem 9.2.10 3. We know that f and g are defined on a punctured neighbourhood of c, which we denote by $N_r^*(c)$. Suppose that (a_n) is a sequence in $N_r^*(c)$ which converges to c. Then, by implication $1\Rightarrow 2$ in Theorem 9.3.1, we conclude that

$$\lim f(a_n) = L \qquad \text{and} \qquad \lim g(a_n) = M.$$

Theorem 9.1.13 implies that

$$\lim[(f \cdot g)(a_n)] = \lim[f(a_n) \cdot g(a_n)] = L \cdot M.$$

Now using the implication $2\Rightarrow 1$ in Theorem 9.3.1, we conclude that $\lim_{x\to c}[f(x) \cdot g(x)] = L \cdot M$, as needed.

The other parts of Theorem 9.3.1 can be proved in a similar way (see Problem 9.13). □

9.4 Continuity and Differentiability

Almost all fundamental notions in analysis rely, in one way or another, on the concept of limits. In this section we demonstrate how continuity and differentiability, two notions you may have encountered before in a calculus course, can be formally defined using limits of functions. These formal definitions allow us to examine more complicated functions and prove general results involving continuity and differentiability.

Continuity

A continuous function $f\colon \mathbb{R} \to \mathbb{R}$ may be described informally as a function whose graph can be drawn "without lifting the pencil off the paper."

Alternatively, one may say that f is continuous if its graph has no holes or jumps. However, neither description can serve as a formal mathematical definition. Having defined rigorously limits of functions, we can now formalize the notion of continuity.

Definition 9.4.1 Let $f(x)$ be a real-valued function, defined on a neighbourhood of $c \in \mathbb{R}$. We say that $f(x)$ is continuous at $x = c$ if $\lim_{x \to c} f(x) = f(c)$.

Remarks.

- Unlike the definition of a limit, in order for a function to be continuous at $x = c$, it must be defined at that point, and not only at nearby points.
- The definition of continuity can also be stated using the epsilon-delta language:

  "We say that $f(x)$ is continuous at $x = c$ if for every $\varepsilon > 0$, there exists a $\delta > 0$, such that for all x in the domain of f, if $|x - c| < \delta$, then
  $$|f(x) - f(c)| < \varepsilon."$$
  Note how we replaced the condition $0 < |x-c| < \delta$ (from Definition 9.2.5) with $|x-c| < \delta$, as for $x = c$ the inequality $|f(x) - f(c)| < \varepsilon$ remains valid.

Example 9.4.2 Let us prove that the function $f\colon \mathbb{R} \to \mathbb{R}$, $f(x) = 3x - 5$ is continuous at every $c \in \mathbb{R}$.

Proof In order to prove continuity at $x = c$, we show that $\lim_{x \to c} f(x) = f(c)$. Let $\varepsilon > 0$. To discover a matching delta, note that

$$|f(x) - f(c)| = |(3x - 5) - (3c - 5)| = 3|x - c|.$$

To make sure $3|x - c| < \varepsilon$, we can choose $\delta = \frac{\varepsilon}{3}$. Now, if $|x - c| < \delta$, then

$$|f(x) - f(c)| = 3|x - c| < 3\delta = 3 \cdot \frac{\varepsilon}{3} = \varepsilon,$$

as needed. We have thus proved that $f(x)$ is continuous at every $c \in \mathbb{R}$. □

Example 9.4.3 Here is a more interesting example. We prove that the following function is continuous at $x = 0$:

$$g\colon \mathbb{R} \to \mathbb{R}, \qquad g(x) = \begin{cases} x \sin\left(\frac{1}{x}\right) & \text{if } x \neq 0 \\ 0 & \text{if } x = 0. \end{cases}$$

Proof Let $\varepsilon > 0$, and choose $\delta = \varepsilon$. Using the fact that $|\sin(t)| \leq 1$ for all $t \in \mathbb{R}$, we conclude that for every $x \neq 0$, if $|x - 0| < \delta$, then

$$|g(x) - g(0)| = |x| \cdot \left|\sin\left(\frac{1}{x}\right)\right| \leq |x| < \delta = \varepsilon.$$

Clearly, the inequality $|g(x) - g(0)| < \varepsilon$ is also valid when $x = 0$, and so we have proved that $\lim_{x\to 0} g(x) = g(0) = 0$, which establishes the continuity of $g(x)$ at $x = 0$.

Note that if we change the value of $g(0)$ to a *non-zero* value, then $g(x)$ will no longer be continuous at $x = 0$. □

Definition 9.4.1 refers to continuity as a pointwise property. That is, a function may be continuous at one point, but not at another. When a function is continuous at every point in its domain, we simply say that it is a continuous function. We need to be careful though when our domain has *endpoints*, as technically we required that our functions are defined on a neighbourhood of a point $x = c$ for the purpose of continuity.

We thus extend the notion of continuity to allow *one-sided continuity* in the case where the domain is a closed interval.

Definition 9.4.4

- A function $f\colon A \to \mathbb{R}$, where A is an open interval or a union of open intervals, is said to be *continuous*, if it is continuous at every $x \in A$.
- A function $g\colon [a, b] \to \mathbb{R}$ is said to be *continuous* if the following conditions hold.
  1. $g(x)$ is continuous at every $x \in (a, b)$.
  2. $g(x)$ is continuous at $x = a$ *from the right*, that is, $\lim_{x\to a^+} g(x) = g(a)$.
  3. $g(x)$ is continuous at $x = b$ *from the left*, that is, $\lim_{x\to b^-} g(x) = g(b)$.

With some effort, one can prove that all elementary functions (such as polynomials, rational, trigonometric and exponential functions, etc.) are continuous on their domain. Some of the proofs are explored in the Problems section.

Here is an example of a general result about continuous functions.

Lemma 9.4.5 *Let f be a real-valued function defined on a neighbourhood of $c \in \mathbb{R}$. Suppose that f is continuous at c. If $f(c) > 0$, then f is positive on some neighbourhood $(c - \delta, c + \delta)$ of c. Similarly, if $f(c) < 0$, then f is negative on some neighbourhood $(c - \delta, c + \delta)$ of c.*

In other words, if a continuous function is positive or negative at a point, it has the same sign in some neighbourhood of that point. Informally, we may say that continuous functions cannot instantly jump from being positive to negative, or vice versa.

Proof Suppose that $f(c) > 0$. As f is continuous at $x = c$, we have $\lim_{x \to c} f(x) = f(c)$. Using Definition 9.2.5 with $\varepsilon = f(c) > 0$, we conclude that there is a $\delta > 0$, such that for every x satisfying $|x - c| < \delta$ (or, equivalently, $c - \delta < x < c + \delta$) we have

$$|f(x) - f(c)| < \varepsilon \quad \Rightarrow \quad -f(c) < f(x) - f(c) < f(c) \quad \Rightarrow \quad 0 < f(x) < 2f(c).$$

We conclude that f is positive on the open interval $(c - \delta, c + \delta)$. The case where $f(c) < 0$ is proved in a similar way, by using $\varepsilon = -f(c)$. □

Exercise 9.4.6

Prove a one-sided version of Lemma 9.4.5. Let f be a real-valued function defined on an interval of the form $[c, t)$, and continuous from the right at $x = c$ (i.e., $\lim_{x \to c^+} f(x) = f(c)$). If $f(c) > 0$ (respectively $f(c) < 0$), then f is positive (respectively negative) on an interval of the form $[c, c + \delta)$ for some $\delta > 0$.

A similar result holds for functions which are continuous from the left at a point $x = c$.

There are several important theorems about continuous real-valued functions whose domain is a closed interval. Here are three such results.

- The Intermediate Value Theorem: A continuous function $f\colon [a, b] \to \mathbb{R}$ achieves every value between $f(a)$ and $f(b)$.
- Weierstrass' First Theorem: A continuous function $f\colon [a, b] \to \mathbb{R}$ is bounded. That is, there exists an $M > 0$ such that $|f(x)| \leq M$ for all $x \in [a, b]$.
- Weierstrass' Second Theorem: A continuous function $f\colon [a, b] \to \mathbb{R}$ achieves a maximum and a minimum value on $[a, b]$. That is, there exist $t, s \in [a, b]$, such that $f(t) \leq f(x) \leq f(s)$ for all $x \in [a, b]$.

We do not discuss the proofs of these theorems as they all rely on a rigorous definition of real numbers, which is beyond the scope of this book. Their proofs are normally discussed in real analysis and calculus textbooks.

Differentiability

The derivative is a central tool in calculus, used to measure rates of change. Given a real-valued function f, we are often interested in finding out how fast the values of the function change around a specific x-value.

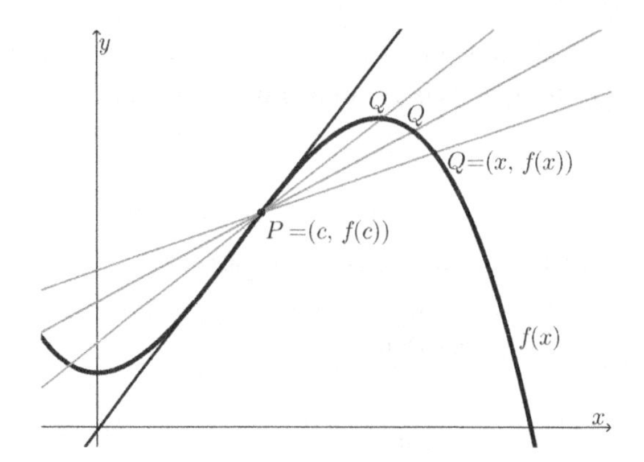

Figure 9.10 The slope of a tangent line is the limit of slopes of secant lines.

Suppose that f is defined on some neighbourhood of a real number c. If $x > c$, and f is defined on the interval $[c,x]$, then $f(x) - f(c)$ indicates the *change in the values of the function* over the interval $[c,x]$. The *rate of change* of f over $[c,x]$ can be described by the quotient

$$\frac{f(x)-f(c)}{x-c}.$$

The same quotient is obtained if $x < c$, and we consider the interval $[x,c]$. If we take the limit of this quotient as x approaches c, the resulting value (assuming the limit exists) represents the rate of change of f *at the point* where $x = c$. This value is called the *derivative* of f at $x = c$, and is often thought of as representing the *instantaneous rate of change* of the function at that point.

From a geometric point of view, the quotient $\frac{f(x)-f(c)}{x-c}$ is the slope of the line passing through the points $P = (c, f(c))$ and $Q = (x, f(x))$. As x approaches c, these secant lines become closer and closer to what appears to be a tangent line to the graph of f at P (see Figure 9.10). We can thus interpret the limit of this quotient as the slope of a tangent line.

Definition 9.4.7 Let f be a real-valued function defined in a neighbourhood of c. We say that f is *differentiable* at c if the limit

$$\lim_{x\to c}\frac{f(x)-f(c)}{x-c}$$

exists. In that case, the value of the limit is called the *derivative* of f at c, and is denoted as $f'(c)$.

Let us examine a few examples.

Example 9.4.8 Calculate the derivative of $g(x) = x^3$ at $x = 2$.

Solution

In order to show that $g(x)$ is differentiable at $x = 2$, and find its derivative, we calculate the limit

$$\lim_{x\to 2} \frac{g(x) - g(2)}{x - 2} = \lim_{x\to 2} \frac{x^3 - 2^3}{x - 2}.$$

We use the identity $a^3 - b^3 = (a-b)(a^2+ab+b^2)$ to simplify our expression:

$$\lim_{x\to 2} \frac{(x-2)(x^2+2x+4)}{x-2} = \lim_{x\to 2}(x^2 + 2x + 4).$$

As polynomials are continuous functions (see Problem 9.17), this limit is equal to

$$2^2 + 2 \cdot 2 + 4 = 12,$$

and so g is differentiable and $g'(2) = 12$.

Example 9.4.9 Prove that the function $h(x) = |x|$ is not differentiable at $x = 0$.

Solution

We use Problem 9.14 to show that h is not differentiable at $x = 0$ by proving that the limit

$$\lim_{x\to 0} \frac{|x| - |0|}{x - 0} = \lim_{x\to 0} \frac{|x|}{x}$$

does not exist. Note that $\frac{|x|}{x}$ equals 1 if $x > 0$, and -1 if $x < 0$. It is quite straightforward to show that $\lim\limits_{x\to 0+} \frac{|x|}{x} = 1$ and $\lim\limits_{x\to 0-} \frac{|x|}{x} = -1$. As the two one-sided limits are distinct, the limit $\lim\limits_{x\to 0} \frac{|x|}{x}$ does not exist, by Problem 9.14. Therefore, h is not differentiable at $x = 0$.

Informally, this result should not be surprising, as the graph of $h(x) = |x|$ has a corner at $x = 0$, and so a tangent line will not exist at that point.

There are several rules that can be used to quickly compute derivatives of commonly used function. Some are mentioned in the Problems section. We will not discuss them here, as we would like to focus on theory rather than computational techniques. Instead, we explore the relation between continuity and differentiability at a point.

Suppose that f is a real-valued function, defined on a neighbourhood of $x = c$. Informally speaking, continuity at c implies that the graph of f does not "tear-up" at that point, meaning that there are no jumps or holes in the graph, and one can draw it without lifting the pencil when $x = c$. On the other hand, if f is differentiable at $x = c$, then its graph has a tangent line, and hence has to appear "smooth" at that point. We cannot have a sharp corner at c.

It thus seems that differentiability is a stronger property than continuity. If a graph looks "smooth" at $x = c$, it cannot "break" at that point and hence must be continuous as well. This observation is formulated in the following theorem, whose proof is actually quite short.

Note that in the proof we rely on the limit laws (Theorem 9.3.1).

Theorem 9.4.10 *Let f be a real-valued function defined on a neighbourhood of $c \in \mathbb{R}$. If f is differentiable at c, then it is also continuous at that point.*

Proof To prove f is continuous at $x = c$, we show that $\lim_{x \to c} f(x) = f(c)$, and in order to do that, we rewrite $f(x)$ as follows (for $x \neq c$):

$$f(x) = \frac{f(x) - f(c)}{x - c} \cdot (x - c) + f(c).$$

We are told that f is differentiable, and hence the limit $\lim_{x \to c} \frac{f(x)-f(c)}{x-c}$ exists and is equal to a real number $f'(c)$. It thus follows, from the limit laws, that

$$\lim_{x \to c} f(x) = \lim_{x \to c} \left[\frac{f(x) - f(c)}{x - c} \cdot (x - c) + f(c) \right] = f'(c) \cdot 0 + f(c) = f(c),$$

and so f is continuous at $x = c$, as needed. □

We have thus showed that functions which are differentiable at a certain point must be continuous there. The converse, however, is not true. As we have seen, the function $h(x) = |x|$ is not differentiable at $x = 0$. It is, however, continuous at $x = 0$, a fact that we ask you to verify below.

Exercise 9.4.11

Prove, using the epsilon-delta definition of continuity, that the function $h(x) = |x|$ is continuous at $x = 0$.

9.5 Problems

9.1 Guess the limits of the following sequences. Then, use Definition 9.1.7 to prove formally that your guess is correct.

a. $a_n = 4 - \frac{2}{n^2}$

b. $b_n = \frac{2n+1}{6n-5}$

c. $c_n = \frac{\cos n + \sin n}{n^2}$

9.2 Use Theorem 9.1.13 to calculate the following limits.

a. $\lim \frac{2n^2-3n+1}{n-10n^2+1}$
b. $\lim \left(2+\frac{n}{n+1}\right) \cdot \left(\frac{1}{5n^2}+3\right)$

9.3 a. Prove, by induction, that $\frac{1}{2^n} \leq \frac{1}{n+1}$ for all $n \in \mathbb{N}$.
b. Use Definition 9.1.7 to prove that $\lim \frac{1}{2^n} = 0$.

9.4 Prove that the following sequences have no limit.

a. $a_n = 4^n$
b. $b_n = \tan\left(\frac{\pi}{2} - \frac{1}{n}\right)$
c. $c_n = (-1)^n + 1$

9.5 Suppose that (a_n) is a bounded sequence. That is, there exists a positive number M such that $|a_n| \leq M$ for all n. Moreover, suppose that (b_n) is a sequence that converges to zero. Prove that $\lim(a_n \cdot b_n) = 0$.

9.6 Prove Part 2 of Theorem 9.1.13.

9.7 Let (b_n) be a sequence that converges to a non-zero real number M.

a. Prove that there exists an $N \in \mathbb{N}$, such that $|b_n| \geq \frac{|M|}{2}$ for all $n \geq N$.
b. Prove that $\lim \frac{1}{b_n} = \frac{1}{M}$.
Hint: Notice that

$$\left|\frac{1}{b_n} - \frac{1}{M}\right| = \frac{|b_n - M|}{|b_n| \cdot |M|} = \frac{1}{|M|} \cdot |b_n - M| \cdot \frac{1}{|b_n|}.$$

Use Part a to argue that $\frac{1}{|b_n|} \leq \frac{2}{|M|}$ when n is large enough.
c. Now prove Part 4 of Theorem 9.1.13.

9.8 We say that a sequence (a_n) *diverges to (positive) infinity*, and write $\lim a_n = \infty$ or $\lim a_n = +\infty$, if for every real number M there exists an $N \in \mathbb{N}$, such that $a_n > M$ whenever $n \geq N$.

a. Use this definition to prove that $\lim(n^2 + 5) = \infty$.
b. Prove that $\lim \left(n + \cos(n) + \frac{1}{n}\right) = \infty$.
c. Formulate a definition for divergence to negative infinity (that is, for $\lim a_n = -\infty$).
d. Prove that if (a_n) and (b_n) are two sequences, such that $\lim a_n = c$ (where $c \in \mathbb{R}$) and $\lim b_n = \infty$, then $\lim(a_n - b_n) = -\infty$.

9.9 Use Definition 9.2.5 to prove the following limits.

a. $\lim_{x \to 5} (7x - 20) = 15$
b. $\lim_{x \to 2} x^2 = 4$
c. $\lim_{x \to 9} \sqrt{x} = 3$

9.10 Let $f(x) = \frac{|x|-2x}{x}$ for $x \neq 0$. Use Definition 9.2.8 to prove that $\lim_{x\to 0^-} f(x) = -3$ and $\lim_{x\to 0^+} f(x) = -1$.

9.11 Definition 9.2.5 can be extended to cases where either c, L or both represent positive or negative infinity rather than a number. For instance, suppose that f is a real-valued function, defined on an interval of the form (a, ∞), and let L be a real number.

We say that the limit of f at $+\infty$ is L, and write $\lim_{x\to+\infty} f(x) = L$, if for every $\varepsilon > 0$, there is a real number M, such that $|f(x) - L| < \varepsilon$ whenever $x > M$.

Informally speaking, the condition $x > M$, which replaces $|x - c| < \delta$, can be thought of representing the idea of "closeness to infinity."

a. Use the above definition to prove that $\lim_{x\to\infty} \frac{3}{x^2} = 0$.
b. Formulate similar definitions for limits of the form $\lim_{x\to-\infty} f(x) = L$, $\lim_{x\to c} f(x) = -\infty$ and $\lim_{x\to+\infty} f(x) = +\infty$.
c. Use the definitions that you formulated, to prove that $\lim_{x\to+\infty} \left(x^2 + \sin x\right) = \infty$.

9.12 Consider the following function $f\colon \mathbb{R} \to \mathbb{R}$:

$$f(x) = \begin{cases} x - 1 & \text{if } x \in \mathbb{Q} \\ 1 - x & \text{if } x \notin \mathbb{Q}. \end{cases}$$

Use Theorem 9.3.1 to prove that the limit $\lim_{x\to 0} f(x)$ does not exist.

9.13 Part 3 of Theorem 9.2.10 is proved in Section 9.3 using sequences. Use a similar strategy to prove Parts 1, 2 and 4 of Theorem 9.2.10.

9.14 Prove the following theorem, relating our original definition of a limit to one-sided limits.

Let f be a real-valued function defined on a punctured neighbourhood of c. Then $\lim_{x\to c} f(x)$ exists if and only if the one-sided limits $\lim_{x\to c^+} f(x)$ and $\lim_{x\to c^-} f(x)$ both exist and are equal to each other.

9.15 Let $f\colon \mathbb{R} \to \mathbb{R}$ be given by $f(x) = a \cdot x + b$, where a, b are constant real numbers. Prove that for every $c \in \mathbb{R}$, $\lim_{x\to c} f(x) = a \cdot c + b$. Distinguish between the cases $a = 0$ and $a \neq 0$.

9.16 Suppose that $n \in \mathbb{N}$ and $f\colon (-1, 1) \to \mathbb{R}$ is a bounded function. Prove that $\lim_{x\to 0} [x^n \cdot f(x)] = 0$.

9.17 Suppose that f and g are two real-valued functions defined on some neighbourhood of c.

a. Explain why if f and g are continuous at c, then $f+g, f-g, f \cdot g$ and $\frac{f}{g}$ (provided that $g(c) \neq 0$) are all continuous at c.
b. Prove that the function $f(x) = x$ is continuous at every point $c \in \mathbb{R}$.
c. Let $K \in \mathbb{R}$. Prove that the constant functions $g(x) = K$ for all $x \in \mathbb{R}$ are continuous at every $c \in \mathbb{R}$.
d. Conclude that polynomials and rational functions in one variable are continuous at every $c \in \mathbb{R}$ in their domain.

9.18 For each statement, decide whether it is necessarily true or could be false. Provide a proof or a counterexample.

a. If f and $f+g$ are continuous at c, then so is g.
b. If f and $f \cdot g$ are continuous at c, then so is g.

9.19 a. Let $f\colon \mathbb{R} \to \mathbb{R}$ be given by

$$f(x) = \begin{cases} 1 & \text{if } x \in \mathbb{Q} \\ -1 & \text{if } x \notin \mathbb{Q}. \end{cases}$$

Prove that $\lim_{x \to 0} f(x)$ does not exist.

b. Show that the function $g\colon \mathbb{R} \to \mathbb{R}$, given by

$$g(x) = \begin{cases} x & \text{if } x \in \mathbb{Q} \\ -x & \text{if } x \notin \mathbb{Q}, \end{cases}$$

is continuous at $x = 0$.

9.20 Calculate the derivative of each function at the given point. Use the definition of the derivative as a limit, and do not use any differentiation rules.

a. $f(x) = x^2$ at $x = 5$.
b. $g(x) = \frac{1}{x}$ at $x = -3$.
c. $h(x) = x^4 + x$ at an arbitrary $c \in \mathbb{R}$.

9.21 Let $f\colon \mathbb{R} \to \mathbb{R}$ be a bounded function, and define $g(x) = x \cdot f(x)$.

a. Prove that g is continuous at $x = 0$.
b. Prove that if f is continuous at $x = 0$, then g is differentiable at $x = 0$, and $g'(0) = f(0)$.

9.22 Let f and g be two real-valued functions, defined on a neighbourhood of a real number c. Prove that if f and g are differentiable at c, then so are $f+g$ and $f-g$, and the following hold. Use Theorem 9.2.10.

- $(f+g)'(c) = f'(c) + g'(c)$
- $(f-g)'(c) = f'(c) - g'(c)$

9.23 Show that if f and g are real-valued functions which are differentiable at c, then their product $f \cdot g$ is also differentiable at c, and the following product rule holds true:

$$(f \cdot g)'(c) = f'(c) \cdot g(c) + f(c) \cdot g'(c).$$

(Hint: Write down the definition of $(f \cdot g)'(c)$ in terms of limits, and then add and subtract the term $f(x) \cdot g(c)$ to the numerator.)

9.24 Is the function $g(x) = x \cdot |x|$ differentiable at $x = 0$? If so, what is $g'(0)$?

9.6 Solutions to Exercises

Solution to Exercise 9.1.6
We solve $|b_n - 1| < \frac{1}{5000}$ and get

$$\frac{10}{n+10} < \frac{1}{5000} \qquad \Leftrightarrow \qquad 49990 < n,$$

and so n must be larger than or equal to 49991.

Solution to Exercise 9.1.11
We know that $a \geq 0$. If $a \neq 0$ then a must be a positive number. Take $\varepsilon = \frac{a}{2} > 0$. It follows that $a < 2\varepsilon = 2 \cdot \frac{a}{2} = a$, which is a contradiction. Therefore, $a = 0$.

Solution to Exercise 9.1.14
We can write

$$a_n = \frac{(3n^2+5)/n^2}{(7n^2-3)/n^2} = \frac{3+\frac{5}{n^2}}{7-\frac{3}{n^2}} = \frac{3+5\cdot\frac{1}{n}\cdot\frac{1}{n}}{7-3\cdot\frac{1}{n}\cdot\frac{1}{n}}.$$

Using Theorem 9.1.13, we get

$$\lim a_n = \frac{3+5\cdot 0\cdot 0}{7-3\cdot 0\cdot 0} = \frac{3}{7}.$$

Solution to Exercise 9.2.2
Note that $\sqrt{8.99} = 2.9983$ and $\sqrt{9.01} = 3.0017$, so we can, for example, use the neighbourhood $N_{0.0015}(3) = (2.9985, 3.0015)$. Every x in this interval will be mapped by $f(x)$ into $(8.99, 9.01)$. There are, of course, many other possible answers.

Solution to Exercise 9.2.4
The given inequality is equivalent to

$$\sqrt{9+\varepsilon} + \sqrt{9-\varepsilon} < 6,$$

and since both sides are positive, we can square and obtain

$$9+\varepsilon+9-\varepsilon+2\sqrt{81-\varepsilon^2} < 36 \qquad \Leftrightarrow \qquad \sqrt{81-\varepsilon^2} < 9.$$

Squaring both sides again gives us the equivalent inequality $81 - \varepsilon^2 < 81$, which is clearly true, and therefore the given inequality holds true as well.

Solution to Exercise 9.2.9

Let us prove that $\lim_{x \to 0^-} h(x) = -1$. The proof of the right-hand limit is similar.
Let $\varepsilon > 0$ and choose $\delta = \varepsilon$. Suppose that x satisfies $0 - \delta < x < 0$. We need to show that $|h(x) - (-1)| < \varepsilon$. Indeed, as $h(x) = x - 1$ for negative x, we get

$$|h(x) - (-1)| = |x - 1 + 1| = |x| < \delta = \varepsilon.$$

The last inequality $|x| < \delta$ follows from the fact that $-\delta < x < 0$. We thus conclude that $\lim_{x \to 0^-} h(x) = -1$.

Solution to Exercise 9.4.6

If $\lim_{x \to c^+} f(x) = f(c) > 0$, then for $\varepsilon = f(c) > 0$, there is a $\delta > 0$, such that whenever $c < x < c + \delta$, we have

$$|f(x) - f(c)| < \varepsilon \quad \Rightarrow \quad -f(c) < f(x) - f(c) < f(c) \quad \Rightarrow \quad 0 < f(x) < 2f(c).$$

This shows that $f > 0$ on $[c, c + \delta)$, as needed. The case where $f(c) < 0$ is proved in a similar way, by choosing $\varepsilon = -f(c)$.

Solution to Exercise 9.4.11

The proof is quite short. The main observation is that the expressions $|x - 0|$ and $|h(x) - h(0)|$ are, in fact, equal. They are both just $|x|$. We thus proceed with our formal proof, and choose δ to be equal to the given ε.

Let $\varepsilon > 0$, and choose $\delta = \varepsilon$. For every real number x, if $|x - 0| = |x| < \delta$, then

$$|h(x) - h(0)| = ||x| - |0|| = |x| < \delta = \varepsilon.$$

This shows that $\lim_{x \to 0} h(x) = h(0)$, and hence h is continuous at $x = 0$.

10 Complex Numbers

In this chapter we introduce the complex number system – an extension of the well-known real numbers. Complex numbers arise naturally in many problems in mathematics and science, and allow to us to study polynomial equations that may not have real solutions (such as $x^2+5=0$). As we will see, many familiar algebraic properties remain valid in the complex number system. In particular, we show that the complex numbers form a field, and that the quadratic formula and the Triangle Inequality can still be used in this new number system.

10.1 Background

We began this book by discussing quadratic equations and how to solve them. The origins of the quadratic formula date back to the Babylonians (~1700 BC), who studied problems that led to quadratic equations. Although their treatment was mainly geometric, the Babylonians came up with a procedure for solving several types of quadratics that is almost equivalent to the modern-day quadratic formula. A general procedure for solving all quadratic equations, similar to the one we use today, was introduced much later by Simon Stevin in 1594.

Not all quadratic equations have solutions that are real numbers. In this chapter, we *extend* our number system, so that *all* quadratic equations have at least one solution. We note, however, that the need to extend the reals arose from attempting to solve *cubic* equations, and *not* quadratics.

In 1545, Girolamo Cardano, an Italian mathematician, published the formula for solving cubic equations of the form $x^3 = px + q$:

$$x = \sqrt[3]{\frac{q}{2} + \sqrt{\left(\frac{q}{2}\right)^2 - \left(\frac{p}{3}\right)^3}} + \sqrt[3]{\frac{q}{2} - \sqrt{\left(\frac{q}{2}\right)^2 - \left(\frac{p}{3}\right)^3}}.$$

Rafael Bombelli, another Italian mathematician who lived at that period, tried to solve the equation $x^3 = 15x + 4$ by setting $p = 15$ and $q = 4$ in Cardano's formula. He knew that $x = 4$ is a solution of the equation, but could not obtain it because of the negative numbers under the square roots:

$$x = \sqrt[3]{2 + \sqrt{-121}} + \sqrt[3]{2 - \sqrt{-121}}.$$

Bombelli assumed the existence of a number he denoted as $\sqrt{-1}$, and was able to obtain the solution $x = 4$ using the above formula and algebraic manipulations. In fact, the equation $x^3 = 15x + 4$ has three distinct real solutions: $x = 4$, $x = -2 + \sqrt{3}$ and $x = -2 - \sqrt{3}$.

In this chapter we extend the real number system to include a new type of numbers called *imaginary* numbers. Elements in the extended system will be called complex numbers. Bombelli was the first to introduce rules of arithmetic for complex numbers, but it took close to 300 years for mathematicians to acknowledge complex numbers and accept them as useful and fundamental to mathematics and science.

Extending a number system is a process you have already encountered in your school years, when you learned about fractions and negative integers. These new numbers allow us to perform calculations and solve equations that would be otherwise impossible.

For instance, fractions are needed to calculate expressions like $12 \div 5$, and negative numbers are required to solve equations such as $x + 9 = 5$. There are other reasons for extending a number system. In geometry, we often need irrational numbers to measure areas and lengths. The diagonal of a unit square has length $\sqrt{2}$, and the area of a circle of radius 1 is π.

Extending a number system is more than just adding new numbers to our current system. We would like to be able to add and multiply elements in the extended system, and as much as possible, preserve rules of arithmetic and other properties that were valid in the original system.

10.2 The Field of Complex Numbers

To extend the real number system, we add a new number i that solves the equation $x^2 + 1 = 0$. That is,

$$i^2 + 1 = 0 \qquad \text{or} \qquad i^2 = -1.$$

Clearly, i cannot be a real number as $a^2 \geq 0$ for every $a \in \mathbb{R}$.

As much as possible, our new number system should preserve the rules of arithmetic for real numbers given by the field axioms (see Definition 2.3.1). In particular, we should be able to add and multiply elements of our new number system. This forces us to consider expressions such as $b \cdot i$ and $a + i$ where a and b are real numbers. In general, our new number system must contain expressions of the form $a + bi$ where $a, b \in \mathbb{R}$. Expressions such as $2 + 3i$, $-0.5 - 13i$ and $\sqrt{2} + \frac{1}{5}i$ will be called *complex numbers*. More formally, we define a complex number as a pair of real numbers.

Definition 10.2.1 A *complex number* z is a pair (a, b) of real numbers, conventionally written as $z = a + bi$. The numbers a and b are called the *real part* and the *imaginary part* of z, respectively. We denote $a = Re(z)$ and $b = Im(z)$. The set of complex numbers is denoted by $\mathbb{C}$.

Remark. A real number a can be identified with the complex number $a + 0i$. This allows us to view the real numbers as a *subset* of the complex numbers. A complex number of the form $0 + bi$, or simply bi, is called an *imaginary number*. Finally, we naturally interpret the expression $a - bi$ as $a + (-b)i$.

Example 10.2.2 Find the real and imaginary parts of the following complex numbers:

$$z = 3 - 5i, \qquad w = 7, \qquad u = \sqrt{2}i.$$

Solution

We write each of the numbers in the standard form $a + bi$ from which the real and imaginary parts can be easily seen.

- $z = 3 + (-5)i$ and so $Re(z) = 3$ and $Im(z) = -5$.
- $w = 7 + 0i$ and therefore $Re(w) = 7$ and $Im(w) = 0$.
- Finally, we have $u = 0 + \sqrt{2}i$ and thus $Re(u) = 0$ and $Im(u) = \sqrt{2}$.

How should we define addition and multiplication of two complex numbers $a + bi$ and $c + di$? In an attempt to preserve commutativity, associativity and distributivity, we obtain

$$(a + bi) + (c + di) = (a + c) + (bi + di) = (a + c) + (b + d)i.$$

Note how commutativity and associativity of addition were used in the first equality, while distributivity was used in the second equality.

Similar considerations, and the fact that $i^2 = -1$ force us to define multiplication of complex numbers in a unique way.

Definition 10.2.3 The *sum* and *product* of two complex numbers $z = a + bi$ and $w = c + di$ are defined as follows:

$$\begin{aligned} z + w &= (a + c) + (b + d)i \\ z \cdot w &= (ac - bd) + (ad + bc)i. \end{aligned}$$

Note that if $z = a$ is a real number, then

$$z \cdot w = a \cdot (c + di) = (ac) + (ad)i$$

as one might expect.

Exercise 10.2.4
Justify the definition of $z \cdot w$ by expanding $(a + bi) \cdot (c + di)$ in the "usual" way, and applying the identity $i^2 = -1$. Which of the field axioms and their consequences did you use in each step?

Example 10.2.5 Find the sum and the product of $z = \frac{1}{3} - 2i$ and $w = 2 + \frac{1}{3}i$.

Solution
From Definition 10.2.3, we get:

$$z + w = \left(\tfrac{1}{3} + 2\right) + \left(-2 + \tfrac{1}{3}\right) i = \tfrac{7}{3} - \tfrac{5}{3}i.$$

To calculate the product, we often expand in the usual way, and use the identity $i^2 = -1$:

$$\begin{aligned} z \cdot w &= \left(\tfrac{1}{3} - 2i\right) \cdot \left(2 + \tfrac{1}{3}i\right) = \tfrac{2}{3} + \tfrac{1}{9}i - 4i - \tfrac{2}{3}i^2 \\ &= \tfrac{2}{3} - \tfrac{35}{9}i - \tfrac{2}{3} \cdot (-1) = \tfrac{4}{3} - \tfrac{35}{9}i. \end{aligned}$$

Example 10.2.6 Let $a \in \mathbb{R}$ and $z = a + ai$. Show that $z^2 = z \cdot z$ is an imaginary number.

Solution
We compute:

$$z^2 = (a + ai) \cdot (a + ai) = a^2 + a^2 i + a^2 i + a^2 i^2 = a^2 + 2a^2 i - a^2 = 2a^2 i.$$

We see that the real part of z^2 is zero, and so this is an imaginary number.

Exercise 10.2.7
Now that multiplication and addition have been defined, the expression $a + bi$, with $a, b \in \mathbb{R}$, can be interpreted in two different ways.

- As a single complex number.
- As the result of adding a to the product $b \cdot i$, that is, as

$$(a + 0i) + (b + 0i) \cdot (0 + 1i).$$

Show that these two interpretations coincide.

Now that we have defined our new number system, addition and multiplication, there are a few natural questions that come to mind.

- Are multiplication and addition in $\mathbb{C}$ commutative and associative? Is the distributive law still valid?

- Is it true that $z + 0 = z$ and $z \cdot 1 = z$ for every $z \in \mathbb{C}$?
- What about the remaining field axioms? Do negatives and reciprocals exist?
- Have we lost any of the properties of real numbers in the extension process?

As we will prove shortly, all the field axioms are satisfied, and so $\mathbb{C}$ is a field. The real numbers 0 and 1, viewed as the complex numbers $0 + 0i$ and $1 + 0i$, continue to be the additive and multiplicative identities in $\mathbb{C}$. The negative of a complex number $a + bi$ will be, naturally, the number $-a + (-b)i = -a - bi$. Finding reciprocals is a little harder.

If $z = a + bi$ is a *non-zero* complex number, which complex number u will satisfy the condition $z \cdot u = 1$? The answer comes from the following observation:

$$(a + bi) \cdot (a - bi) = (a^2 + b^2) + (-ab + ab)i = a^2 + b^2.$$

We can now divide these equalities by the *positive* number $a^2 + b^2$ to get:

$$(a + bi) \cdot \left(\frac{a}{a^2 + b^2} - \frac{b}{a^2 + b^2} i \right) = 1.$$

This gives us a clear candidate for the multiplicative inverse of z, namely $u = \frac{a}{a^2+b^2} - \frac{b}{a^2+b^2} i$.

Example 10.2.8 In practice, to find reciprocals, we often multiply and divide z by $a - bi$, instead of using the formula derived above. For instance, let us find the reciprocal of $z = -2 + 3i$.

Solution

To find

$$z^{-1} = \frac{1}{z} = \frac{1}{-2 + 3i},$$

we multiply both the numerator and denominator by $-2 - 3i$, and get

$$\frac{-2 - 3i}{(-2 + 3i) \cdot (-2 - 3i)} = \frac{-2 - 3i}{2^2 + 3^2} = -\frac{2}{13} - \frac{3}{13} i.$$

One can now verify that

$$(-2 + 3i) \cdot \left(-\frac{2}{13} - \frac{3}{13} i \right) = 1.$$

We are now ready to prove the following theorem.

Theorem 10.2.9 *The set of complex numbers, with addition and multiplication defined above, forms a field.*

Proof To prove that $\mathbb{C}$ is a field, we must verify all the field axioms (see Definition 2.3.1). As this process is quite straightforward, we check only some of the axioms, and ask that you check the remaining ones yourself in the exercise following the proof.

Closure. If $z, w \in \mathbb{C}$, then $z + w \in \mathbb{C}$ and $z \cdot w \in \mathbb{C}$. This follows directly from Definition 10.2.3.

Associativity of Addition. If $z = a + bi$, $w = c + di$ and $u = p + qi$ are three arbitrary complex numbers, we show that $z + (w + u) = (z + w) + u$ by carefully computing each side:

$$\begin{aligned} z + (w + u) &= (a + bi) + [(c + di) + (p + qi)] \\ &= (a + bi) + [(c + p) + (d + q)i] \\ &= [a + (c + p)] + [b + (d + q)]i \end{aligned}$$

and

$$\begin{aligned} (z + w) + u &= [(a + bi) + (c + di)] + (p + qi) \\ &= [(a + c) + (b + d)i] + (p + qi) \\ &= [(a + c) + p] + [(b + d) + q]i. \end{aligned}$$

From associativity of real numbers, we know that $a+(c+p)=(a+c)+p$ and $b+(d+q) = (b+d)+q$, and so the expressions obtained for $z+(w+u)$ and for $(z + w) + u$ represent the same complex number, as expected.

Commutativity of Multiplication. If $z = a + bi$ and $w = c + di$ are complex numbers, then

$$\begin{aligned} z \cdot w &= (a + bi) \cdot (c + di) = (ac - bd) + (ad + bc)i \\ w \cdot z &= (c + di) \cdot (a + bi) = (ca - db) + (cb + da)i. \end{aligned}$$

Multiplication and addition of real numbers are commutative, and so $ac - bd = ca - db$ and $ad + bc = cb + da$. We conclude that $z \cdot w = w \cdot z$, as needed.

The Additive Identity Axiom. For every $z = a + bi$, we have, according to Definition 10.2.3,

$$z + 0 = (a + bi) + (0 + 0i) = (a + 0) + (b + 0)i = a + bi = z,$$

and we see that $z + 0 = z$ as needed.

Existence of Negatives and Reciprocals. For every $z = a + bi \in \mathbb{C}$, set $w = (-a) + (-b)i = -a - bi$, and compute

$$z + w = [a + (-a)] + [b + (-b)]i = 0 + 0i = 0.$$

Moreover, if $z \neq 0$ then $a^2 + b^2 > 0$. We define $u = \frac{a}{a^2+b^2} - \frac{b}{a^2+b^2}i$ and check:

$$\begin{aligned} z \cdot u &= (a+bi) \cdot \left(\frac{a}{a^2+b^2} - \frac{b}{a^2+b^2} i \right) \\ &= \left(\frac{a^2}{a^2+b^2} + \frac{b^2}{a^2+b^2} \right) + \left(\frac{a \cdot (-b)}{a^2+b^2} + \frac{ba}{a^2+b^2} \right) i = 1 + 0i = 1. \end{aligned}$$

We conclude that negatives and reciprocals exist, and are given by:

$$-z = -a - bi \quad \text{and} \quad z^{-1} = \frac{a}{a^2+b^2} - \frac{b}{a^2+b^2} i \quad (\text{for } z \neq 0).$$

There are a few more axioms that we ask you to check in the exercise below. Once that is done, we have shown that $\mathbb{C}$ is a field. □

Exercise 10.2.10

Complete the proof of Theorem 10.2.9 by proving the following remaining field axioms.

1. **Associativity of Multiplication**: For every $z, w, u \in \mathbb{C}$, we have $z \cdot (w \cdot u) = (z \cdot w) \cdot u$.
2. **Commutativity of Addition**: For every $z, w \in \mathbb{C}$, we have $z + w = w + z$.
3. **The Multiplicative Identity Axiom**: For every $z \in \mathbb{C}$, we have $z \cdot 1 = z$.
4. **The Distributivity Axiom**: For any $z, w, u \in \mathbb{C}$, we have $z \cdot (w + u) = z \cdot w + z \cdot u$.

In any field, subtraction and division can be defined in terms of addition and multiplication.

Definition 10.2.11 Let z and w be two complex numbers. We define:

$$z - w = z + (-w) \quad \text{and} \quad \frac{z}{w} = z \cdot w^{-1} \quad (\text{assuming } w \neq 0).$$

Example 10.2.12 Solve the equation $2z + 2iz - 3 = 5i$.

Solution

Since $\mathbb{C}$ is a field, we perform the "usual" algebraic manipulations and solve for z. First, we factor z and add 3 to both sides:

$$(2 + 2i)z = 3 + 5i.$$

Then, we divide both sides by $2 + 2i$:

$$z = \frac{3+5i}{2+2i}.$$

In order to provide a final answer in the standard form $a + bi$, we use the strategy from Example 10.2.8, and multiply the numerator and denominator by $2 - 2i$:

$$z = \frac{(3+5i)\cdot(2-2i)}{(2+2i)\cdot(2-2i)} = \frac{6-6i+10i+10}{2^2+2^2} = \frac{16+4i}{8} = 2 + \tfrac{1}{2}i.$$

Therefore, the equation has the unique solution $z = 2 + \frac{1}{2}i$.

Remark. Even though we were successful in preserving all the field axioms, we do pay a small price. The complex numbers cannot be *ordered* in a way that is compatible with the field axioms. That is, there is no way to make sense of statements such as $z < w$, where z and w are complex numbers, in a satisfactory way. This is discussed in the following exercise.

Exercise 10.2.13

Suppose for a moment, that we managed to order the complex numbers in a way that preserves basic properties of inequalities. Would the imaginary number i be greater or smaller than zero?

1. Show that if $i > 0$, and we multiply both sides by i, we get a contradiction.
2. Show that if we assume that $i < 0$, we get the same contradiction.

These contradictions show that the complex numbers cannot be ordered in a satisfactory way.

We end this section by discussing conjugates and absolute values of complex numbers.

Definition 10.2.14 For every complex number $z = a + bi$, we define:

$$\bar{z} = a - bi \qquad \text{and} \qquad |z| = \sqrt{a^2 + b^2}.$$

We call $\bar{z}$ the *conjugate* of z, and $|z|$ the *absolute value* (or *modulus*) of z.

Note that, in general, $\bar{z}$ is a complex number, while $|z|$ is a non-negative real number. In the special case where $z = a + 0i$ is a *real* number, we get

$$\bar{z} = a - 0i = z \qquad \text{and} \qquad |z| = \sqrt{a^2 + 0^2} = |a|.$$

Conjugates and absolute values are often used to simplify calculations with complex numbers, but can also be interpreted geometrically, as the following exercise illustrates.

Exercise 10.2.15

We can represent a complex number $z = a + bi$ as a point (a, b) in the plane $\mathbb{R}^2$. Using this representation, give a geometrical interpretation of $\bar{z}$ and $|z|$.

Example 10.2.16 Let $z = -12 - 5i$.

a. Calculate the number $w = \frac{\bar{z}}{|z|^2}$.
b. Calculate $z \cdot w$. What can we conclude about z and w?

Solution

a. According to Definition 10.2.14, we get:

$$\bar{z} = -12 + 5i \qquad \text{and} \qquad |z| = \sqrt{(-12)^2 + 5^2} = 13,$$

and so $w = \frac{-12+5i}{13^2} = -\frac{12}{169} + \frac{5}{169}i$.

b. We calculate:

$$z \cdot w = (-12 - 5i) \cdot \left(-\frac{12}{169} + \frac{5}{169}i\right) = \frac{144}{169} + \frac{60}{169}i - \frac{60}{169}i + \frac{25}{169} = \frac{169}{169} = 1.$$

We conclude that w is the reciprocal of z. That is, $w = \frac{1}{z}$.

The conclusion that $w = \frac{1}{z}$ in the last example is not a coincidence.

Claim 10.2.17 *For every $z \in \mathbb{C}$, we have $z \cdot \bar{z} = |z|^2$. It follows that if $z \neq 0$, then*

$$z^{-1} = \frac{\bar{z}}{|z|^2}.$$

Proof Let $z = a + bi$. Then

$$z \cdot \bar{z} = (a+bi) \cdot (a-bi) = [a \cdot a - b \cdot (-b)] + [a \cdot (-b) + ba]i = a^2 + b^2 = |z|^2.$$

In fact, we already proved this identity while discussing reciprocals on page 270. If $z \neq 0$, we get

$$z \cdot \frac{\bar{z}}{|z|^2} = z \cdot \bar{z} \cdot \frac{1}{|z|^2} = |z|^2 \cdot \frac{1}{|z|^2} = 1,$$

from which it follows that $z^{-1} = \frac{\bar{z}}{|z|^2}$. □

Exercise 10.2.18
Verify the following statements.

1. A complex number z is *real* if and only if $\bar{z} = z$.
2. A complex number z is *imaginary* if and only if $\bar{z} = -z$.

The following proposition shows that conjugates behave nicely with respect to sums, differences, products and quotients of complex numbers.

Proposition 10.2.19 (Properties of Conjugates) *For every $z, w \in \mathbb{C}$, we have the following.*

1. $\overline{z+w} = \overline{z} + \overline{w}$.
2. $\overline{z-w} = \overline{z} - \overline{w}$.
3. $\overline{z \cdot w} = \overline{z} \cdot \overline{w}$.
4. $\overline{\left(\frac{z}{w}\right)} = \frac{\overline{z}}{\overline{w}}$ *(for* $w \neq 0$*).*

Proof Suppose $z = a + bi$ and $w = c + di$ are two complex numbers. We prove Parts 1 and 3 and leave the rest as an exercise.

1 By carefully applying the definitions of conjugates and sums, we get

$$\overline{z+w} = \overline{(a+bi)+(c+di)} = \overline{(a+c)+(b+d)i} = (a+c)-(b+d)i$$

and

$$\overline{z}+\overline{w} = \overline{(a+bi)} + \overline{(c+di)} = (a-bi)+(c-di) = (a+c)-(b+d)i.$$

Therefore, $\overline{z+w} = \overline{z} + \overline{w}$.

3 Again, we compute:

$$\overline{z \cdot w} = \overline{(a+bi)\cdot(c+di)} = \overline{(ac-bd)+(ad+bc)i} = (ac-bd)-(ad+bc)i$$

and

$$\overline{z} \cdot \overline{w} = \overline{(a+bi)} \cdot \overline{(c+di)} = (a-bi)\cdot(c-di) = (ac-bd)-(ad+bc)i.$$

We conclude that $\overline{z \cdot w} = \overline{z} \cdot \overline{w}$. □

Exercise 10.2.20
Complete the proof of Proposition 10.2.19. For Part 4, start by proving that $\overline{w^{-1}} = (\overline{w})^{-1}$.

Many properties satisfied by absolute values of real numbers extend to complex numbers. In particular, we have the following.

Proposition 10.2.21 *For every two complex numbers z and w, we have*

$$|z \cdot w| = |z| \cdot |w| \qquad \textit{and} \qquad \left|\frac{z}{w}\right| = \frac{|z|}{|w|} \qquad (\textit{if } w \neq 0).$$

Proof To prove the first identity, we use Claim 10.2.17 and Proposition 10.2.19 to calculate $|z \cdot w|^2$:

$$|z \cdot w|^2 = (z \cdot w) \cdot \overline{z \cdot w} = z \cdot w \cdot \overline{z} \cdot \overline{w} = (z \cdot \overline{z}) \cdot (w \cdot \overline{w}) = |z|^2 \cdot |w|^2 = (|z| \cdot |w|)^2.$$

We get $|z \cdot w|^2 = (|z| \cdot |w|)^2$ which implies $|z \cdot w| = |z| \cdot |w|$ as needed.

In particular, if $w \neq 0$, we have

$$1 = \left|w \cdot \frac{1}{w}\right| = |w| \cdot \left|\frac{1}{w}\right| \qquad \Rightarrow \qquad \left|\frac{1}{w}\right| = \frac{1}{|w|}.$$

We can now prove the second identity as follows:

$$\left|\frac{z}{w}\right| = \left|z \cdot \frac{1}{w}\right| = |z| \cdot \left|\frac{1}{w}\right| = |z| \cdot \frac{1}{|w|} = \frac{|z|}{|w|}.$$

□

If z and w are two complex numbers, there is no reason to expect that $|z+w|$ and $|z| + |w|$ will be equal to each other. After all, these quantities can be different even when z and w are real numbers. In the next section we show that the Triangle Inequality in $\mathbb{R}$ remains valid for complex numbers.

Exercise 10.2.22

1. Find two complex numbers $z = a + bi$ and $w = c + di$, with a, b, c, d all non-zero, such that $|z + w| \neq |z| + |w|$.
2. Show that if $z = 6 - 9i$ and $w = 4 - 6i$, then $|z + w| = |z| + |w|$.

10.3 The Complex Plane and the Triangle Inequality

Every real number can be represented as a point on a number line. Complex numbers, which we defined as *pairs of real numbers*, can be represented as points on a plane equipped with a standard xy-coordinate system (Figure 10.1).

Every complex number $z = a + bi$ is identified with the point (a, b) on that plane. The horizontal axis is called the *real axis*, as its points represent real numbers. Similarly, points on the vertical axis represent imaginary numbers, which is why we call it the *imaginary axis*.

This *complex plane* allows us to visualize elements and subsets of $\mathbb{C}$. Operations and properties involving complex numbers can often be interpreted geometrically through the complex plane. For instance, conjugates and absolute values can be interpreted geometrically (see Exercise 10.2.15).

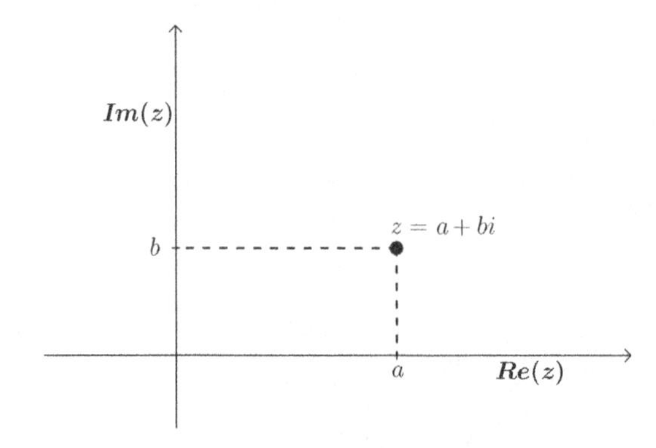

Figure 10.1 The complex plane.

Exercise 10.3.1
Given two complex numbers z and w, describe the geometrical meaning of $|z - w|$.

Example 10.3.2 Here are two equations in the complex variable z. Describe geometrically the solution sets on the complex plane.

a. $z \cdot \overline{z} = 9$.
b. $|z - i| = |z + 1|$.

Solution

a. Using the identity $z \cdot \overline{z} = |z|^2$ the equation can be written as

$$|z|^2 = 9 \qquad \text{or} \qquad |z| = 3,$$

which represents a circle of radius 3 centered at the origin.

b. This equation is a little more complicated. We write $z = x + yi$ and simplify the resulting equation:

$$\begin{aligned} |(x + yi) - i| &= |(x + yi) + 1| \\ |x + (y - 1)i|^2 &= |(x + 1) + yi|^2 \\ x^2 + (y - 1)^2 &= (x + 1)^2 + y^2 \\ x^2 + y^2 - 2y + 1 &= x^2 + 2x + 1 + y^2 \\ -2y &= 2x \\ y &= -x. \end{aligned}$$

We conclude that the line $y = -x$ represents the solutions of the equation.

Exercise 10.3.3
Identify the regions in the complex plane determined by the following inequalities:

a. $|z - 1 - i| \leq 1$,
b. $Re(z) < 3$.

Sums of Complex Numbers

There is a simple geometric description for the addition operation of complex numbers.

Theorem 10.3.4 *Let z and w be two complex numbers. The points corresponding to $z, 0, w$ and $z + w$ either lie on the same straight line, or are the vertices of a parallelogram.*

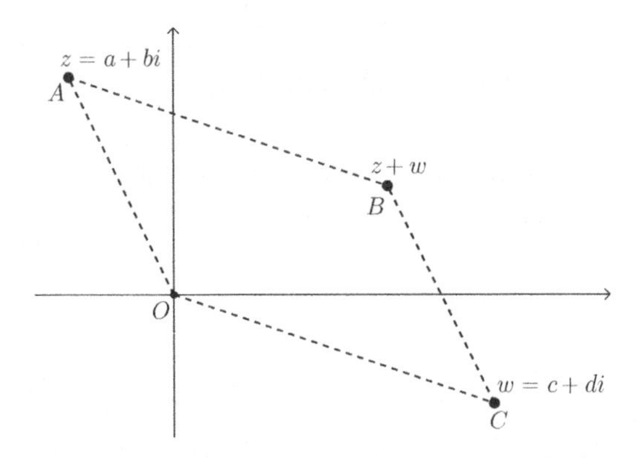

Figure 10.2 Addition of complex numbers.

Proof Let $z = a + bi$ and $w = c + di$ be two arbitrary complex numbers, and denote by A, B and C the points corresponding to $z, z + w$ and w, respectively (Figure 10.2). Also denote by O the origin, and suppose that $a \neq 0$ and $c \neq 0$.

As $z + w = (a + c) + (b + d)i$, the coordinates of B are $(a + c, b + d)$. We now compute the slope m of the line segments OA, OC, CB and AB:

$$m_{OA} = \frac{b - 0}{a - 0} = \frac{b}{a} \qquad m_{OC} = \frac{d - 0}{c - 0} = \frac{d}{c}$$

$$m_{CB} = \frac{b + d - d}{a + c - c} = \frac{b}{a} \qquad m_{AB} = \frac{b + d - b}{a + c - a} = \frac{d}{c}.$$

OA and CB have the same slope, and so are parallel to each other. Similarly, OC and AB are parallel.

If the slopes $\frac{b}{a}$ and $\frac{d}{c}$ are different, then $OABC$ is a parallelogram.

If $\frac{b}{a} = \frac{d}{c} = m$, then all four points lie on the line $y = mx$.

To complete the proof, we must check the case where a or c is zero, which is left as an exercise. □

The geometrical interpretation of products is discussed in Section 10.5.

Exercise 10.3.5

Complete the proof for the case where a or c is zero.

The Triangle Inequality

In Section 1.3 we proved the Triangle Inequality for real numbers. We now extend this result to the field of complex numbers.

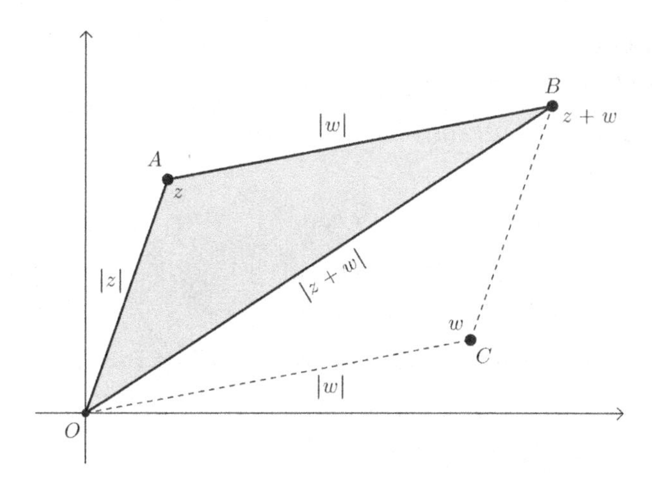

Figure 10.3 The Triangle Inequality.

Theorem 10.3.6 (The Triangle Inequality) *For every two complex numbers z and w, we have*

$$|z + w| \leq |z| + |w|.$$

The points corresponding to z, w and $z + w$ are plotted in Figure 10.3. We already know that the quadrilateral $OABC$ is a parallelogram, and thus triangle OAB has sides of length $|z|$, $|w|$ and $|z + w|$.

The inequality expresses the fact that a side of a triangle is shorter than the sum of the lengths of the remaining two sides. The possibility of equality is included to capture cases where the points O, A and B lie on a straight line and do not form a triangle.

Our goal is to provide an algebraic proof of the Triangle Inequality, which relies solely on our definitions and previous results. To do so, we can use the *rough work* strategy from Chapter 1. That is, we try to manipulate the Triangle Inequality and make it simpler. However, we can no longer use identities such as $|a|^2 = a^2$ as they may not be valid in $\mathbb{C}$.

Exercise 10.3.7

Show that the inequality $|z + w| \leq |z| + |w|$, for $z, w \in \mathbb{C}$, can be simplified to

$$z\overline{w} + \overline{z}w \leq 2|z||w|.$$

To do so, square both sides and use Claim 10.2.17.

The previous exercise suggests that we should first attempt to prove the inequality $z\overline{w} + \overline{z}w \leq 2|z||w|$. This inequality, as we will shortly see, is equivalent to $Re(z \cdot \overline{w}) \leq |z||w|$, which we prove using consequences from the following exercise.

Exercise 10.3.8
Show that for every $u \in \mathbb{C}$ we have

1. $\overline{\overline{u}} = u$ and $u + \overline{u} = 2Re(u)$,
2. $Re(u) \leq |u|$.

Claim 10.3.9 *For every $z, w \in \mathbb{C}$, we have*

$$Re(z \cdot \overline{w}) \leq |z| \cdot |w|.$$

Proof The proof follows directly by using Part 2 of Exercise 10.3.8 with $u = z \cdot \overline{w}$:

$$Re(z \cdot \overline{w}) \leq |z \cdot \overline{w}| = |z| \cdot |\overline{w}| = |z| \cdot |w|.$$

Note that we also used the fact that $|\overline{w}| = |w|$. □

We are finally ready to prove the Triangle Inequality for complex numbers.

Proof of the Triangle Inequality We use previous results to expand $|z + w|^2$:

$$|z + w|^2 = (z + w) \cdot \overline{(z + w)} = (z + w) \cdot (\overline{z} + \overline{w}) = z\overline{z} + z\overline{w} + \overline{z}w + w\overline{w}.$$

Note that $\overline{z}w$ is the conjugate of $z\overline{w}$, since

$$\overline{z\overline{w}} = \overline{z} \cdot \overline{\overline{w}} = \overline{z}w,$$

and so we have $z\overline{w} + \overline{z}w = 2Re(z\overline{w})$ by Part 1 of Exercise 10.3.8. Using Part 2 of that exercise, we get

$$|z + w|^2 = z\overline{z} + 2Re(z\overline{w}) + w\overline{w} \leq |z|^2 + 2|z||w| + |w|^2 = (|z| + |w|)^2.$$

We conclude that $|z + w|^2 \leq (|z| + |w|)^2$, which implies $|z + w| \leq |z| + |w|$ as needed. □

10.4 Square Roots and Quadratic Equations

The square root of a real number $a \geq 0$ is defined as the *non-negative* solution to the equation $x^2 = a$, for example,

$$\sqrt{9} = 3.$$

Now that we have extended our number system to include imaginary numbers, we can solve such equations even if $a < 0$.

Example 10.4.1 Find the complex solutions of the equations $z^2 = -1$ and $z^2 = -9$.

Solution
The equation $z^2 = -1$ is equivalent to

$$z^2 + 1 = 0 \qquad \Rightarrow \qquad (z+i)(z-i) = 0,$$

and thus has two complex solutions: $z = i$ and $z = -i$. Here, we use the fact that in a any field, if $a \cdot b = 0$ then $a = 0$ or $b = 0$.

Similarly, the equation $z^2 = -9$ has solutions $z = 3i$ and $z = -3i$, since

$$(3i)^2 = 9i^2 = -9 \qquad \text{and} \qquad (-3i)^2 = 9i^2 = -9.$$

Exercise 10.4.2
For $a \in \mathbb{R}$, describe all the complex solutions of $z^2 = a$. Distinguish between the cases $a > 0$, $a = 0$ and $a < 0$.

Square roots of arbitrary complex numbers can be defined in a similar way, with one important difference: the complex numbers are *not ordered*, and so the notion of being non-negative is meaningless. We therefore write

$$\sqrt{-9} = \pm 3i$$

and refer to *both* $3i$ and $-3i$ as square roots of -9.

Definition 10.4.3 Let $w \in \mathbb{C}$. A *square root* of w is a complex number z that satisfies the equation $z^2 = w$.

Due to the lack of order in $\mathbb{C}$, there is no preferred square root. We can no longer refer to *the* square root in the context of complex numbers.

Does every complex number have a square root, and if so, how many? Let us look at the following example.

Example 10.4.4 Find all complex square roots, if any, of $w = 9 - 40i$.

Solution
We need to find all complex solutions to the equation

$$z^2 = 9 - 40i.$$

If we write $z = x + yi$, then $z^2 = (x^2 - y^2) + 2xyi$, and we get

$$(x^2 - y^2) + 2xyi = 9 - 40i.$$

This leads to the following two equations in the *real* variables x and y:

$$\begin{aligned} x^2 - y^2 &= 9 \\ 2xy &= -40. \end{aligned}$$

The second equation implies that both x and y are non-zero, and so we solve it for y and substitute $y = -\frac{20}{x}$ in the first equation:

$$x^2 - \left(-\frac{20}{x}\right)^2 = 9 \qquad \Rightarrow \qquad x^4 - 9x^2 - 400 = 0.$$

We let $t = x^2$, and get the quadratic equation $t^2 - 9t - 400 = 0$. Its solutions are

$$t = -16 \qquad \text{and} \qquad t = 25.$$

As $t = x^2$, it cannot be negative, and therefore

$$t = x^2 = 25 \qquad \Rightarrow \qquad x = \pm 5 \qquad \Rightarrow \qquad y = -\frac{20}{x} = \mp 4.$$

We conclude that

$$\sqrt{9 - 40i} = \pm(5 - 4i).$$

The last example can be generalized.

Theorem 10.4.5 *Every non-zero complex number has exactly two complex square roots. These two roots are negative to each other.*

Proof Suppose that $w = a + bi$ is a non-zero complex number. If $b = 0$ then w is a real number, and the theorem follows from Exercise 10.4.2. We thus assume that $b \neq 0$.

The equation $z^2 = w$, where $z = x + yi$, leads to the following two equations in x and y:

$$\begin{aligned} x^2 - y^2 &= a \\ 2xy &= b. \end{aligned}$$

We use the second equation to get $y = \frac{b}{2x}$ (the assumption $b \neq 0$ implies that $x \neq 0$). The first equation then becomes

$$x^2 - \frac{b^2}{4x^2} = a \qquad \Rightarrow \qquad 4x^4 - 4ax^2 - b^2 = 0.$$

By setting $t = x^2$ we get the quadratic equation $4t^2 - 4at - b^2 = 0$, whose solutions are

$$t = \frac{4a \pm \sqrt{16a^2 + 16b^2}}{8} \qquad \Rightarrow \qquad t = \frac{a \pm \sqrt{a^2 + b^2}}{2}.$$

As $t = x^2 \geq 0$, we conclude that

$$t = x^2 = \frac{a + \sqrt{a^2 + b^2}}{2} > 0.$$

We obtain two solutions for x, namely $x = \sqrt{t}$ and $x = -\sqrt{t}$, and each is paired with a solution for $y = \frac{b}{2x}$.

The complex number w has therefore two square roots, given by

$$\sqrt{t} + \frac{b}{2\sqrt{t}}i \qquad \text{and} \qquad -\sqrt{t} - \frac{b}{2\sqrt{t}}i.$$

Note that these two roots are indeed negatives of each other. □

Remark. We emphasize once more the difference between the square root operation for real and complex numbers.

- For real numbers, the square root operation is a *function*, producing a single output for every non-negative input:

$$f\colon [0, \infty) \to \mathbb{R}, \qquad f(x) = \sqrt{x}.$$

- In the context of complex numbers, the square root operation produces *two* outputs (unless the argument is 0), and hence the square root operation is *not* a function.

Complex Quadratic Equations

In Chapter 1 we proved the quadratic formula. In the proof, we used the four basic operations and square roots. We can now repeat this process and show that the formula remains valid in the complex case.

Moreover, as square roots exist for all complex numbers, we conclude that *every* quadratic equation has at least one solution.

Theorem 10.4.6 *Let $a, b, c \in \mathbb{C}$ with $a \neq 0$. The equation $az^2 + bz + c = 0$ has the following.*

1. *A unique complex solution if $b^2 - 4ac = 0$, given by $z = -\frac{b}{2a}$.*
2. *Two distinct complex solutions if $b^2 - 4ac \neq 0$, given by*

$$z = \frac{-b + \sqrt{b^2 - 4ac}}{2a}.$$

Note that there is no need to use the $\pm$ sign in the quadratic formula. The square root of a non-zero complex number always produces two values.

Proof The proof is almost identical to the proof of the real case. We repeat it here briefly. Start by multiplying both sides of the equation by $4a$ and complete the square:

$$\begin{aligned} 4a^2z^2 + 4abz + 4ac &= 0 \\ 4a^2z^2 + 4abz + b^2 - b^2 + 4ac &= 0 \\ (2az + b)^2 &= b^2 - 4ac. \end{aligned}$$

If $b^2 - 4ac = 0$, we get

$$(2az + b)^2 = 0 \quad \Rightarrow \quad 2az + b = 0 \quad \Rightarrow \quad z = -\frac{b}{2a}.$$

On the other hand, if $b^2 - 4ac \neq 0$, then $\sqrt{b^2 - 4ac}$ provides two distinct square roots which are negative to each other (Theorem 10.4.5). In that case, we can solve for z and obtain the familiar quadratic formula:

$$2az + b = \sqrt{b^2 - 4ac} \quad \Rightarrow \quad z = \frac{-b + \sqrt{b^2 - 4ac}}{2a}.$$ □

Example 10.4.7 Solve the following quadratic equation:

$$2z^2 - 7z + 5 + 5i = 0.$$

Solution

This equation is in the form $az^2 + bz + c = 0$ with $a = 2$, $b = -7$ and $c = 5 + 5i$, and the discriminant is given by

$$b^2 - 4ac = (-7)^2 - 4 \cdot 2 \cdot (5 + 5i) = 49 - 40 - 40i = 9 - 40i.$$

From the quadratic formula we get

$$z = \frac{7 + \sqrt{9 - 40i}}{4}.$$

In Example 10.4.4 we showed that

$$\sqrt{9 - 40i} = \pm(5 - 4i),$$

which we can use to obtain our solutions:

$$z = \frac{7 \pm (5 - 4i)}{4} \quad \Rightarrow \quad z = 3 - i \quad \text{or} \quad z = \frac{1}{2} + i.$$

Therefore, the given equation has two distinct complex solutions.

In Section 10.6 we describe another method for computing square roots of complex numbers through polar representations.

10.5 Polar Representation of Complex Numbers

A complex number has been defined as an expression of the form $z = a + bi$ where a and b are real numbers. The point (a, b) represents z in the complex plane, and is often referred to as the *rectangular* or *Cartesian* representation of z. We now discuss another important and useful way of representing complex numbers.

Suppose that $z = a + bi$ is a *non-zero* complex number. We denote by $r = |z| = \sqrt{a^2 + b^2}$ the distance of the point $P = (a, b)$ from the origin O and

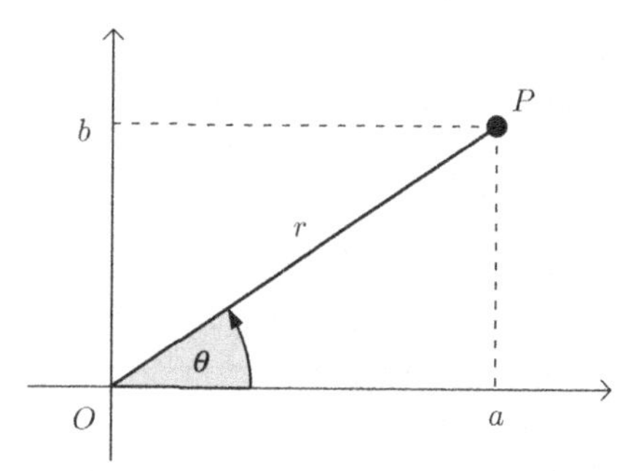

Figure 10.4 Polar representation of a complex number.

by θ the angle between the positive x-axis and the line segment OP, measured counter-clockwise (Figure 10.4).

It follows that

$$a = r\cos\theta \qquad \text{and} \qquad b = r\sin\theta$$

which implies that

$$z = (r\cos\theta) + (r\sin\theta)i = r(\cos\theta + i\sin\theta).$$

Remark. Note that the relations $a = r\cos\theta$ and $b = r\sin\theta$ remain valid when P lies on one of the axes and in any quadrant.

Definition 10.5.1 Let $z = a + bi$ be a complex number. We say that

$$r(\cos\theta + i\sin\theta)$$

is a *polar representation* of z if $r \geq 0$ and θ are real numbers, such that

$$a = r\cos\theta \qquad \text{and} \qquad b = r\sin\theta.$$

We call θ the *argument* of z.

Remarks.

- Some abbreviate and write $\operatorname{cis}\theta$ for $\cos\theta + i\sin\theta$. In that case, a polar representation can be written, in short, as $z = r\operatorname{cis}\theta$.
- For convenience, we allowed the argument θ to be *any* real number. Therefore, θ is defined up to an integer multiple of 2π (in radians), and consequently, complex numbers have multiple polar representations. For instance,

$$4\left[\cos\left(\tfrac{\pi}{3}\right) + \sin\left(\tfrac{\pi}{3}\right)i\right],$$
$$4\left[\cos\left(\tfrac{7\pi}{3}\right) + \sin\left(\tfrac{7\pi}{3}\right)i\right], \qquad \text{and}$$
$$4\left[\cos\left(-\tfrac{5\pi}{3}\right) + \sin\left(-\tfrac{5\pi}{3}\right)i\right]$$

are all polar representations of $z = 2 + 2\sqrt{3}i$.

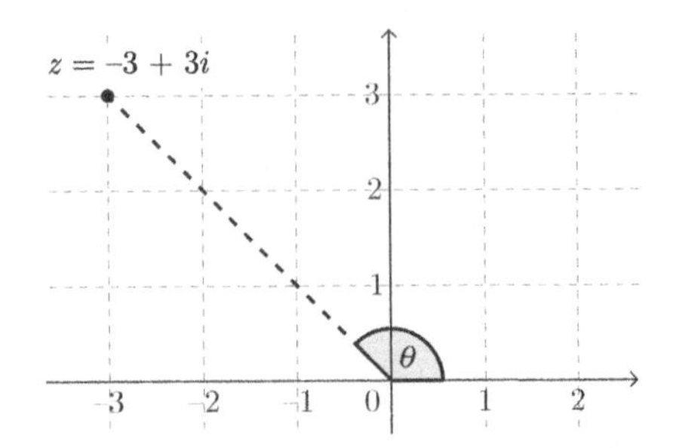

Figure 10.5 Polar representation of $z = -3 + 3i$.

- When $z = 0$ we can use *any* real number as the argument θ. If $z \neq 0$ we often choose the argument to be in the interval $[0, 2\pi)$, so that it is determined uniquely. This choice is called the argument's *principal value*.

Example 10.5.2 Find all polar representations of $z = -3 + 3i$ (see Figure 10.5).

Solution

The absolute value of z is $r = \sqrt{18} = 3\sqrt{2}$, and so any polar representation has the form $z = 3\sqrt{2}(\cos\theta + i\sin\theta) = 3\sqrt{2}\operatorname{cis}\theta$. To find θ, note that the equations

$$a = r\cos\theta \qquad \text{and} \qquad b = r\sin\theta$$

imply that

$$\tan\theta = \frac{b}{a},$$

and so we solve

$$\tan\theta = \frac{3}{-3} = -1 \qquad \Rightarrow \qquad \theta = \frac{3\pi}{4} + \pi k \qquad (\text{for } k \in \mathbb{Z}).$$

The point $(-3, 3)$ corresponding to z is in the second quadrant, so only solutions with an *even* k are suitable.

We conclude that all polar representations of z are given by

$$z = 3\sqrt{2}\operatorname{cis}\left(\tfrac{3\pi}{4} + 2\pi k\right) \qquad (k \in \mathbb{Z}).$$

If we use the principal value $\theta = \frac{3\pi}{4}$, we obtain the representation

$$z = 3\sqrt{2}\operatorname{cis}\left(\tfrac{3\pi}{4}\right) = 3\sqrt{2}\left[\cos\left(\tfrac{3\pi}{4}\right) + i\sin\left(\tfrac{3\pi}{4}\right)\right].$$

Geometric Interpretation of Products

In Section 10.3 we discussed the geometrical representation of sums of complex numbers. Now, using polar representations, we provide a geometrical interpretation of products.

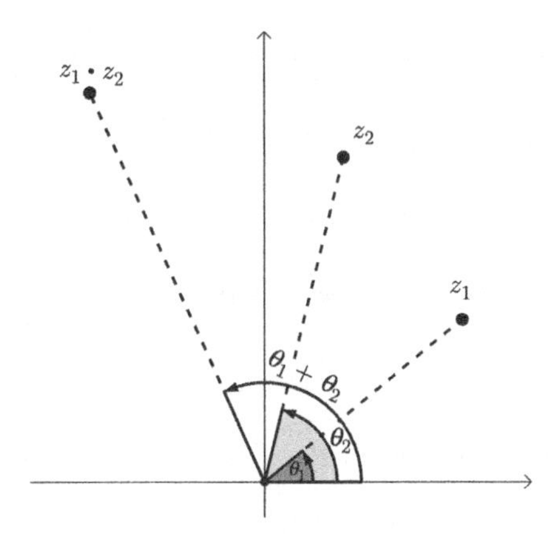

Figure 10.6 Multiplication of complex numbers.

Suppose that z_1 and z_2 are two complex numbers with polar representations (see Figure 10.6)

$$z_1 = r_1(\cos\theta_1 + i\sin\theta_1) \qquad \text{and} \qquad z_2 = r_2(\cos\theta_2 + i\sin\theta_2).$$

Their product is given by

$$\begin{aligned} z_1 \cdot z_2 &= (r_1 \cos\theta_1 + ir_1 \sin\theta_1) \cdot (r_2 \cos\theta_2 + ir_2 \sin\theta_2) \\ &= r_1r_2(\cos\theta_1 \cos\theta_2 - \sin\theta_1 \sin\theta_2) + r_1r_2(\cos\theta_1 \sin\theta_2 + \sin\theta_1 \cos\theta_2)i. \end{aligned}$$

Using trigonometric identities for $\cos(\alpha + \beta)$ and $\sin(\alpha + \beta)$, we get a polar representation for the product

$$\begin{aligned} z_1 \cdot z_2 &= r_1r_2 \cos(\theta_1 + \theta_2) + r_1r_2 \sin(\theta_1 + \theta_2)i \\ &= r_1r_2[\cos(\theta_1 + \theta_2) + i\sin(\theta_1 + \theta_2)]. \end{aligned}$$

Conclusion *If* $z_1 = r_1(\cos\theta_1 + i\sin\theta_1)$ *and* $z_2 = r_2(\cos\theta_2 + i\sin\theta_2)$ *then*

$$z_1 \cdot z_2 = r_1r_2[\cos(\theta_1 + \theta_2) + i\sin(\theta_1 + \theta_2)]. \tag{10.1}$$

That is, a polar representation for $z_1 \cdot z_2$ *is obtained by multiplying the absolute values and adding up the arguments.*

In a similar way we can obtain a polar representation of quotients. But first, try out the following exercise.

Exercise 10.5.3
Suppose $z = r(\cos\theta + i\sin\theta)$ is a non-zero complex number.

1. Explain why $-r(\cos\theta + i\sin\theta)$ is *not* a polar representation of $-z$. Similarly, $r(\cos\theta - i\sin\theta)$ is *not* a polar representation of $\overline{z}$.
2. Find a polar representation of $-z, \overline{z}$ and z^{-1}.

Now suppose that z_1 and z_2 are two complex numbers, with $z_2 \neq 0$, and with polar representations

$$z_1 = r_1(\cos\theta_1 + i\sin\theta_1) \qquad \text{and} \qquad z_2 = r_2(\cos\theta_2 + i\sin\theta_2).$$

Their quotient is then given by

$$\begin{aligned}\frac{z_1}{z_2} &= z_1 \cdot (z_2)^{-1} = [r_1(\cos\theta_1 + i\sin\theta_2)] \cdot \left[\frac{1}{r_2}(\cos(-\theta_2) + i\sin(-\theta_2))\right] \\ &= \frac{r_1}{r_2}[\cos(\theta_1 - \theta_2) + i\sin(\theta_1 - \theta_2)],\end{aligned}$$

which is a polar representation for $\frac{z_1}{z_2}$.

10.6 De Moivre's Theorem and Roots

Integer powers of a complex number z can be defined in a natural way. For $n \in \mathbb{N}$, we simply define z^n as repeated multiplication

$$z^n = z \cdot z \cdot z \cdots z.$$

If $z \neq 0$, we define

$$z^{-n} = (z^{-1})^n = \frac{1}{z^n}$$

and also $z^0 = 1$.

Remark. This above definition applies to *any* field. Usual exponent rules, such as

$$z^{m+n} = z^m \cdot z^n, \qquad z^{m-n} = \frac{z^m}{z^n} \qquad \text{and} \qquad z^{n \cdot m} = \left(z^n\right)^m \qquad (\text{for } n, m \in \mathbb{Z})$$

are valid in every field and, in particular, for complex numbers.

In this section we see how polar representations can be used to calculate powers and roots of complex numbers.

Suppose that $z = \cos\theta + i\sin\theta \in \mathbb{C}$. Formula (10.1) implies that

$$\begin{aligned}z^2 &= z \cdot z = \cos(2\theta) + i\sin(2\theta), \\ z^3 &= z \cdot z^2 = \cos(3\theta) + i\sin(3\theta),\end{aligned}$$

and so on. In general, we obtain the following theorem, whose proof is left as an exercise.

Theorem 10.6.1 (De Moivre's Theorem) *For every integer n and a real number θ, we have*

$$(\cos\theta + i\sin\theta)^n = \cos(n\theta) + i\sin(n\theta).$$

Exercise 10.6.2

1. Use induction on n to prove De Moivre's Theorem in the case where $n > 0$.
2. Complete the proof of the theorem for $n \leq 0$.

De Moivre's Theorem implies that for an arbitrary complex number $z = r(\cos\theta + i\sin\theta)$ we have

$$z^n = r^n[\cos(n\theta) + i\sin(n\theta)]$$

for all $n \in \mathbb{Z}$ (assuming that $z \neq 0$ if $n \leq 0$). This observation is useful for computing high powers of complex numbers.

Example 10.6.3 Calculate $(-3+3i)^{11}$.

Solution
In Example 10.5.2 we saw that

$$3\sqrt{2}\left[\cos\left(\tfrac{3\pi}{4}\right) + \sin\left(\tfrac{3\pi}{4}\right) i\right]$$

is a polar representation of $-3+3i$. Using De Moive's Theorem, we get

$$\begin{aligned}(-3+3i)^{11} &= (3\sqrt{2})^{11}\left[\cos\left(11\cdot\tfrac{3\pi}{4}\right) + i\sin\left(11\cdot\tfrac{3\pi}{4}\right)\right] \\ &= (3\sqrt{2})^{11}\left[\cos\left(\tfrac{33\pi}{4}\right) + i\sin\left(\tfrac{33\pi}{4}\right)\right] \\ &= (3\sqrt{2})^{11}\cdot\left(\tfrac{1}{\sqrt{2}} + \tfrac{1}{\sqrt{2}}i\right) = 3^{11}2^5(1+i) = 3^{11}2^5 + 3^{11}2^5 i.\end{aligned}$$

Arbitrary Roots of Complex Numbers

In Section 10.4 we learned how to solve equations of the form $z^2 = w$, but the method we used does not work well for higher powers of z. Fortunately, polar representations provide a simpler way.

Theorem 10.6.4 *Let $w = \rho(\cos\alpha + i\sin\alpha)$ be a polar representation of a complex number, and $n \in \mathbb{N}$. The equation $z^n = w$ has solutions given by*

$$z = \sqrt[n]{\rho}\cdot\left[\cos\left(\frac{\alpha}{n} + \frac{2\pi k}{n}\right) + i\sin\left(\frac{\alpha}{n} + \frac{2\pi k}{n}\right)\right] \quad \textit{where} \quad k = 0, 1, \ldots, n-1.$$

These solutions are distinct if $w \neq 0$.

Proof If $w = 0$, the only solution to $z^n = w$ is $z = 0$, which is consistent with the given formula. Assume, for the remainder of the proof, that $w \neq 0$.

Let $z = r(\cos\theta + i\sin\theta)$ be a polar representation of a solution to $z^n = w$. De Moivre's Theorem implies that

$$r^n[\cos(n\theta) + i\sin(n\theta)] = \rho(\cos\alpha + i\sin\alpha). \tag{10.2}$$

By taking absolute values of both sides, we get

$$r^n = \rho \qquad \Rightarrow \qquad r = \sqrt[n]{\rho}.$$

Equation (10.2) simplifies to

$$\cos(n\theta) + i\sin(n\theta) = \cos\alpha + i\sin\alpha \qquad \Rightarrow \qquad \begin{cases} \cos(n\theta) = \cos\alpha \\ \sin(n\theta) = \sin\alpha \end{cases}$$

and we conclude that

$$n\theta = \alpha + 2\pi k \qquad \Rightarrow \qquad \theta = \frac{\alpha}{n} + \frac{2\pi k}{n} \quad \text{where } k \in \mathbb{Z}.$$

A solution of $z^n = w$ must, therefore, have the form

$$z = \sqrt[n]{\rho} \cdot \left[\cos\left(\frac{\alpha}{n} + \frac{2\pi k}{n}\right) + i\sin\left(\frac{\alpha}{n} + \frac{2\pi k}{n}\right)\right].$$

As $\cos x$ and $\sin x$ are periodic functions with period 2π, solutions coincide if and only if their arguments differ by an integer multiple of 2π. Consequently, we get exactly n solutions, corresponding to $k = 0, 1, 2, \ldots, n-1$. □

Note that we could have chosen a different set of ks to produce the n solutions of $z^n = w$.

Exercise 10.6.5
If $w \neq 0$, what do the solutions of $z^n = w$ represent in the complex plane?

Example 10.6.6 Find all solutions of $z^4 = \sqrt{3} + i$.

Solution
A polar representation of $\sqrt{3}+i$ is $2[\cos\left(\frac{\pi}{6}\right) + i\sin\left(\frac{\pi}{6}\right)]$ (check!). By Theorem 10.6.4, we get

$$z = \sqrt[4]{2}\left[\cos\left(\frac{\pi}{24} + \frac{2\pi k}{4}\right) + i\sin\left(\frac{\pi}{24} + \frac{2\pi k}{4}\right)\right] \quad \text{for} \quad k = 0, 1, 2, 3.$$

More explicitly, we obtain four solutions:

$$z_0 = z = \sqrt[4]{2}\left[\cos\left(\frac{\pi}{24}\right) + i\sin\left(\frac{\pi}{24}\right)\right]$$
$$z_1 = z = \sqrt[4]{2}\left[\cos\left(\frac{13\pi}{24}\right) + i\sin\left(\frac{13\pi}{24}\right)\right]$$
$$z_2 = z = \sqrt[4]{2}\left[\cos\left(\frac{25\pi}{24}\right) + i\sin\left(\frac{25\pi}{24}\right)\right]$$
$$z_3 = z = \sqrt[4]{2}\left[\cos\left(\frac{37\pi}{24}\right) + i\sin\left(\frac{37\pi}{24}\right)\right].$$

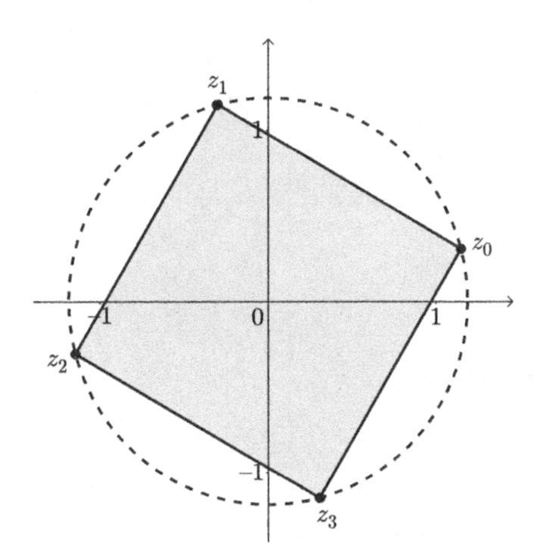

Figure 10.7 The solutions of the equation $z^4 = \sqrt{3} + i$.

These solutions represent vertices of a square, lying on a circle of radius $r = \sqrt[4]{2} \approx 1.19$ (Figure 10.7).

Example 10.6.7 Let w be a complex number. One of the solutions to the equation $z^6 = w$ is $u = \frac{\sqrt{3}+1}{2} + \frac{\sqrt{3}-1}{2}i$. Find the value of w, and all other solutions of the equation in polar coordinates.

Solution

As u solves the equation $z^6 = w$, we can find w by computing u^6. First, let us find a polar representation for u. If $u = r(\cos\theta + i\sin\theta)$, then

$$r = \sqrt{\left(\frac{\sqrt{3}+1}{2}\right)^2 + \left(\frac{\sqrt{3}-1}{2}\right)^2} = \sqrt{\frac{4+2\sqrt{3}+4-2\sqrt{3}}{4}} = \sqrt{2}$$

$$\tan\theta = \frac{\sqrt{3}-1}{\sqrt{3}+1} = \frac{(\sqrt{3}-1)(\sqrt{3}-1)}{(\sqrt{3}+1)(\sqrt{3}-1)} = \frac{4-2\sqrt{3}}{2} = 2-\sqrt{3}.$$

The number u is in the first quadrant, and so θ must be between 0 and $\frac{\pi}{2}$. As $\tan\left(\frac{\pi}{12}\right) = 2-\sqrt{3}$, we conclude that $\theta = \frac{\pi}{12} = 15°$, and so a polar representation for u is $\sqrt{2}\left[\cos\left(\frac{\pi}{12}\right) + i\sin\left(\frac{\pi}{12}\right)\right]$.

Now, we compute u^6:

$$u^6 = (\sqrt{2})^6 \cdot \left[\cos\left(6\cdot\frac{\pi}{12}\right) + i\sin\left(6\cdot\frac{\pi}{12}\right)\right] = 8\cdot\left[\cos\left(\frac{\pi}{2}\right) + i\sin\left(\frac{\pi}{2}\right)\right] = 8i.$$

Therefore, $w = 8i$.

Using Theorem 10.6.4 we can describe all the solutions to $z^6 = 8i$:

$$z_k = \sqrt{2}\cdot\left[\cos\left(\frac{\pi}{12} + \frac{\pi}{3}k\right) + i\sin\left(\frac{\pi}{12} + \frac{\pi}{3}k\right)\right] \qquad \text{for } k = 0, 1, 2, 3, 4, 5.$$

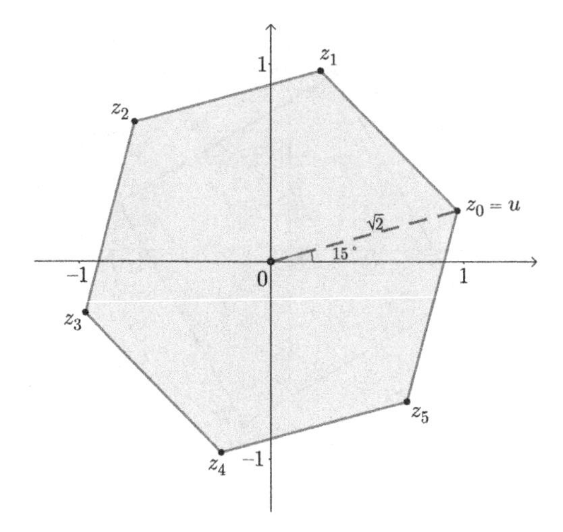

Figure 10.8 The solutions of the equation $z^6 = w$.

Note that $z_0 = u$, and that the solutions are the vertices of a regular hexagon in the complex plane, centered at the origin (Figure 10.8).

Solutions to the equation $z^n = w$ are called *nth roots* of w. However, as there is no preferred solution, we avoid the notation $\sqrt[n]{w}$.

General Polynomial Equations

As we have seen, not all quadratic equations have real solutions. However, every quadratic equation with complex coefficients has *at least one complex solution*.

A natural question arises: what about polynomial equations of higher degrees? Do all cubic and quartic equations with complex coefficients have at least one complex solution? What about equations of higher degree? Do we need to further extend $\mathbb{C}$ to ensure existence of solutions?

The answer is given by the famous Fundamental Theorem of Algebra.

Theorem 10.6.8 (The Fundamental Theorem of Algebra) *Let $p(z)$ be a non-constant polynomial with complex coefficients, that is,*

$$p(z) = a_n z^n + a_{n-1} z^{n-1} + \cdots + a_2 z^2 + a_1 z + a_0,$$

where $n \in \mathbb{N}$, $a_0, \ldots, a_n \in \mathbb{C}$ and $a_n \neq 0$. Then $p(z_0) = 0$ for some $z_0 \in \mathbb{C}$.

The theorem implies that there is no need to further extend $\mathbb{C}$ in order to include solutions to higher-degree polynomial equations. For that reason, we say that $\mathbb{C}$ is an *algebraically closed field*.

There are several proofs of this theorem, but all require results from analysis which are beyond the scope of this book. We therefore skip the proof of the theorem. It can be found in most books on complex analysis.

10.7 The Exponential Function

In this section we extend the exponential function $y = e^x$ from the real numbers to the field of complex numbers. As a result, we obtain an interesting and important connection between exponential and trigonometric functions, and the famous Euler Identity.

Our goal is to define a function $f(z) = e^z$, from $\mathbb{C}$ to $\mathbb{C}$, that coincides with the real exponential function when $z \in \mathbb{R}$. Moreover, we would like to preserve, as much as possible, properties of the real exponential function.

For instance, if x and y are real numbers, then

$$e^{x+y} = e^x \cdot e^y.$$

Consequently, for every $z = a + bi \in \mathbb{C}$ we would like to have

$$e^z = e^{a+bi} = e^a \cdot e^{bi}.$$

As a is a real number, the quantity e^a is already well defined, and so it is enough to focus on finding a satisfying definition for e^{bi}, where b is a real number.

Recall from Section 10.5 that if $b_1, b_2 \in \mathbb{R}$, then

$$(\cos b_1 + i \sin b_1) \cdot (\cos b_2 + i \sin b_2) = \cos(b_1 + b_2) + i \sin(b_1 + b_2).$$

To preserve exponent rules, we would like to have

$$e^{ib_1} \cdot e^{ib_2} = e^{i(b_1+b_2)},$$

which suggests defining e^{ib} as $\cos b + i \sin b$. This definition is often called *Euler's Formula* and, as it turns out, is consistent with other exponent rules.

Definition 10.7.1 For every complex number $z = a + bi$, define

$$e^z = e^a \cdot (\cos b + i \sin b).$$

A more convincing and rigorous justification for the above definition can be given using tools from real analysis. We provide a brief overview of such an argument.

We wish to define a function $f\colon \mathbb{R} \to \mathbb{C}, f(\theta) = e^{i\theta}$, satisfying the following differentiation rule:

$$(e^{i\theta})' = i \cdot e^{i\theta} \qquad \text{or} \qquad f'(\theta) = i \cdot f(\theta) \qquad \text{for all } \theta \in \mathbb{R}.$$

This is a reasonable requirement, as it extends the rule $(e^{cx})' = c \cdot e^{cx}$ for the real-valued exponential function.

The derivative of f is computed component-wise. That is, if $u(\theta)$ and $v(\theta)$ are the real and imaginary parts of $f(\theta)$, then

$$f'(\theta) = [u(\theta) + i \cdot v(\theta)]' = u'(\theta) + i \cdot v'(\theta).$$

Moreover, we have that

$$f'(\theta) = i \cdot f(\theta) = i \cdot [u(\theta) + i \cdot v(\theta)] = i \cdot u(\theta) - v(\theta).$$

By comparing the real and imaginary parts of the two expressions for $f'(\theta)$, we get:

$$u'(\theta) = -v(\theta) \qquad \text{and} \qquad v'(\theta) = u(\theta).$$

We differentiate each of these equations and conclude that

$$u''(\theta) = -v'(\theta) = -u(\theta) \qquad \text{and} \qquad v''(\theta) = u'(\theta) = -v(\theta).$$

In other words, the real-valued functions $u(\theta)$ and $v(\theta)$ must satisfy the following conditions

$$u'' = -u \qquad \text{and} \qquad v'' = -v.$$

In addition, we expect that $f(0) = e^0 = 1$, and so $u(0) = 1$ and $v(0) = 0$. It can be shown, using tools from real analysis and differential equations, that the only functions satisfying these conditions are $u(\theta) = \cos(\theta)$ and $v(\theta) = \sin(\theta)$. Consequently, we see that we must define $f(\theta)$ as $\cos(\theta) + i \cdot \sin(\theta)$, which is consistent with Definition 10.7.1.

The notation $e^{i\theta}$ for $\cos\theta + i\sin\theta$ can simplify many of the expressions and results obtained in previous sections. Here are a few examples.

- Polar representations of a complex number $z = a + bi$ can be written as $z = r \cdot e^{i\theta}$. In particular, complex numbers of the form $e^{i\theta}$ represent points on the unit circle $|z| = 1$.
- If $z_1 = r_1 e^{\theta_1}$ and $z_2 = r_2 e^{i\theta_2}$ are two complex numbers, then

$$z_1 \cdot z_2 = r_1 r_2 e^{i(\theta_1+\theta_2)} \qquad \text{and} \qquad \frac{z_1}{z_2} = \frac{r_1}{r_2} e^{i(\theta_1-\theta_2)} \quad \text{if } z_2 \neq 0.$$

- De Moivre's Theorem (Theorem 10.6.1) can be restated as follows. For every $n \in \mathbb{Z}$ and $\theta \in \mathbb{R}$, we have $(e^{i\theta})^n = e^{in\theta}$.
- Theorem 10.6.4 can also be restated. Let $w = \rho e^{i\alpha} \in \mathbb{C}$, where $\rho \geq 0$ and α are real numbers, and $n \in \mathbb{N}$. The equation $z^n = w$ has solutions given by

$$z = \sqrt[n]{\rho} \cdot e^{\frac{i\cdot(\alpha+2\pi k)}{n}} \qquad \text{where} \qquad k = 0, 1, \ldots, n-1.$$

Euler's Identity

Setting $\theta = \pi$ in Euler's Formula gives the following unexpected relation between the imaginary number i and two other important mathematical constants, e and π:

$$e^{\pi i} = \cos \pi + i \sin \pi = -1.$$

10.8 Problems

10.1 Calculate the following expressions. Make sure your final answer is in the standard form $a + bi$.

a. $(5 + 2i) - i \cdot (2 + 3i)$
b. $(1 + i)^3$
c. $\frac{\sqrt{3}+i}{1+\sqrt{3}i}$
d. $\frac{3}{4-i}$
e. $\frac{(5+\sqrt{2}i)^2}{i}$

10.2 Solve the following equations for z.

a. $|z|^2 + 2i = 2z$
b. $-4 = 3iz + |z|$
c. $z^2 + 1 = |z|^2$

10.3 Use induction to show that for every $z \in \mathbb{C}$ and $n \in \mathbb{N}$, we have $\overline{z^n} = \overline{z}^n$ and $|z^n| = |z|^n$.

10.4 a. Show that if $p(z) = \sum_{k=0}^{n} a_k z^k$ is a polynomial with real coefficients, and $p(w) = 0$, then $p(\overline{w}) = 0$.
b. Find a quartic polynomial $p(z)$ (that is, a polynomial of degree 4), with real coefficients, such that $p(1 + i) = p(2 + i) = 0$.

10.5 Which of the following statements are true for every $z \in \mathbb{C}$? Justify your answer with a short proof or a counterexample.

a. $|z| = |\overline{z}|$
b. $z^2 = |z|^2$
c. If $iz = \overline{iz}$, then $Re(z) = 0$.
d. $(z + i) \cdot (\overline{z} - i)$ is a real number.
e. $z^2 + \overline{z}^2 = 2|z|^2$

10.6 Let $z, w \in \mathbb{C}$ such that both $z + w$ and $z \cdot w$ are real numbers. Prove that either z and w are both real numbers or $z = \overline{w}$.

10.7 Let a, b be two integers that can be expressed as sums of two squares, that is,

$$a = m^2 + n^2 \qquad \text{and} \qquad b = k^2 + l^2$$

for some integers m, n, k and l. Prove that $a \cdot b$ can also be expressed as the sum of two squares.
(Hint: If $z = m + ni$, then $a = |z|^2$.)

10.8 Draw the solution sets to the following in the complex plane.

a. $|z + 1 - 2i| = 5$
b. $z + i\bar{z} = 3 + 3i$
c. $z \cdot \bar{z} \leq \sqrt{20} \cdot |2 - i|$

10.9 Let z be a complex number such that $Im(z) \neq 0$. Prove that z corresponds to a point on the unit circle if and only if $z + \frac{1}{z}$ is a real number.

10.10 The numbers $0, 1 + i$ and $-2 + 3i$ represent three vertices of a parallelogram in the complex plane. Where is the fourth vertex? Find all possible answers.

10.11 a. Let $u \in \mathbb{C}$. Show that $Re(u) = |u|$ if and only if u is a non-negative real number.
b. Let $z, w \in \mathbb{C}$. Use Part a to prove that $Re(z \cdot \overline{w}) = |z| \cdot |w|$ if and only if $w = \alpha \cdot z$ or $z = \alpha \cdot w$ for some non-negative real number α.
c. Conclude that in the Triangle Inequality, equality occurs if and only if either $z = \alpha w$ or $w = \alpha z$ for some non-negative real number α. What is the geometrical meaning of this condition?

10.12 Prove the reverse Triangle Inequality for complex numbers. For every $z, w \in \mathbb{C}$, we have

$$||z| - |w|| \leq |z - w|.$$

10.13 a. Prove the *Parallelogram Identity*. For all $z, w \in \mathbb{C}$, we have $|z + w|^2 + |z - w|^2 = 2(|z|^2 + |w|^2)$.
Hint: Use Claim 10.2.17.
b. If $0, z, w, z+w$ are the vertices of a parallelogram, then $|z|, |w|, |z+w|$ and $|z - w|$ represent lengths of the parallelogram's sides and diagonals. Use these observations to re-state the Parallelogram Identity in geometrical terms.

10.14 Calculate the following square roots. Provide your answer in the form $a + bi$.

a. $\sqrt{i}$
b. $\sqrt{-i}$
c. $\sqrt{-3 + 2\sqrt{10}i}$
d. $\sqrt{9a^2 - 1 + 6ai}$ (where $a \in \mathbb{R}$ is a constant).

10.15 Find all complex solutions of the following equations.

a. $6z^2 - (14+3i)z + 4 + i = 0$
b. $z^2 - 2i\sqrt{2}z - 2 = 0$
c. $4z^3 - 4z^2 + 5z = 0$

10.16 Let $z \in \mathbb{C}$ such that $Im(z) \neq 0$ or $Re(z) > 0$. Prove the following.

a. $z + |z| \neq 0$
b. $\sqrt{z} = \pm\sqrt{|z|} \cdot \frac{z+|z|}{|z+|z||}$

10.17 a. Suppose that $a, b, c \in \mathbb{C}$ with $a \neq 0$. Show that if z_1 and z_2 are two distinct solutions of $az^2 + bz + c = 0$, then

$$z_1 + z_2 = -\frac{b}{a} \quad \text{and} \quad z_1 \cdot z_2 = \frac{c}{a}.$$

We see that Vieta's Formulas, discussed in Problem 1.1 in the context of real numbers, remain valid in the complex case.

b. Suppose that z_1 and z_2 are the solutions of

$$z^2 + (3+2i)z + (5+7i) = 0.$$

Without finding z_1 and z_2, calculate $z_1^2 + z_2^2$.

10.18 Let $w \in \mathbb{C}$. One of the solutions to $z^2 - w \cdot z + 8 + 2i = 0$ is $z = 1 + i$. Use Vieta's Formulas (Problem 10.17) to find the other solution and the value of w.

10.19 a. Show that if $a, b \in \mathbb{R}$, then $(a^2 + abi) \cdot (ab - b^2 i)$ is a real number.
b. Show that if z is a non-zero complex number, then $\frac{1}{\bar{z}} - z^{-1}$ is an imaginary number.

10.20 Find a polar representation of the following complex numbers (θ represents an arbitrary real number).

a. $-1 + i$
b. $3 - \sqrt{3}i$
c. $-2 - \sqrt{12}i$
d. $\sin\theta + i\cos\theta$
e. $-\sin\theta + i\cos\theta$

10.21 Let $z = r(\cos\theta + i\sin\theta)$ be a polar representation of a non-zero complex number. Find a polar representation of the following numbers (in terms of r and θ):

$$z - \bar{z}, \qquad \frac{i}{z}, \qquad \left(\frac{z}{|z|}\right)^3.$$

10.22 The vertices of a regular pentagon lie on the unit circle. If one of the vertices corresponds to $z = \frac{\sqrt{3}}{2} - \frac{1}{2}i$, find the other four vertices.

10.23 Use De Moivre's Theorem to calculate the following. Provide your final answer in the form $a + bi$:

$$(-1+i)^4, \qquad (1-\sqrt{3}i)^5, \qquad \left(\sqrt{\frac{3}{2}} + \frac{1}{\sqrt{2}}i\right)^6.$$

10.24 Find all solutions of the following equations.

a. $z^6 = 8i$
b. $z^3 = -8 - 8\sqrt{3}i$
c. $z^5 = -32$

10.25 Prove, by induction, that for every complex number $z \neq 1$ and $n \in \mathbb{N}$, we have

$$1 + z + z^2 + \cdots + z^{n-1} = \frac{z^n - 1}{z - 1}.$$

10.26 Let w be a non-zero complex number and $n \geq 2$ a natural number.

a. Show that the n distinct solutions to the equation $z^n = w$ can be written as $z_0, z_0u, z_0u^2, \ldots, z_0u^{n-1}$, where $z_0, u \in \mathbb{C}$ and $u^n = 1$.
b. Use Problem 10.25 to show that the sum of all solutions to $z^n = w$ is zero.

10.27 Let $n \in \mathbb{N}$. Show that the product of all complex solutions to $z^n = 1$ is $(-1)^{n+1}$.

10.28 a. Let $z = \cos\theta + i\sin\theta$. Compute z^3 in two ways: by expanding $z \cdot z \cdot z$ using the definition of multiplication, and by applying De Moivre's Theorem.
b. Obtain formulas for $\cos(3\theta)$ and $\sin(3\theta)$ in terms of $\cos\theta$ and $\sin\theta$.

10.29 Let $z = \cos\theta + i\sin\theta$ be a complex number on the unit circle, and $n \in \mathbb{Z}$. Use De Moivre's Theorem to express

$$z^n + \frac{1}{z^n} \qquad \text{and} \qquad z^n - \frac{1}{z^n}$$

in terms of θ.

10.30 In this exercise we use Problem 10.25 and De Moivre's Theorem to obtain formulas for the trigonometric sums

$$\sum_{k=0}^{n-1} \cos(k\theta) \qquad \text{and} \qquad \sum_{k=0}^{n-1} \sin(k\theta).$$

We assume that θ is a real number, which is *not* a multiple of 2π. Define $z = \cos\theta + i\sin\theta$ and $w = \cos\left(\frac{\theta}{2}\right) + i\sin\left(\frac{\theta}{2}\right)$.

a. Verify that

$$w^2 = z, \qquad z = \overline{z} \qquad \text{and} \qquad w = \overline{w}.$$

b. Problem 10.25 implies that

$$\sum_{k=0}^{n-1} z^k = \frac{z^n - 1}{z - 1}. \tag{10.3}$$

Use De Moivre's Theorem to show that the real and imaginary parts of the left-hand side are precisely the trigonometric sums we are trying to compute.

c. To calculate the real and imaginary parts of the right-hand side in Equation (10.3), we write:

$$\frac{z^n - 1}{z - 1} = \frac{w^{2n} - 1}{w^2 - 1} = \frac{w^n \cdot (w^n - \overline{w}^n)}{w \cdot (w - \overline{w})} = w^{n-1} \cdot \frac{w^n - \overline{w}^n}{w - \overline{w}}.$$

Justify each step in this computation.

d. Show that both $w^n - \overline{w}^n$ and $w - \overline{w}$ are purely imaginary, and express them in terms of θ.

e. Express

$$w^{n-1} \cdot \frac{w^n - \overline{w}^n}{w - \overline{w}}$$

in terms of θ, in such a way that the real and imaginary parts can be clearly identified.

f. Compare the real and imaginary parts of (10.3) to obtain the formulas for the two trigonometric sums.

10.31 Is the function $f\colon \mathbb{C} \to \mathbb{C}$, $f(z) = e^z$ injective? What is the image of this function?

10.32 Use Euler's Formula $e^{i\theta} = \cos\theta + i\sin\theta$ to derive expressions for $\cos\theta$ and $\sin\theta$ in terms of exponents.

10.9 Solutions to Exercises

Solution to Exercise 10.2.4

Using familiar algebraic manipulations we get:

$$\begin{aligned}
(a + bi) \cdot (c + di) &= ac + adi + bci + bdi^2 && (10.4)\\
&= ac - bd + adi + bci && (10.5)\\
&= (ac - bd) + (ad + bc)i. && (10.6)
\end{aligned}$$

In (10.4), commutativity of multiplication and distributivity were used. In (10.5), we used commutativity of addition, the identity $i^2 = -1$, and the fact that $(-1)x = -x$, which follows from the field axioms. Finally, in (10.6) we used distributivity to factor i from the last two terms. Note that associativity of addition and multiplication are also used implicitly in all three steps.

Solution to Exercise 10.2.7

We carefully apply Definition 10.2.3 to calculate $(a+0i)+(b+0i)\cdot(0+1i)$. Multiplying the complex numbers $b+0i$ and $0+1i$ results in $0+bi$, or simply bi.

Adding the numbers $a+0i$ and $0+bi$ gives

$$(a+0)+(0+b)i = a+bi,$$

which shows that the two interpretations coincide.

Solution to Exercise 10.2.10

1. Let $z = a+bi$, $w = c+di$ and $u = p+qi$ be three complex numbers. Then $w \cdot u = (cp-dq)+(cq+dp)i$, and so

$$\begin{aligned} z\cdot(w\cdot u) &= (a+bi)\cdot[(cp-dq)+(cq+dp)i] \\ &= [a\cdot(cp-dq)-b\cdot(cq+dp)] \\ &\quad +[a\cdot(cq+dp)+b\cdot(cp-dq)]i. \end{aligned}$$

   Similarly, we get

$$\begin{aligned} (z\cdot w)\cdot u &= [(ac-bd)+(ad+bc)i]\cdot(p+qi) \\ &= [(ac-bd)\cdot p-(ad+bc)\cdot q] \\ &\quad +[(ac-bd)\cdot q+(ad+bc)\cdot p]i. \end{aligned}$$

   Using properties of real numbers, we can easily verify that the real and imaginary parts of $z\cdot(w\cdot u)$ match those of $(z\cdot w)\cdot u$. This proves the associativity of multiplication axiom.
2. If $z = a+bi$ and $w = c+di$, then

$$z+w = (a+c)+(b+d)i \qquad \text{and} \qquad w+z = (c+a)+(d+b)i.$$

   Addition of real numbers is commutative, and thus $z+w = w+z$.
3. For every $z = a+bi \in \mathbb{C}$, we have

$$z\cdot 1 = (a+bi)\cdot(1+0i) = (a\cdot 1-b\cdot 0)+(a\cdot 0+b\cdot 1)i = a+bi = z.$$

4. Again, set $z = a+bi$, $w = c+di$ and $u = p+qi$. Then

$$\begin{aligned} z\cdot(w+u) &= (a+bi)\cdot[(c+p)+(d+q)i] \\ &= [a\cdot(c+p)-b\cdot(d+q)]+[a\cdot(d+q)+b\cdot(c+p)]i \end{aligned}$$

   and

$$\begin{aligned} z\cdot w+z\cdot u &= [(ac-bd)+(ad+bc)i]+[(ap-bq)+(aq+bp)i] \\ &= (ac-bd+ap-bq)+(ad+bc+aq+bp)i. \end{aligned}$$

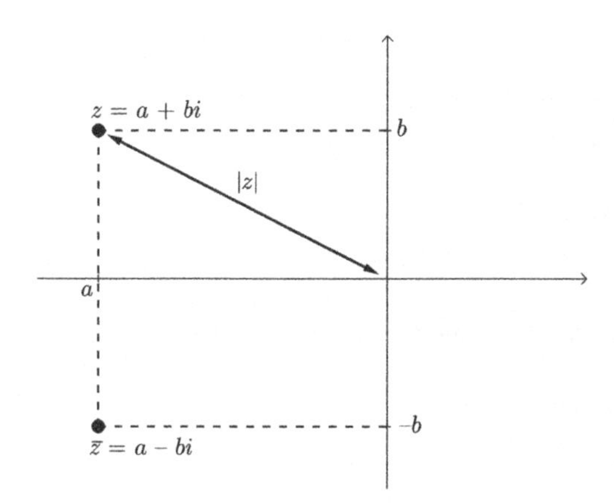

Figure 10.9 The geometrical representation of $\bar{z}$ and $|z|$.

We can now easily verify that the real and imaginary parts of $z \cdot (w + u)$ and $z \cdot w + z \cdot u$ match, which proves the distributive law for complex numbers.

Solution to Exercise 10.2.13
If we assume that $i > 0$, and multiply both sides by i, we get

$$i^2 > 0 \qquad \Rightarrow \qquad -1 > 0,$$

which is false. If we suppose that $i < 0$, and multiply both sides by i, we get $-1 > 0$ again, as i is now assumed to be negative and so we must switch the inequality sign.

This argument shows that complex numbers cannot be ordered in a "reasonable" way.

Solution to Exercise 10.2.15
If z is represented by the point (a, b), then $\bar{z}$ is represented by the point $(a, -b)$, which is obtained by *reflecting* (a, b) about the x-axis.

The absolute value of z is the *distance* of the point (a, b) from the origin (Figure 10.9).

Solution to Exercise 10.2.18
Let $z = a + bi$ be a complex number.

1. The equality $\bar{z} = z$ is equivalent to $a - bi = a + bi$, which is valid if and only if $b = 0$. It follows that $\bar{z} = z$ if and only if $z = a + 0i$ is real.
2. Similarly, $\bar{z} = -z$ is equivalent to $a - bi = -(a + bi)$, which is true if and only if $a = 0$. In this case $z = 0 + bi$ is an imaginary number.

Solution to Exercise 10.2.20

2. From

$$\overline{z - w} = \overline{(a + bi) - (c + di)} = \overline{(a - c) + (b - d)i} = (a - c) - (b - d)i$$

and

$$\overline{z} - \overline{w} = \overline{(a+bi)} - \overline{(c+di)} = (a-bi) - (c-di) = (a-c) - (b-d)i$$

we see that $\overline{z-w} = \overline{z} - \overline{w}$.

4. According to Part 3,

$$\overline{w} \cdot \overline{w^{-1}} = \overline{w \cdot w^{-1}} = \overline{1} = 1.$$

This means that $\overline{w^{-1}}$ is the reciprocal of $\overline{w}$. That is, $\overline{w^{-1}} = (\overline{w})^{-1}$.
We conclude that

$$\overline{\left(\frac{z}{w}\right)} = \overline{z \cdot w^{-1}} = \overline{z} \cdot \overline{w^{-1}} = \overline{z} \cdot (\overline{w})^{-1} = \frac{\overline{z}}{\overline{w}},$$

as needed.

Solution to Exercise 10.2.22

1. We can take $z = 1 + 3i$ and $w = 1 - 3i$, then

$$|z| = \sqrt{10}, \qquad |w| = \sqrt{10} \qquad \text{and} \qquad |z+w| = |2+0i| = 2.$$

Clearly, $2 \neq \sqrt{10} + \sqrt{10}$.

2. In this case, we have:

$$|z| = \sqrt{117}, \qquad |w| = \sqrt{52} \qquad \text{and} \qquad |z+w| = |10 - 15i| = \sqrt{325}.$$

Note that $\sqrt{117} + \sqrt{52} = 3\sqrt{13} + 2\sqrt{13} = 5\sqrt{13} = \sqrt{325}$, and hence $|z+w| = |z| + |w|$.

Solution to Exercise 10.3.1
If $z = x_1 + y_1 i$ and $w = x_2 + y_2 i$, then

$$|z - w| = |(x_1 - x_2) + (y_1 - y_2)i| = \sqrt{(x_1 - x_2)^2 + (y_1 - y_2)^2}.$$

This is the *distance* between the points representing z and w in the complex plane.

Solution to Exercise 10.3.3

a. Write the equation as $|z - (1+i)| \leq 1$. The left-hand side is the distance between z and $1+i$, and so the inequality represents a disk of radius 1 centered at $1+i$, including points on its boundary (Figure 10.10).
b. If $z = x + yi \in \mathbb{C}$, then the inequality $Re(z) < 3$ becomes $x < 3$. The corresponding region is shown in Figure 10.11 (note that the vertical line $x = 3$ is not included).

Solution to Exercise 10.3.5
If $a = 0$ and $b \neq 0$, then A lies on the y-axis and so OA is vertical. In that case BC is also vertical as B and C have the same x-coordinate. We conclude that OA and CB are parallel. Similarly, if $c = 0$ and $d \neq 0$, OC and AB will

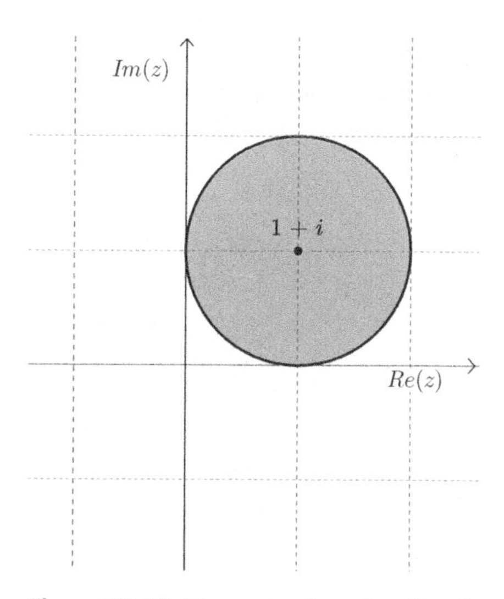

Figure 10.10 The region $|z - 1 - i| \leq 1$.

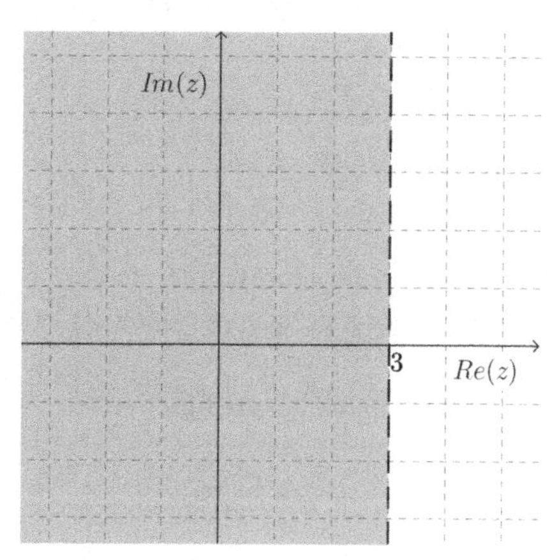

Figure 10.11 The region $Re(z) < 3$.

be vertical and hence parallel. Finally, if $a = b = 0$ or $c = d = 0$ then either $z = 0$ or $w = 0$, and the theorem can be easily verified.

Solution to Exercise 10.3.7

Following the suggestion, we square both sides and use the claim:

$$\begin{aligned}
|z+w|^2 &\leq |z|^2 + 2|z||w| + |w|^2 \\
(z+w)\cdot\overline{(z+w)} &\leq z\overline{z} + 2|z||w| + w\overline{w} \\
(z+w)\cdot(\overline{z}+\overline{w}) &\leq z\overline{z} + 2|z||w| + w\overline{w} \\
z\overline{z} + z\overline{w} + \overline{z}w + w\overline{w} &\leq z\overline{z} + 2|z||w| + w\overline{w} \\
z\overline{w} + \overline{z}w &\leq 2|z||w|.
\end{aligned}$$

Solution to Exercise 10.3.8

Let $u = x + yi$ be a complex number. To prove Part 1, we simply calculate

$$\overline{\overline{u}} = \overline{x - yi} = x + yi = u$$

and

$$u + \overline{u} = (x + yi) + (x - yi) = 2x = 2Re(u).$$

Part 2 can be proved as follows:

$$Re(u) = x \leq |x| = \sqrt{x^2} \leq \sqrt{x^2 + y^2} = |u|.$$

Solution to Exercise 10.4.2

If $a > 0$, the equation $z^2 = a$ has two real solutions: $z = \sqrt{a}$ and $z = -\sqrt{a}$.
If $a = 0$, the equation $z^2 = 0$ has a unique solution $z = 0$.
If $a < 0$, then $-a > 0$, and the equation $z^2 = a$ has two imaginary solutions: $z = i\sqrt{-a}$ and $z = -i\sqrt{-a}$, as

$$(i\sqrt{-a}) = i^2 \cdot (\sqrt{-a})^2 = (-1) \cdot (-a) = a$$

and

$$(-i\sqrt{-a}) = (-i)^2 \cdot (\sqrt{-a})^2 = (-1) \cdot (-a) = a.$$

Solution to Exercise 10.5.3

1. $-r(\cos\theta + i\sin\theta)$ is not a polar representation as $-r$ is a negative number. Also, $r(\cos\theta - i\sin\theta)$ is not a polar representation as we should have a $+$ sign between $\cos\theta$ and $i\sin\theta$.
2. We use properties of the sine and cosine functions to obtain polar representations of $-z$ and $\overline{z}$:

$$-z = r[-\cos\theta + i(-\sin\theta)] = r[\cos(\theta + \pi) + i\sin(\theta + \pi)]$$
$$\overline{z} = r(\cos\theta - i\sin\theta) = r[\cos(-\theta) + i\sin(-\theta)].$$

For z^{-1}, we use Claim 10.2.17:

$$z^{-1} = \frac{\overline{z}}{|z|^2} = \frac{r[\cos(-\theta) + i\sin(-\theta)]}{r^2} = \frac{1}{r} \cdot [\cos(-\theta) + i\sin(-\theta)].$$

Solution to Exercise 10.6.2

1. For $n = 1$, both sides evaluate to $\cos\theta + i\sin\theta$, which confirms the base case. Suppose the formula is valid for some $k \in \mathbb{N}$ and compute:

$$\begin{aligned}(\cos\theta + i\sin\theta)^{k+1} &= (\cos\theta + i\sin\theta)^k \cdot (\cos\theta + i\sin\theta) \\ &= [\cos(k\theta) + i\sin(k\theta)] \cdot (\cos\theta + i\sin\theta) \qquad (10.7) \\ &= \cos[(k+1)\theta] + i\sin[(k+1)\theta] \qquad (10.8)\end{aligned}$$

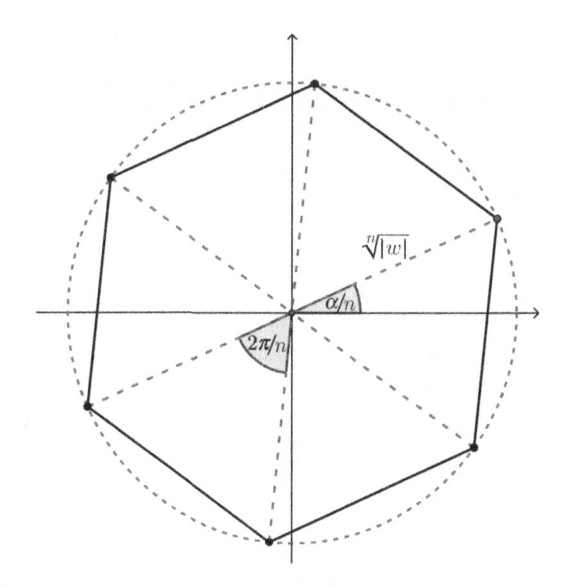

Figure 10.12 The solutions of the equation $z^n = w$.

which verifies the case $n = k + 1$. This proves the theorem for $n > 0$. The induction hypothesis was used in (10.7).

2. The case $n = 0$ can be verified directly (both sides evaluate to 1). For $n < 0$, denote by $z = \cos\theta + i\sin\theta$, and note that $|z| = \cos^2\theta + \sin^2\theta = 1$. Consequently,

$$z^{-1} = \frac{\bar{z}}{|z|^2} = \bar{z} = \cos\theta - i\sin\theta = \cos(-\theta) + i\sin(-\theta).$$

Using Part 1 and the fact that $-n > 0$, we get:

$$z^n = (z^{-1})^{-n} = [\cos(-\theta) + i\sin(-\theta)]^{-n} = \cos(n\theta) + i\sin(n\theta).$$

This completes the proof of De Moivre's Theorem.

Solution to Exercise 10.6.5

Theorem 10.6.4 implies that $z^n = w$ has n distinct solutions, all lying on a circle centered at the origin, and with radius $\sqrt[n]{|w|}$ (Figure 10.12). The arguments of consecutive solutions defer by $\frac{2\pi}{n}$, and hence represent the vertices of a regular polygon with n sides.

11 Preview of Linear Algebra

Linear algebra is a branch of mathematics that deals with linear equations, systems of linear equations, and their representation as functions between algebraic structures called vector spaces. Linear algebra is an essential tool in many disciplines such as engineering, statistics and computer science, and is also central within mathematics, in areas such as analysis and geometry.

Much like how a field is defined, a vector space is defined through a list of axioms, motivated by concrete observations in familiar spaces, such as the standard two-dimensional plane and three-dimensional space. We begin this chapter by taking a closer look at real n-dimensional spaces and vectors, and then move on to discussing abstract vector spaces and linear maps. Our experience with sets and functions, developed in previous chapters, as well as certain proof techniques will prove to be useful in our discussion.

11.1 The Spaces $\mathbb{R}^n$ and their Properties

Given sets $A_1, A_2, \ldots, A_n$, we defined their Cartesian product as the collection of all n-tuples $(x_1, x_2, \ldots, x_n)$ where each x_i is in A_i (see Definition 2.1.11).

In particular, for each $n \in \mathbb{N}$, elements in

$$\mathbb{R}^n = \mathbb{R} \times \mathbb{R} \times \cdots \times \mathbb{R}$$

are n-tuples of real numbers.

For example, elements of $\mathbb{R}^2$ are pairs of real numbers, which we can view as points in a two-dimensional coordinate system. Elements in $\mathbb{R}^3$ can be thought of as points in a three-dimensional coordinate system (see Figure 11.1). As we will shortly see, $\mathbb{R}^n$ is more than just a set. Addition and multiplication in $\mathbb{R}$ allow us to introduce, quite naturally, algebraic operations in $\mathbb{R}^n$.

Moving forward, we often refer to elements in $\mathbb{R}^n$ as *vectors*, rather than points, and to real numbers as *scalars*. This terminology will be justified later.

Definition 11.1.1 Let $n \in \mathbb{N}$. An element $\underline{v} = (v_1, v_2, \ldots, v_n)$ in $\mathbb{R}^n$ is called an *n-dimensional algebraic vector*, or simply a *vector*. The numbers $v_1, \ldots, v_n$ are the *coordinates* of $\underline{v}$.

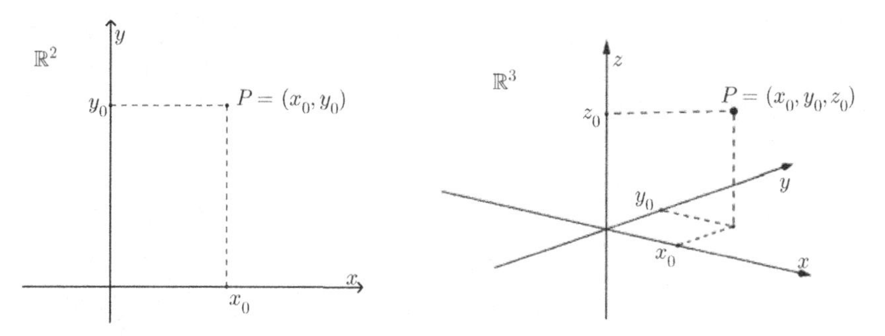

Figure 11.1 Two-dimensional and three-dimensional coordinate systems.

Remarks.

- We will often *underline* vectors in $\mathbb{R}^n$ to distinguish them from real numbers.
- Naturally, two vectors $\underline{v} = (v_1, \ldots, v_n)$ and $\underline{u} = (u_1, \ldots, u_n)$ are equal if their coordinates match. That is, if $v_i = u_i$ for all $1 \leq i \leq n$.
- The vector $\underline{0} = (0, \ldots, 0)$ is called the *zero vector*.

Next, we introduce two important operations on vectors.

Definition 11.1.2 For every $\underline{v} = (v_1, \ldots, v_n)$ and $\underline{u} = (u_1, \ldots, u_n)$ in $\mathbb{R}^n$, and $c \in \mathbb{R}$, we define:

$$\underline{v} + \underline{u} = (v_1 + u_1, \ldots, v_n + u_n) \qquad \text{(vector addition)}$$

and

$$c \cdot \underline{v} = (c \cdot v_1, \ldots, c \cdot v_n) \qquad \text{(scalar multiplication).}$$

Example 11.1.3 Consider the vectors $\underline{u} = (-1, 3)$ and $\underline{v} = (5, 7)$ in $\mathbb{R}^2$. Calculate the vectors $\frac{1}{2}\underline{u}$ and $\underline{u} + \underline{v}$.

Solution

According to Definition 11.1.2, we get

$$\frac{1}{2}\underline{u} = \left(\frac{1}{2} \cdot (-1), \frac{1}{2} \cdot 3\right) = \left(-\frac{1}{2}, \frac{3}{2}\right)$$

and

$$\underline{u} + \underline{v} = (-1 + 5, 3 + 7) = (4, 10).$$

Exercise 11.1.4

Let $\underline{u} = (0, -4, 0.5)$ and $\underline{v} = (1, 0, -1)$ be two vectors in $\mathbb{R}^3$. Calculate the vector $(3\underline{u}) + (2\underline{v})$.

Vector addition and scalar multiplication are new operations on vectors in $\mathbb{R}^n$. The use of the words *addition* and *multiplication* is no coincidence. The following theorem shows that these operations satisfy properties similar to addition and multiplication of real numbers.

Theorem 11.1.5 *Properties of vector addition.*

1. $\underline{v} + \underline{u} = \underline{u} + \underline{v}$ *for all* $\underline{v}, \underline{u} \in \mathbb{R}^n$ *(commutativity).*
2. $(\underline{v} + \underline{u}) + \underline{w} = \underline{v} + (\underline{u} + \underline{w})$ *for all* $\underline{v}, \underline{u}, \underline{w} \in \mathbb{R}^n$ *(associativity).*
3. $\underline{v} + \underline{0} = \underline{v}$ *for all* $\underline{v} \in \mathbb{R}^n$ *(existence of an additive identity).*
4. *For every* $\underline{v} \in \mathbb{R}^n$*, there is a vector* $\underline{u} \in \mathbb{R}^n$*, such that* $\underline{v} + \underline{u} = \underline{0}$*. We denote* $\underline{u} = -\underline{v}$ *(existence of negatives).*

Properties of scalar multiplication.

5. $1 \cdot \underline{v} = \underline{v}$ *for all* $\underline{v} \in \mathbb{R}^n$.
6. $(c \cdot d) \cdot \underline{v} = c \cdot (d \cdot \underline{v})$ *for every* $c, d \in \mathbb{R}$ *and* $\underline{v} \in \mathbb{R}^n$.

Distributivity.

7. $c \cdot (\underline{v} + \underline{u}) = c \cdot \underline{v} + c \cdot \underline{u}$ *for all* $c \in \mathbb{R}$ *and* $\underline{v}, \underline{u} \in \mathbb{R}^n$.
8. $(c + d) \cdot \underline{v} = c \cdot \underline{v} + d \cdot \underline{v}$ *for all* $c, d \in \mathbb{R}$ *and* $\underline{v} \in \mathbb{R}^n$.

Remarks.

- We will see that the vector $\underline{u}$ in property 4 is unique. This allows us to declare it as *the* negative of $\underline{v}$ and use the notation $-\underline{v}$. Moreover, vector subtraction can be naturally defined as follows:

$$\underline{u} - \underline{v} = \underline{u} + (-\underline{v}) \qquad \text{for } \underline{u}, \underline{v} \in \mathbb{R}^n.$$

- Now that we have introduced vector operations, the symbols $+$ and $\cdot$ can be interpreted in more than one way, which can be confusing at first. For instance, in an expression such as $(c \cdot d) \cdot \underline{v}$, the first dot represents multiplication in $\mathbb{R}$, while the second dot represents scalar multiplication.

Proof The proof of the theorem is straightforward. We prove properties 4, 6 and 8, and leave the rest as an exercise.

4. For every $\underline{v} = (v_1, v_2, \ldots, v_n) \in \mathbb{R}^n$, define $\underline{u} = (-v_1, -v_2, \ldots, -v_n)$. Then, by Definition 11.1.2, we have

$$\underline{v} + \underline{u} = (v_1 + (-v_1), v_2 + (-v_2), \ldots, v_n + (-v_n)) = (0, 0, \ldots, 0) = \underline{0}.$$

This shows that the negative of a vector $\underline{v}$ is $-\underline{v} = (-v_1, -v_2, \ldots, -v_n)$. Note that such $\underline{u}$ is unique. That is, the only way to guarantee $\underline{v} + \underline{u} = \underline{0}$ is to choose each coordinate of $\underline{u}$ to be the negative of the corresponding coordinate of $\underline{v}$.

6. Suppose $\underline{v} = (v_1, v_2, \ldots, v_n) \in \mathbb{R}^n$ and $c, d \in \mathbb{R}$. We compute each side of the identity:

$$(c \cdot d) \cdot \underline{v} = (c \cdot d) \cdot (v_1, v_2, \ldots, v_n) = ((c \cdot d)v_1, (c \cdot d)v_2, \ldots, (c \cdot d)v_n)$$
$$c \cdot (d \cdot \underline{v}) = c \cdot (d \cdot v_1, d \cdot v_2, \ldots, d \cdot v_n) = (c(d \cdot v_1), c(d \cdot v_2), \ldots, c(d \cdot v_n)).$$

Due to associativity of multiplication in $\mathbb{R}$, we see that $(c \cdot d) \cdot \underline{v} = c \cdot (d \cdot \underline{v})$ as needed.

8. Again, assume that c, d and $\underline{v}$ are as in the proof of property 6, and compute:

$$\begin{aligned}(c+d) \cdot \underline{v} &= (c+d) \cdot (v_1, v_2, \ldots, v_n) \\ &= ((c+d)v_1, (c+d)v_2, \ldots, (c+d)v_n)\end{aligned}$$

and

$$\begin{aligned}c \cdot \underline{v} + d \cdot \underline{v} &= (cv_1, cv_2, \ldots, cv_n) + (dv_1, dv_2, \ldots, dv_n) \\ &= (cv_1 + dv_1, cv_2 + dv_2, \ldots, cv_n + dv_n).\end{aligned}$$

Distributivity in $\mathbb{R}$ implies that $(c+d)v_i = cv_i + dv_i$ for all $1 \leq i \leq n$, from which it follows that $(c+d) \cdot \underline{v} = c \cdot \underline{v} + d \cdot \underline{v}$ as needed. □

Exercise 11.1.6
Complete the proof of Theorem 11.1.5 by proving the remaining properties.

Exercise 11.1.7
There are many other properties satisfied by vectors in $\mathbb{R}^n$. We will soon see the reason for including these specific eight properties in Theorem 11.1.5. For now, prove the following (here, t represents a real number and $\underline{v}$ a vector in $\mathbb{R}^n$).

1. If $t\underline{v} = \underline{0}$, then $t = 0$ or $\underline{v} = \underline{0}$.
2. $(-1)\underline{v} = -\underline{v}$.
3. $-(t\underline{v}) = (-t)\underline{v}$.

11.2 Geometric Vectors

Historically, the notion of a vector was introduced in physics to describe quantities that have both magnitude and direction, such as forces and velocities. These quantities are often represented as arrows to indicate a direction. The length of the arrow represents the magnitude of the quantity.

We call such arrows *geometric vectors* and define them formally as pairs of points in $\mathbb{R}^n$.

Definition 11.2.1 A *geometric vector* in $\mathbb{R}^n$ is an ordered pair of points (A, B) in $\mathbb{R}^n$, denoted as $\vec{AB}$. We interpret this pair as representing an arrow from A, its *initial point*, to B, its *terminal point* (Figure 11.2).

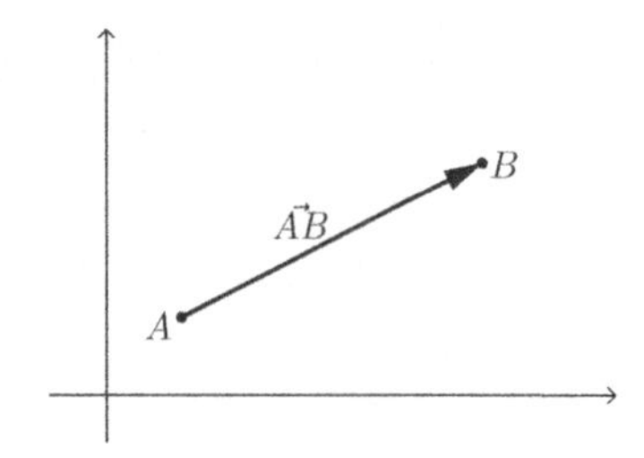

Figure 11.2 A geometric vector.

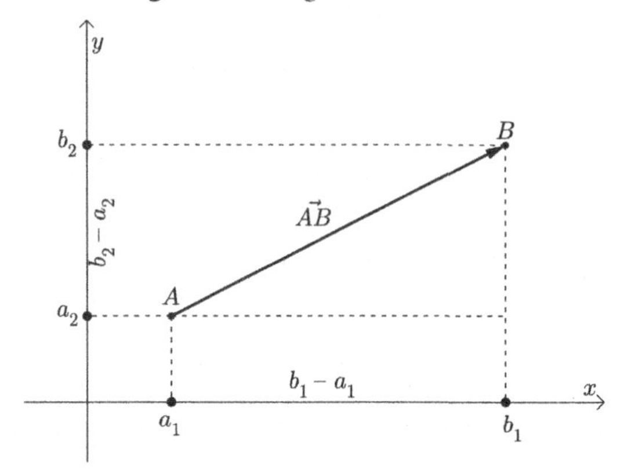

Figure 11.3 The coordinates of $\vec{AB}$ are $b_1 - a_1$ and $b_2 - a_2$.

Geometric vectors are used to represent algebraic vectors. The following definition allows us to attach an algebraic vector to a given geometric vector.

Definition 11.2.2 Let $A = (a_1, \ldots, a_n)$ and $B = (b_1, \ldots, b_n)$ be two points, and $\underline{u} = (u_1, \ldots, u_n)$ an algebraic vector in $\mathbb{R}^n$. We say that $\underline{u}$ *represents*, or *corresponds* to, the geometric vector $\vec{AB}$ if $u_i = b_i - a_i$ for all $1 \leq i \leq n$, and write $\underline{u} = \vec{AB}$.

Remarks.

- The quantity $b_i - a_i$ is the projection of $\vec{AB}$ onto the ith coordinate. For instance, in the two-dimensional case, $b_1 - a_1$ and $b_2 - a_2$ are the projections onto the x-axis and y-axis, respectively (Figure 11.3).
- Note how we refer to elements of $\mathbb{R}^n$ both as points and as vectors. There is no formal mathematical difference between the two terms, only in the way we think about them. *Algebraic vectors* are thought of as representing arrows, while *points* are viewed as dots in an n-dimensional coordinate system.

- What happens if the two points A and B coincide, and our "arrow" shrinks to a single point? In this case, $b_i = a_i$ and so $\underline{u}$ becomes the zero vector. In fact, the converse is true as well. The vector $\vec{AB}$ is represented by the zero vector if and only if $A = B$.
- If the initial point of $\vec{AB}$ is the origin (that is, $A = (0,\ldots,0)$), then $u_i = b_i$ for all $1 \leq i \leq n$. In other words, whenever the initial point of $\vec{AB}$ is the origin, the coordinates of $\underline{u}$ coincide with those of the terminal point B.

Given a geometric vector $\vec{AB}$, the condition $u_i = b_i - a_i$ determines *uniquely* the corresponding algebraic vector $\underline{u}$. However, there are many geometric vectors represented by a given algebraic vector $\underline{u}$. Here is an example.

Example 11.2.3 Let $\underline{u} = (-1, 2, 5)$ be a vector in $\mathbb{R}^3$.

1. Let $A = (1, 2, 3), B = (0, 4, 8)$ and $C = (-1, 6, 13)$. Show that $\underline{u}$ represents both $\vec{AB}$ and $\vec{BC}$.
2. Find another pair of points D and E in $\mathbb{R}^3$, such that $\underline{u} = \vec{DE}$.

Solution

1. We subtract the coordinate of A from B and get
$$\vec{AB} = (0 - 1, 4 - 2, 8 - 3) = (-1, 2, 5) = \underline{u}.$$
Similarly, $\vec{BC} = (-1 - 0, 6 - 4, 13 - 8) = (-1, 2, 5) = \underline{u}$.
2. We can choose arbitrary coordinates for D, say $D = (9, -7, 12)$, and then determine the coordinates of $E = (x, y, z)$ so that $\vec{DE} = \underline{u}$. Definition 11.2.2 implies that
$$\begin{aligned} x - 9 &= -1 \\ y - (-7) &= 2 \\ z - 12 &= 5 \end{aligned}$$
and so $E = (x, y, z) = (8, -5, 17)$.

Exercise 11.2.4
Suppose A, B are points in $\mathbb{R}^n$. If $\vec{AB}$ is represented by $\underline{u}$, which algebraic vector represents $\vec{BA}$?

Suppose that $\vec{AB}$ and $\vec{CD}$ are two vectors in $\mathbb{R}^2$, represented by the same algebraic vector $\underline{u} = (u_1, u_2)$. Denote by a_i, b_i, c_i and d_i the coordinates of A, B, C and D, respectively. Definition 11.2.2 implies that the respective projections

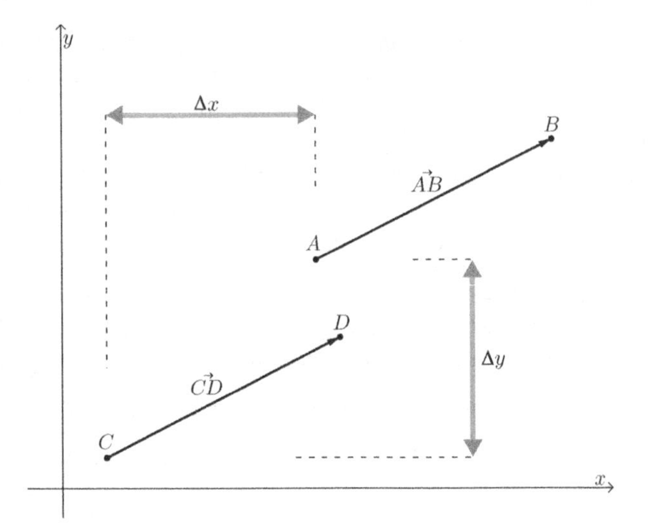

Figure 11.4 $\vec{AB}$ and $\vec{CD}$ represent the same algebraic vector.

onto the coordinate axes are equal:

$$\begin{aligned} b_1 - a_1 &= d_1 - c_1 = u_1 \\ b_2 - a_2 &= d_2 - c_2 = u_2. \end{aligned}$$

We can also argue that the two vectors have the same length, and that they point in the same direction. To see why, denote by Δx and Δy the difference between the x- and y-coordinates of A and C:

$$\begin{aligned} \Delta x &= a_1 - c_1 \\ \Delta y &= a_2 - c_2. \end{aligned}$$

Now translate the vector $\vec{CD}$ by Δx units horizontally and by Δy units vertically (Figure 11.4). By doing so, we shift the point $C = (c_1, c_2)$ to

$$(c_1 + \Delta x, c_2 + \Delta y) = (a_1, a_2) = A,$$

and the point $D = (d_1, d_2)$ to

$$(d_1 + \Delta x, d_2 + \Delta y) = (d_1 + a_1 - c_1, d_2 + a_2 - c_2) = (b_1, b_2) = B.$$

Consequently, the translated vector coincides with $\vec{AB}$. As lengths and directions are preserved by translations, the two vectors have the same length and direction.[1]

[1] This is an informal argument, as we have not defined formally the notions of length and direction. We are also relying on the fact that translations preserve length and direction.

This conclusion remains valid in $\mathbb{R}^3$ and, with the proper definitions, can be generalized to $\mathbb{R}^n$. We summarize as follows.

Conclusion If two geometric vectors in $\mathbb{R}^n$ correspond to the same algebraic vector, then they have the same length and direction.

Exercise 11.2.5
The converse is also true: if two geometric vectors have the same length and direction, they correspond to the same algebraic vector. Prove this statement in $\mathbb{R}^2$.
(Hint: If two vectors in $\mathbb{R}^2$ have the same length and direction, then one vector can be translated so that it coincides with the other vector.)

Addition

We now discuss the geometric interpretation of vector addition in $\mathbb{R}^n$. Most of the discussion will be restricted to $\mathbb{R}^2$ and $\mathbb{R}^3$, but the results can be generalized to higher dimensions.

Theorem 11.2.6 (The Parallelogram Law) *Let $\underline{u}$ and $\underline{v}$ be vectors in $\mathbb{R}^2$ (or $\mathbb{R}^3$), represented by $\vec{AB}$ and $\vec{AD}$, respectively. Suppose that the points A, B and D do not lie on the same straight line. The vector $\underline{u}+\underline{v}$ is represented by $\vec{AC}$ if and only if $ABCD$ is a parallelogram.*

Proof We prove the theorem in $\mathbb{R}^2$, as the proof in $\mathbb{R}^3$ is nearly identical.
First, we note that

$$\vec{AB}+\vec{BC}=(b_1-a_1,b_2-a_2)+(c_1-b_1,c_2-b_2)=(c_1-a_1,c_2-a_2)=\vec{AC}.$$

If $ABCD$ is a parallelogram, then $\vec{BC} = \underline{v}$, as it has the same length and direction as $\vec{AD}$, and the equality $\vec{AB}+\vec{BC}=\vec{AC}$ becomes $\underline{u}+\underline{v}=\vec{AC}$.
Conversely, if $\vec{AC}=\underline{u}+\underline{v}$, we get

$$\underline{u}+\vec{BC}=\underline{u}+\underline{v}$$

from which it follows that $\vec{BC}=\underline{v}$. We conclude that $\vec{BC}=\vec{AD}$, and so the line segments BC and AD are parallel. A similar argument shows that AB and CD are parallel, and so the quadrilateral $ABCD$ is a parallelogram (see Figure 11.5). □

The theorem allows us to interpret sums of algebraic vectors geometrically. If $\underline{u}$ and $\underline{v}$ are represented by non-parallel arrows with the same initial point A, then their sum is represented by the diagonal through A of the parallelogram determined by $\underline{u}$ and $\underline{v}$.

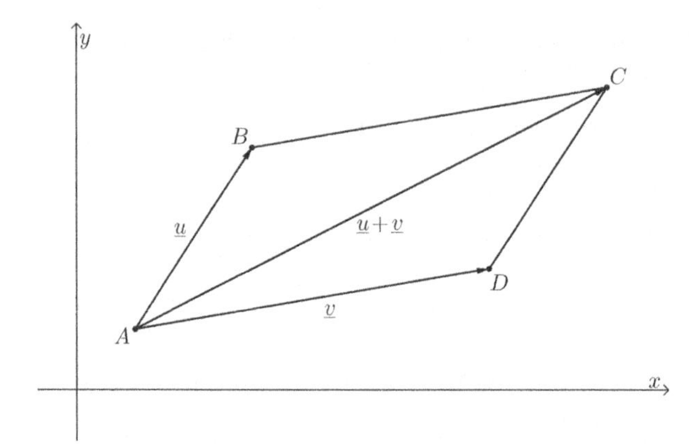

Figure 11.5 The Parallelogram Law.

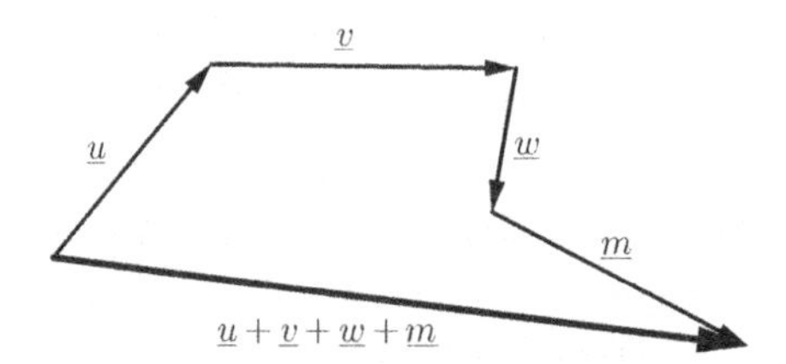

Figure 11.6 The tip-to-tail rule.

In the course of the proof of Theorem 11.2.6, we showed that $\vec{AB} + \vec{BC} = \vec{AC}$, which provided another strategy for adding vectors geometrically. If $\underline{u}$ and $\underline{v}$ are represented by $\vec{AB}$ and $\vec{BC}$, then their sum is represented by $\vec{AC}$. This is often called the *tip-to-tail rule*, and can be used with more than two vectors, as indicated in Figure 11.6. As always, we use subscripts to denote the coordinates of vectors and points in $\mathbb{R}^2$: $\underline{u} = (u_1, u_2), A = (a_1, a_2)$, etc.

Scalar Multiplication

We now turn to the geometrical interpretation of scalar multiplication.

Theorem 11.2.7 *Let $\underline{u}$ and $\underline{v}$ be two non-zero vectors in $\mathbb{R}^2$ (or $\mathbb{R}^3$), represented by $\vec{AB}$ and $\vec{CD}$. Then $\underline{v} = t\underline{u}$ for some $t \neq 0$ if and only if $\vec{AB}$ and $\vec{CD}$ are parallel, or lie on the same straight line.*

Proof Again, we prove the theorem in $\mathbb{R}^2$ only. We can translate the vectors $\vec{AB}$ and $\vec{CD}$ so that their initial point is at the origin, as this will not change the directions of the vectors (Figure 11.7). We also assume that both $\vec{AB}$ and $\vec{CD}$ lie in the first quadrant.

Note that, under these assumptions, we have

$$\begin{aligned} B &= \underline{u} = (b_1, b_2) \\ D &= \underline{v} = (d_1, d_2) \end{aligned}$$

for some real numbers b_1, b_2, d_1, d_2. If $\underline{v} = t\underline{u}$ for some $t \neq 0$, then $d_1 = tb_1$ and $d_2 = tb_1$. Equivalently, we have

$$\frac{DF}{BE} = \frac{OF}{OE} = t$$

which means that the two triangles OBE and ODF are similar, and hence their corresponding angles are equal. We conclude that the points O, B and D must be on the same line. It follows that the original vectors $\vec{AB}$ and $\vec{CD}$ are parallel or lie on the same straight line.

Conversely, if $\vec{AB}$ and $\vec{CD}$ are parallel or lie on the same straight line, then O, B and D lie on the same line. In that case, the triangles OBE and ODF are similar, and so the ratios $\frac{DF}{BE}$ and $\frac{OF}{OE}$ are the same. Set $t = \frac{DF}{BE} = \frac{OF}{OE}$.

In terms of the coordinates of B and D we have $t = \frac{d_2}{b_2} = \frac{d_1}{b_1}$, and

$$\underline{v} = \vec{OD} = (d_1, d_2) = (tb_1, tb_2) = t(b_1, b_2) = t\underline{u},$$

as needed. This has proved the theorem under the assumption that $\vec{AC}$ and $\vec{BD}$, when moved so that their initial point is the origin, lie in the first quadrant. If both vectors lie in one of the other quadrants the proof is similar. It is also quite straightforward to prove the theorem when the vectors lie on the coordinate axes. We do ask that you attempt to prove the theorem when the vectors lie in different quadrants. □

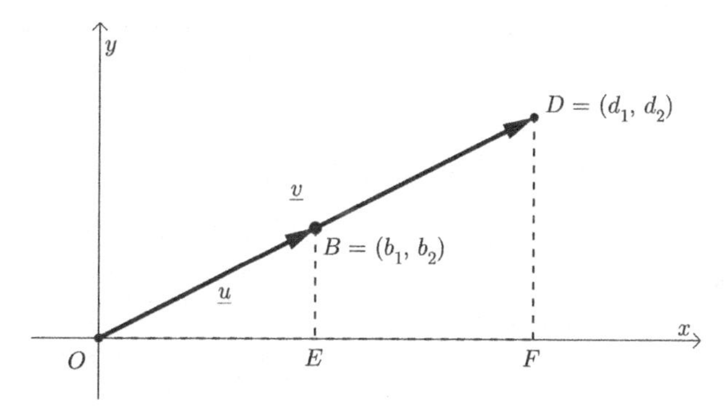

Figure 11.7 The vector $\underline{v}$ is a scalar multiple of $\underline{u}$.

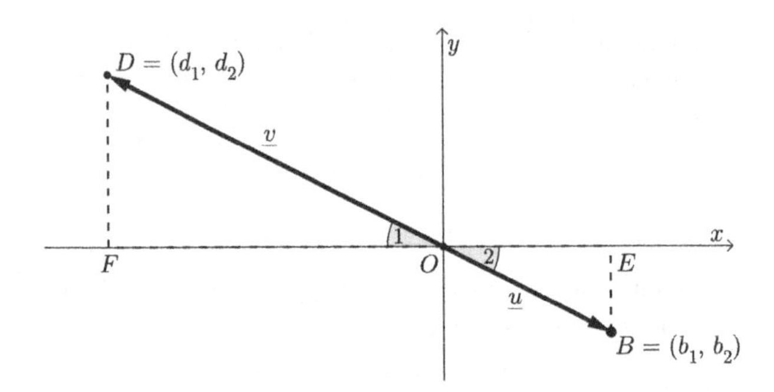

Figure 11.8 The vector $\underline{v}$ is a scalar multiple of $\underline{u}$.

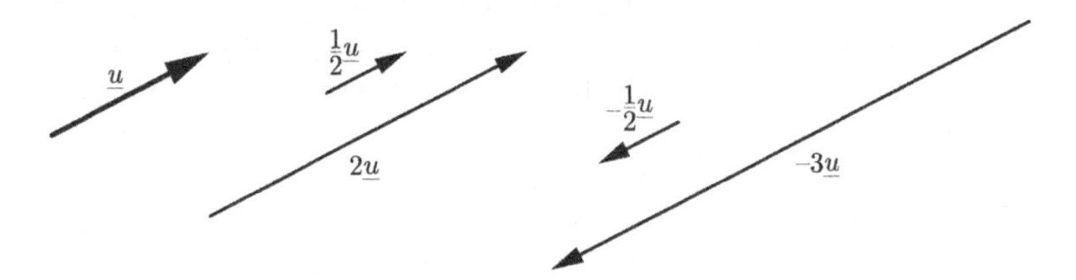

Figure 11.9 Scalar multiplication of a vector.

Exercise 11.2.8
Prove Theorem 11.2.7 in the case where $\vec{AB}$ and $\vec{CD}$, once translated, lie in the second and fourth quadrants. Use Figure 11.8 as a guide.

We see that multiplying a vector $\underline{u} \neq \underline{0}$ by a $t \neq 0$ can change its length,[2] while its direction either remains the same or is changed to the opposite direction. If $t > 0$, then $\underline{u}$ and $t\underline{u}$ have the same direction. If $t < 0$, the directions are opposite to each other. Moreover, the vector $t\underline{u}$ is longer than $\underline{u}$ when $|t| > 1$, and shorter when $|t| < 1$ (Figure 11.9).

Example 11.2.9 In the triangle in Figure 11.10, M is the midpoint of BC. Let $\underline{u} = \vec{AB}$ and $\underline{v} = \vec{AC}$. Express the median $\vec{AM}$ in terms of $\underline{u}$ and $\underline{v}$.

Solution
Using the tip-to-tail rule, we have

$$\vec{BC} = -\underline{u} + \underline{v}.$$

Also, since M is the midpoint of BC, the vectors $\vec{BM}$ and $\vec{MC}$ are equal (same length and direction). We conclude that

$$\vec{BC} = \vec{BM} + \vec{MC} = 2\vec{BM} \quad \Rightarrow \quad 2\vec{BM} = -\underline{u} + \underline{v} \quad \Rightarrow \quad \vec{BM} = -\tfrac{1}{2}\underline{u} + \tfrac{1}{2}\underline{v}.$$

[2] That is, the length of a geometric vector representing $\underline{u}$.

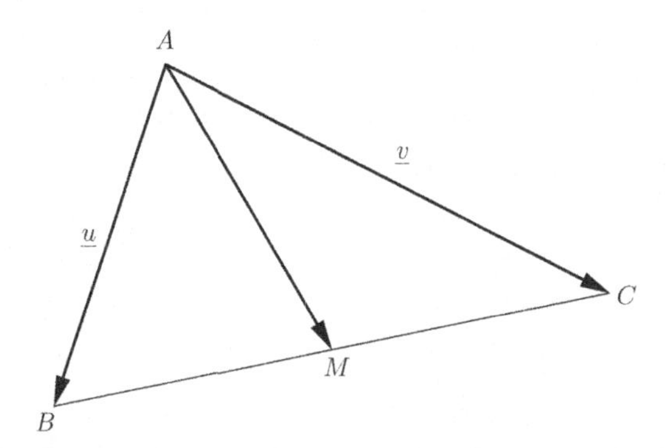

Figure 11.10 In triangle ABC, AM is the median to BC.

Finally, we use the tip-to-tail rule one more time to obtain the median:

$$\vec{AM} = \underline{u} + \vec{BM} = \underline{u} - \tfrac{1}{2}\underline{u} + \tfrac{1}{2}\underline{v} = \tfrac{1}{2}\underline{u} + \tfrac{1}{2}\underline{v}.$$

11.3 Abstract Vector Spaces

In Section 11.1 we saw that vector addition and scalar multiplication in $\mathbb{R}^n$ satisfy several algebraic properties. As it turns out, there are many other mathematical systems in which addition and scalar multiplication can be naturally defined, and the properties outlined in Theorem 11.1.5 remain valid.

For instance, functions from $\mathbb{R}$ to $\mathbb{R}$ can be added and multiplied by a real number, and these operations behave much like addition and scalar multiplication in $\mathbb{R}^n$. Such observations led mathematicians to generalize the notion of an (algebraic) vector, and consider spaces other than $\mathbb{R}^n$. In such spaces, elements may no longer be n-tuples of real numbers, but other objects, such as functions, matrices, sequences, etc.

The abstract definition of a vector space, provided below, is motivated by observations made in $\mathbb{R}^n$. Its structure will remind you of the definition of field, from Chapter 2. Informally speaking, a vector space is a set V of objects (the vectors) in which elements can be added and multiplied by scalars, such that the eight properties in Theorem 11.1.5 are valid. Moreover, instead of restricting ourselves to $\mathbb{R}$, we allow our scalars to come from an arbitrary field $\mathbb{F}$.

Definition 11.3.1 A *vector space* V over a field $\mathbb{F}$ is a set, with a distinguished element 0, and two operations

$$+ : V \times V \to V \qquad \text{(vector addition)}$$

and

$$\cdot : \mathbb{F} \times V \to V \qquad \text{(scalar multiplication)},^3$$

such that the following axioms hold.

Vector addition.

1. $v + u = u + v$ for all $v, u \in V$ (commutativity).
2. $(v + u) + w = v + (u + w)$ for all $v, u, w \in V$ (associativity).
3. $v + 0 = v$ for all $v \in V$ (identity axiom).
4. For every $v \in V$, there is a $u \in V$ such that $v + u = 0$. We write $u = -v$ (existence of negatives).

Scalar multiplication.

5. $1 \cdot v = v$ for all $v \in V$.
6. $(c \cdot d) \cdot v = c \cdot (d \cdot v)$ for all $c, d \in \mathbb{F}$ and $v \in V$.

Distributivity.

7. $c \cdot (v + u) = c \cdot v + c \cdot u$ for all $c \in \mathbb{F}$ and $v, u \in V$.
8. $(c + d) \cdot v = c \cdot v + d \cdot v$ for all $c, d \in \mathbb{F}$ and $v \in V$.

Remarks.

- To maintain consistency with our terminology in $\mathbb{R}^n$, elements in V and $\mathbb{F}$ are called *vectors* and *scalars*, respectively.
- You might have noticed that we are no longer underlining our vectors, and write u, v, w instead of $\underline{u}, \underline{v}, \underline{w}$. From now on, vectors need not be elements of $\mathbb{R}^n$, and hence can no longer be represented by arrows. We removed the underline to emphasize that fact. We keep writing $\underline{u}, \underline{v}, \underline{w}$ etc. for vectors in $\mathbb{R}^n$, but without the underline in other settings and in general discussions.
- We emphasize one more time the multiple uses of symbols in algebra. Addition of vectors, real numbers and other type of numbers are all denoted by $+$, and we must be careful in interpreting this symbol based on the context.
- We often omit the dot symbol for scalar multiplication and write cv instead of $c \cdot v$.
- A vector space V comes with a distinguished element $0 \in V$ satisfying $v + 0 = v$ for all $v \in V$. It is natural to wonder whether we can have another

[3] That is, addition is a function that assigns to each pair $(v, u) \in V \times V$ their sum $v + u \in V$. Similarly, scalar multiplication is a function that assigns to each pair $(c, v) \in \mathbb{F} \times V$ an element $c \cdot v \in V$.

element, say $0' \in V$, that also satisfies $v + 0' = v$ for all $v \in V$. It turns out that this cannot happen, for the following reason.

If $v + 0 = v$ for all $v \in V$, then by taking $v = 0'$ we get $0' + 0 = 0'$. Similarly, by setting $v = 0$ in $v + 0' = v$ we get $0 + 0' = 0$. Commutativity of addition implies that $0' = 0' + 0 = 0 + 0' = 0$.

We conclude that $0 = 0'$. In other words, 0 is the only element in V satisfying $v + 0 = v$ for all $v \in V$.

- We can also wonder whether negative elements are unique. If $v \in V$, then Axiom 4 guarantees that $v + u = 0$ for some $u \in V$. Can we have another element $u' \in V$ that also satisfies $v + u' = 0$? No. And here is why. If such u' exists, then

$$u' = u' + 0 = 0 + u' = (u + v) + u' = u + (v + u') = u + 0 = u.$$

Each step can be justified by one of the vector space axioms (check!), and so $u = u'$ and negatives are indeed unique. That is the reason for referring to *the* negative of v, and for the notation $-v$.

- Subtraction and scalar division are defined in the obvious way.

If $v, u \in V$, we define $v - u = v + (-u)$.
If $v \in V$ and $c \neq 0$ is a scalar, we define $\frac{v}{c} = \frac{1}{c}v$.

Example 11.3.2 The first example of a vector space that comes to mind is $\mathbb{R}^n$. Theorem 11.1.5 asserts that all the vector space axioms in $\mathbb{R}^n$ are satisfied, and hence, it is a vector space over $\mathbb{R}$.

Example 11.3.3 Let $\mathcal{F}$ be the set of all functions from $[0, 1]$ to $\mathbb{R}$. If $f, g \in \mathcal{F}$ and $c \in \mathbb{R}$, we naturally define:

$$\begin{aligned} (f + g)(x) &= f(x) + g(x) \\ (cf)(x) &= c \cdot f(x). \end{aligned}$$

The zero element is the zero function $f(x) = 0$ for all $x \in [0, 1]$. It is straightforward to verify all the vector space axioms, and so $\mathcal{F}$ is a vector space over $\mathbb{R}$.

Example 11.3.4 Let ℓ denote the set of all infinite bounded sequences of real numbers. Each element $a \in \ell$ is a bounded sequence

$$a = (a_1, a_2, \ldots).$$

That is, for some $M \in \mathbb{R}$, $|a_n| \leq M$ for all $n \in \mathbb{N}$. Addition and scalar multiplication are defined as follows.

If $a = (a_1, a_2, \ldots), b = (b_1, b_2, \ldots) \in \ell$, we set $a + b = (a_1 + b_1, a_2 + b_2, \ldots)$.
If $a = (a_1, a_2, \ldots) \in \ell$ and $c \in \mathbb{R}$, we define $ca = (ca_1, ca_2, \ldots)$.

We must check that the resulting vectors $a + b$ and ca are in ℓ. Indeed, if M_1 and M_2 are bounds for a and b, respectively, the Triangle Inequality implies that

$$|a_n + b_n| \leq |a_n| + |b_n| \leq M_1 + M_2 \qquad \text{for all } n \in \mathbb{N}.$$

Similarly, if $|a_n| \leq M$ for all $n \in \mathbb{N}$, then

$$|ca_n| \leq |c| \cdot M.$$

We conclude that both $a + b$ and ca are in ℓ, and thus addition and scalar multiplication are defined properly.

Again, it is straightforward to check the vector space axioms. The zero vector is, in this case, the zero sequence $(0, 0, \ldots)$, and if $a = (a_1, a_2, \ldots) \in \ell$, its negative is the sequence $-a = (-a_1, -a_2, \ldots)$.

Example 11.3.5 Our motivating example $\mathbb{R}^n$ can be generalized. For any field $\mathbb{F}$ and $n \in \mathbb{N}$, the set

$$\mathbb{F}^n = \mathbb{F} \times \mathbb{F} \times \cdots \times \mathbb{F}$$

can be turned into a vector space, by mimicking the definition of addition and scalar multiplication in $\mathbb{R}^n$. The field axioms can be used to easily verify the vector space axioms in $\mathbb{F}^n$.

Exercise 11.3.6

1. Which of the following are vector spaces over $\mathbb{R}$ (with respect to the natural addition and scalar multiplication)?
   a. All polynomials with real coefficients of degree at most 3.
   b. The set of all functions $f\colon \mathbb{R} \to \mathbb{R}$, satisfying $f(x) \geq -1$ for all $x \in \mathbb{R}$.
2. Let $\mathbb{F} = \{0, 1, a\}$ be a field with three elements. In the vector space $\mathbb{F}^2$, calculate the following.
   a. $a \cdot (1, a)$
   b. $(a, 1) + (0, 1)$
   c. $-(1, 1) + a(0, a)$

The vector space axioms, listed in Definition 11.3.1, include some of the basic properties of vectors observed in $\mathbb{R}^n$, but definitely not all of them. In mathematics, we try to keep our lists of axioms as short as possible, and omit statements that can be proved from the existing ones. We now prove a few basic consequences of the vector space axioms.

Theorem 11.3.7 *Let V be a vector space over a field $\mathbb{F}$. Then we have the following.*

1. $0 \cdot v = 0$ *for all* $v \in V$.
2. $c \cdot 0 = 0$ *for all* $c \in \mathbb{F}$.
3. *For all* $c \in \mathbb{F}$ *and* $v \in V$, *if* $cv = 0$, *then* $c = 0$ *or* $v = 0$.

Exercise 11.3.8

The zero symbol appears several times in Theorem 11.3.7. Which occurrences refer to the scalar $0 \in \mathbb{F}$, and which refer to the zero vector $0 \in V$?

Proof of Theorem 11.3.7

1. We prove that $0 \cdot v = 0$ through the following chain of equalities. The numbers in parentheses indicate the vector space axiom used in each step:

$$\begin{aligned} 0 \cdot v &\overset{(3)}{=} 0 \cdot v + 0 \overset{(4)}{=} 0 \cdot v + [0 \cdot v + (-0 \cdot v)] \overset{(2)}{=} [0 \cdot v + 0 \cdot v] + (-0 \cdot v) \\ &\overset{(8)}{=} (0+0) \cdot v + (-0 \cdot v) = 0 \cdot v + (-0 \cdot v) \overset{(4)}{=} 0. \end{aligned}$$

   Note that in the second last step we used one of the field axioms to replace $0+0$ by 0.

2. The proof of $c \cdot 0 = 0$ is very similar to Part 1, and is left as an exercise.
3. We show that if $cv = 0$ and $c \neq 0$, then $v = 0$:

$$v \overset{(5)}{=} 1 \cdot v = (c^{-1} \cdot c) \cdot v \overset{(2)}{=} c^{-1} \cdot (cv) = c^{-1} \cdot 0 = 0.$$

   In the second step we used $c^{-1} \cdot c = 1$, which can be justified by the field axioms, and the last step follows from Part 2 of the theorem. □

Exercise 11.3.9

Complete the proof of Part 2 of Theorem 11.3.7.

We ask that you prove a few more important consequences in Problem 11.11 on page 332.

11.4 Subspaces

In Exercise 11.3.6 we saw that the set of all real polynomials of degree at most 3 is a vector space with respect to the usual addition and multiplication of functions. We denote this space by $\mathcal{P}_3$:

$$\mathcal{P}_3 = \{ax^3 + bx^2 + cx + d \colon a, b, c, d \in \mathbb{R}\}.$$

This space, however, is a subset of another vector space $\mathcal{P}$, containing *all* real polynomials (of *any* degree). We invite you to verify that $\mathcal{P}$ is indeed a vector space over $\mathbb{R}$.

In mathematics, we are often interested in studying *sub-structures*. That is, subsets that possess the same properties as the underlying set. In the context of vector spaces, such sub-structures are called *subspaces*. We thus say that $\mathcal{P}_3$ is a *subspace* of $\mathcal{P}$.

Definition 11.4.1 Let V be a vector space over a field $\mathbb{F}$. A subset $U \subseteq V$ is a *(vector) subspace* of V if under the operations of vector addition and scalar multiplication in V, U is a vector space over $\mathbb{F}$.

Example 11.4.2 A vector space V can be regarded as a subspace of itself. Moreover, the set $\{0\}$, containing only the zero vector, is a subspace (check this!). We refer to $\{0\}$ and V as the *trivial subspaces* of V, and view them as the smallest and largest subspaces of V. As you can imagine, we will usually be more interested in subspaces other than those trivial ones.

Given a subset U of a vector space V, how would one prove that it is a subspace? At first, it may seem that we should attempt to verify all the vector space axioms in U. Fortunately, there is a shortcut. Most of the vector space axioms clearly remain valid in U. For instance, if commutativity of addition holds in V, then it clearly must hold in any subset of V. In the following proposition we show that it is enough to verify closure under addition and scalar multiplication in order to prove that a non-empty subset of V is a subspace.

Proposition 11.4.3 *Let V be a vector space over a field $\mathbb{F}$, and U a non-empty subset of V. Then U is a subspace of V if and only if it is closed under vector addition and scalar multiplication. That is,*

$$u_1 + u_2 \in U \text{ for all } u_1, u_2 \in U$$

and

$$cu \in U \text{ for all } c \in \mathbb{F} \text{ and } u \in U.$$

Proof First, if U is a subspace of V, then Definition 11.3.1 must hold in U. In particular, U is closed under vector addition and scalar multiplication: $u_1 + u_2 \in U$ for all $u_1, u_2 \in U$ and $cu \in U$ for all $c \in \mathbb{F}$ and $u \in U$.

Conversely, assume that U is closed under the operations of V. In order for U to be a vector space, it must contain the zero vector. As $U \neq \varnothing$, there exists some $u \in U$. From our assumptions, we conclude that $(-1)u = -u \in U$, and hence $u + (-u) = 0 \in U$. Thus, the zero vector is indeed in U.[4]

Axioms 1, 2, 3, 5, 6, 7 and 8 clearly remain valid in U, as they are valid in V. To confirm Axiom 4, we note again that for every $u \in U$ we have $(-1)u = -u \in U$, as U is closed under scalar multiplication. Thus, negatives of vectors in U also lie in U. The condition $u + (-u) = 0$ remains valid in U, and hence Axiom 4 also holds true.

We conclude that U is a vector space, as needed. □

[4] Note how we used the fact that $(-1)u = -u$, which is true in every vector space. See Problem 11.11.

Proposition 11.4.3 is a useful tool that can be used to quickly verify whether a subset is a vector subspace. Let us look at a few examples.

Example 11.4.4 Prove that the set

$$U = \{(x, y, z) \in \mathbb{R}^3 : x = 2y + 3z\}$$

is a subspace of $\mathbb{R}^3$.

Solution

The condition $x = 2y + 3z$ implies that if $y = z = 1$, then $x = 5$, and so the triple $(5, 1, 1)$ is an element of U. This shows that U is non-empty.

To show closure under addition, let $\underline{u}_1 = (x_1, y_1, z_1)$ and $\underline{u}_2 = (x_2, y_2, z_2)$ be two vectors in U. This means that

$$x_1 = 2y_1 + 3z_1 \qquad \text{and} \qquad x_2 = 2y_2 + 3z_2,$$

and by adding the two equalities, we get

$$x_1 + x_2 = 2(y_1 + y_2) + 3(z_1 + z_2).$$

It follows that $\underline{u}_1 + \underline{u}_2 \in U$, as needed.

Finally, we verify closure under scalar multiplication. Let $\underline{u} = (x, y, z) \in U$ and $c \in \mathbb{R}$. As $\underline{u}$ is an element of U, we have

$$x = 2y + 3z \qquad \Rightarrow \qquad cx = 2cy + 3cz,$$

and so $c\underline{u} = (cx, cy, cz) \in U$.

By Proposition 11.4.3 we conclude that U is a subspace of $\mathbb{R}^3$.

In the example above, the subspace U can be viewed as a subset of three-dimensional space. Geometrically, U is a plane in $\mathbb{R}^3$ that passes through the origin (see Figure 11.11). With some effort, one can show that, other than the trivial subspaces $\{(0, 0, 0)\}$ and $\mathbb{R}^3$ itself, the only subspaces of $\mathbb{R}^3$ are lines and planes through the origin.

Example 11.4.5 Bounded Functions. Let $\mathcal{F}$ be the set of all functions from $\mathbb{R}$ to $\mathbb{R}$. It is straightforward to show that $\mathcal{F}$ is a vector space over $\mathbb{R}$ with respect to addition and scalar multiplication of functions. Show that $\mathcal{L}$, the set of all bounded functions $f : \mathbb{R} \to \mathbb{R}$ forms a subspace of $\mathcal{F}$.

Solution

We use Proposition 11.4.3. Clearly $\mathcal{L}$ is non-empty. If $f, g \in \mathcal{L}$, then there exist real numbers M and N such that

$$|f(x)| \leq M \qquad \text{and} \qquad |g(x)| \leq N \qquad \text{for all } x \in \mathbb{R}.$$

The Triangle Inequality implies that

$$|f(x) + g(x)| \leq |f(x)| + |g(x)| \leq M + N$$

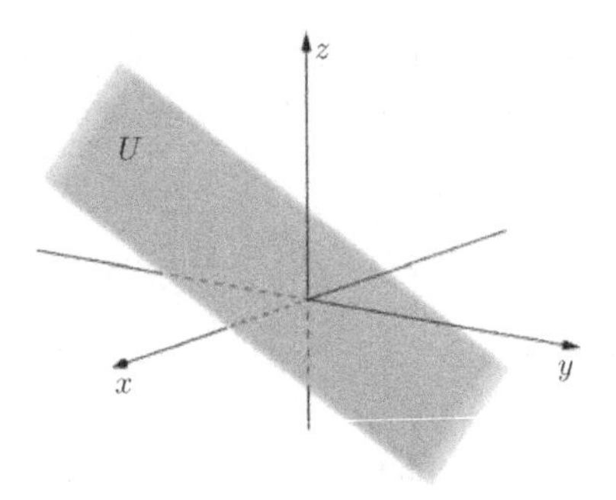

Figure 11.11 The plane U is a subspace of $\mathbb{R}^3$.

for all $x \in \mathbb{R}$, from which it follows that $f+g$ is bounded. Therefore, $f+g \in \mathcal{L}$ and so $\mathcal{L}$ is closed under addition.

Moreover, if $c \in \mathbb{R}$, then

$$|c \cdot f(x)| = |c| \cdot |f(x)| \leq |c| \cdot M,$$

which shows that cf is a bounded function. We conclude that $cf \in \mathcal{L}$, which proves closure under scalar multiplication.

We have thus showed that $\mathcal{L}$ is a vector subspace of $\mathcal{F}$.

Example 11.4.6 Polynomials. We already mentioned the space $\mathcal{P}$ of all polynomials with real coefficients. For each $n \in \mathbb{N}$, we define $\mathcal{P}_n$ to be the set of all real polynomials of degree at most n:

$$\mathcal{P}_n = \{a_n x^n + \cdots + a_1 x + a_0 \colon a_0, a_1, \ldots, a_n \in \mathbb{R}\}.$$

Addition and scalar multiplication in $\mathcal{P}_n$ can never produce polynomials of degree higher than n, and hence $\mathcal{P}_n$ is a subspace of $\mathcal{P}$ according to Proposition 11.4.3.

Example 11.4.7 Sequences. The space of all real infinite sequences, which we denote by V, is a vector space over $\mathbb{R}$. The vector space operations are the usual (component-wise) addition and scalar multiplication. Let U be the set of all sequences in V which are *eventually zero*, that is, sequences of the form

$$a_1, a_2, \ldots, a_m, 0, 0, 0, \ldots.$$

Then U is a subspace of V as it is non-empty, and closed under addition and scalar multiplication (check!).

We end this section by discussing how to form new subspaces from existing ones. If U are W are subspaces, would their union and intersection also be a subspace? How else can we try to create new subspaces from U and W?

Exercise 11.4.8
Find an example showing that the union of two subspaces need not be a subspace.

The following theorem asserts that the intersection of two subspace is a subspace, and provides another way of forming subspaces.

Theorem 11.4.9 *Let V be a vector space over a field $\mathbb{F}$, and $U, W \subseteq V$ two subspaces. Then the following holds.*

1. *The intersection $U \cap W$ is a subspace of V.*
2. *The set*

$$U + W = \{v \in V \colon v = u + w \text{ for some } u \in U \text{ and } w \in W\},$$

called the sum of U and W, is a subspace of V.

Proof 1. Clearly $U \cap W \neq \varnothing$, since both U and V contain the zero vector. If $v_1, v_2 \in U \cap W$, then $v_1 + v_2 \in U \cap W$, as both U and W are closed under addition. Similarly, if $v \in U \cap W$ and $c \in \mathbb{F}$, then $cv \in U \cap W$ as both U and W are closed under scalar multiplication. By Proposition 11.4.3, $U \cap W$ is a subspace of V.
2. Again, we use Proposition 11.4.3 in our proof. $U + W \neq \varnothing$ as it contains the zero vector. If $v_1, v_2 \in U + W$, then according to the definition of $U + W$,

$$v_1 = u_1 + w_1 \quad \text{and} \quad v_2 = u_2 + w_2 \quad \text{for } u_1, u_2 \in U \quad \text{and} \quad w_1, w_2 \in W.$$

The vector space axioms imply that

$$v_1 + v_2 = (u_1 + w_1) + (u_2 + w_2) = (u_1 + u_2) + (w_1 + w_2) \in U + W,$$

and so $U + W$ is closed under addition. Now suppose that $v = u + w \in U + W$, where $u \in U$ and $w \in W$, and $c \in \mathbb{F}$. As U and W are closed under scalar multiplication, we conclude that $cu \in U$ and $cw \in W$. Consequently, $cv = cu + cw \in U + W$, which shows closure under scalar multiplication. We have thus shown that $U + W$ is a subspace of V. □

11.5 Linear Maps and Isomorphisms

In previous chapters we discussed functions between two sets. When the underlying sets are equipped with additional structure, we are often interested in studying functions which preserve that structure. In particular, functions between two vector spaces that preserve addition and scalar multiplication are called *linear maps*, and are fundamental in science and mathematics. In this section, we begin with the formal definition of a linear map and look at some examples and a few basic properties of such functions.

Throughout this section, V and W denote vector spaces over the same field $\mathbb{F}$.

Definition 11.5.1 A function $T\colon V \to W$ is called a *linear map* (or a *linear transformation*) from V to W if the following conditions hold.

1. $T(v_1 + v_2) = T(v_1) + T(v_2)$ for all $v_1, v_2 \in V$.
2. $T(cv) = cT(v)$ for all $v \in V$ and $c \in \mathbb{F}$.

Proving that a function is a linear map is often a routine check of the definition.

Example 11.5.2 Prove that the function $T\colon \mathbb{R}^3 \to \mathbb{R}^2$, given by

$$T(x, y, z) = (x + y, y - z)$$

is a linear map.

Solution

If $\underline{v}_1 = (x_1, y_1, z_1)$ and $\underline{v}_2 = (x_2, y_2, z_2)$ are vectors in $\mathbb{R}^3$, then

$$\begin{aligned} T(\underline{v}_1 + \underline{v}_2) &= T(x_1 + x_2, y_1 + y_2, z_1 + z_2) \\ &= (x_1 + x_2 + y_1 + y_2, y_1 + y_2 - z_1 - z_2) \end{aligned}$$

and

$$\begin{aligned} T(\underline{v}_1) + T(\underline{v}_2) &= (x_1 + y_1, y_1 - z_1) + (x_2 + y_2, y_2 - z_2) \\ &= (x_1 + x_2 + y_1 + y_2, y_1 + y_2 - z_1 - z_2), \end{aligned}$$

which shows that $T(\underline{v}_1 + \underline{v}_2) = T(\underline{v}_1) + T(\underline{v}_2)$.

Similarly, if $\underline{v} = (x, y, z) \in \mathbb{R}^3$ and $c \in \mathbb{R}$, then

$$T(c\underline{v}) = T(cx, cy, cz) = (cx + cy, cy - cz)$$

and

$$cT(\underline{v}) = c(x + y, y - z) = (cx + cy, cy - cz),$$

and so $T(c\underline{v}) = cT(\underline{v})$. We have thus proved that T is a linear map from $\mathbb{R}^3$ to $\mathbb{R}^2$.

A basic property of linear maps is that they map the zero vector in the domain to the zero vector in the codomain. We ask that you attempt to prove this fact before proceeding.

Exercise 11.5.3

Show that if $T\colon V \to W$ is a linear map, then $T(0) = 0$.

We now define two sets that can be associated with a given linear map, the kernel and the image. We then show that these sets are, in fact, subspaces of the domain and the codomain.

Definition 11.5.4 Let $T\colon V \to W$ be a linear map.

1. The *kernel* (or *null space*) of T, denoted by $Ker(T)$, is the set of all $v \in V$ whose image under T is the zero vector in W:
$$Ker(T) = \{v \in V\colon Tv = 0\} \subseteq V.$$
2. The *image* (or *range*) of T, denoted by $Im(T)$, is the image of T as a function:
$$Im(T) = \{Tv\colon v \in V\} \subseteq W.$$

Example 11.5.5 Find the kernel and the image of the linear map in Example 11.5.2.

Solution

By definition, an element $\underline{v} = (x, y, z) \in \mathbb{R}^3$ is in $Ker(T)$ if and only if $T(\underline{v}) = \underline{0}$. This condition is equivalent to

$$T(x, y, z) = (0, 0) \quad \Leftrightarrow \quad \begin{cases} x + y = 0 \\ y - z = 0 \end{cases} \quad \Leftrightarrow \quad \begin{cases} x = -y \\ z = y. \end{cases}$$

That is, the kernel of T is the set of all vectors (x, y, z) in $\mathbb{R}^3$, for which $x = -y$ and $z = y$. Hence we write

$$Ker(T) = \{(-y, y, y)\colon y \in \mathbb{R}\}.$$

For the image, note that for each $(a, b) \in \mathbb{R}^2$, we have

$$T(a, 0, -b) = (a + 0, 0 - (-b)) = (a, b).$$

Consequently, it follows that $Im(T) = \mathbb{R}^2$.

Theorem 11.5.6 *Let $T\colon V \to W$ be a linear map. Then $Ker(T)$ is a subspace of V and $Im(T)$ is a subspace of W.*

Proof We use Proposition 11.4.3 to prove that both $Ker(T)$ and $Im(T)$ are subspaces.

- Clearly, $Ker(T) \neq \varnothing$ as $T(0) = 0$ and so $0 \in Ker(T)$. Let $v_1, v_2 \in Ker(T)$. Then $T(v_1) = T(v_2) = 0$. As T is a linear map, we get
$$T(v_1 + v_2) = T(v_1) + T(v_2) = 0 + 0 = 0$$

and so $v_1 + v_2 \in Ker(T)$, which has proved that $Ker(T)$ is closed under addition.

Moreover, if $c \in \mathbb{F}$ and $v \in Ker(T)$, then

$$T(cv) = cT(v) = c \cdot 0 = 0$$

from which we conclude that $cv \in Ker(T)$ and thus the kernel is closed under scalar multiplication.

By Proposition 11.4.3, $Ker(T)$ is a subspace of V.

- We now turn to proving that $Im(T)$ is a subspace of W. First, $Im(T) \neq \varnothing$ as $T(0) = 0 \in Im(T)$. Let w_1, w_2 be two vectors in $Im(T)$. This means that

$$w_1 = T(v_1) \quad \text{and} \quad w_2 = T(v_2) \quad \text{for some} \quad v_1, v_2 \in V.$$

But then

$$w_1 + w_2 = T(v_1) + T(v_2) = T(v_1 + v_2)$$

and hence $w_1 + w_2 \in Im(T)$.

Finally, if $c \in \mathbb{F}$ and $w \in Im(T)$, then $w = T(v)$ for some $v \in V$, and we have

$$cw = cT(v) = T(cv).$$

Therefore, $Im(T)$ is closed under addition and scalar multiplication, and hence is a subspace of W. □

Example 11.5.7 The function $P\colon \mathbb{R}^2 \to \mathbb{R}^2$, given by

$$P(x, y) = \left(\frac{x+y}{2}, \frac{x+y}{2}\right)$$

is a linear map (check!). Geometrically, it represents the projection of a vector $\underline{v} = (x, y) \in \mathbb{R}^2$ onto the line $y = x$ (see Figure 11.12). Find the kernel and the image of P.

Solution

The kernel of P is

$$Ker(P) = \left\{(x, y)\colon \frac{x+y}{2} = 0\right\} = \{(x, y)\colon y = -x\},$$

which aligns with the geometrical interpretation. The projection of a vector lying on (or parallel to) the line $y = -x$ is a single point, representing the zero vector, while other vectors in $\mathbb{R}^2$ have a non-zero projection.

We can also see from the diagram that the image of P will be simply the set of all points on the line $y = x$. Here is an algebraic proof of this observation. For every $(a, b) \in \mathbb{R}^2$ we have $P(a, b) = \left(\frac{a+b}{2}, \frac{a+b}{2}\right)$, which is a point on the

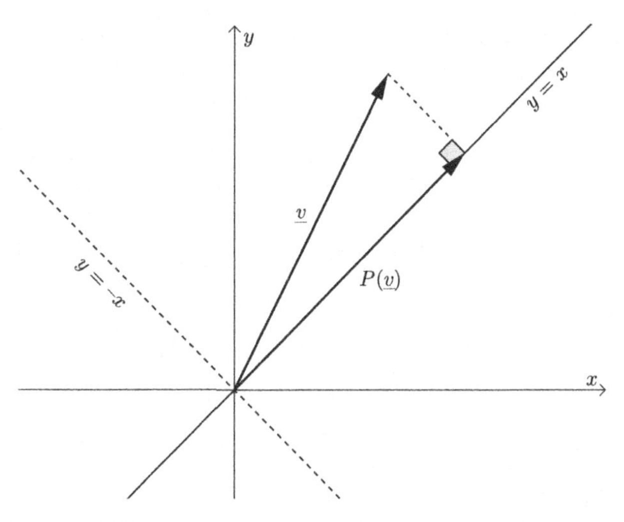

Figure 11.12 The map P is the projection onto the line $y = x$.

line $y = x$, as both coordinates are the same. On the other hand, take any point (x, x) on this line, and note that $P(0, 2x) = (x, x)$, which shows that (x, x) is in the image of P. In conclusion:

$$Im(P) = \{(x, y) \colon y = x\} = \{(x, x) \colon x \in \mathbb{R}\}.$$

One of the interesting properties of the kernel, is that it can be used to check whether a linear map is injective.

Theorem 11.5.8 *A linear map $T \colon V \to W$ is injective if and only if $Ker(T) = \{0\}$.*

Proof Assume T is injective. Then there is at most one vector in V which is mapped to the zero vector in W. As $T(0) = 0$ we conclude that $Ker(T) = \{0\}$.

Conversely, suppose that $Ker(T) = \{0\}$, and that v_1 and v_2 are two vectors in V such that $T(v_1) = T(v_2)$. As T is a linear map, we get

$$T(v_1) - T(v_2) = 0 \quad \Rightarrow \quad T(v_1 - v_2) = 0 \quad \Rightarrow \quad v_1 - v_2 \in Ker(T).$$

But $Ker(T) = \{0\}$, from which it follows that $v_1 - v_2 = 0$ or $v_1 = v_2$. We thus have proved that T is an injective function. □

Take a look at Example 11.5.7. In that case, $Ker(P) \neq \{0\}$, and indeed, the projection map P is not injective, as different points in $\mathbb{R}^2$ may have the same image.

If a linear map between the two vector spaces is a bijection, would its inverse be a linear map as well? The following proposition answers this question.

Proposition 11.5.9 *If $T\colon V \to W$ is a bijective linear map, then $T^{-1}\colon W \to V$ is also a linear map.*

Proof Let $w_1, w_2 \in W$. Then $w_1 = T(v_1)$ and $w_2 = T(v_2)$ for some $v_1, v_2 \in V$. We get:

$$T(v_1) + T(v_2) = w_1 + w_2 \quad \Rightarrow \quad T(v_1 + v_2) = w_1 + w_2$$

and thus

$$T^{-1}(w_1 + w_2) = v_1 + v_2 = T^{-1}(w_1) + T^{-1}(w_2).$$

Also, if $w \in W$ and $c \in \mathbb{F}$, then $w = T(v)$ for some $v \in V$, and we get

$$T(cv) = cT(v) = cw \quad \Rightarrow \quad cv = T^{-1}(cw) \quad \Rightarrow \quad cT^{-1}(w) = T^{-1}(cw).$$

According to Definition 11.5.1, T^{-1} is a linear map. □

Bijective linear maps are of great importance in algebra and other areas of mathematics. They allow us to identify vector spaces that may initially look different, as being essentially the same.

Definition 11.5.10 A bijective linear map $T\colon V \to W$ is called an *isomorphism*. Two vector spaces (over the same field) are said to be *isomorphic*, if there is an isomorphism between them.

Example 11.5.11 Show that the space $\mathcal{P}_2$ of all real polynomials of degree at most two is isomorphic to $\mathbb{R}^3$.

Solution

The key observation here is that a polynomial of degree at most 2 has the form $ax^2 + bx + c$, and hence is determined by three real numbers. Therefore, we define a function

$$T\colon \mathcal{P}_2 \to \mathbb{R}^3 \qquad \text{by} \qquad T(ax^2 + bx + c) = (a, b, c).$$

It is straightforward to check that T is a linear map. Moreover, if $f(x) = ax^2 + bx + c$ is an element of $\mathcal{P}_2$, and $T(f) = 0$, then $a = b = c = 0$, and so f is the zero polynomial. By Theorem 11.5.8, T is injective. T is also surjective, as given any vector $(a, b, c) \in \mathbb{R}^3$, the polynomial $ax^2 + bx + c$ is mapped by T to that triple:

$$T(ax^2 + bx + c) = (a, b, c).$$

Therefore, T is an isomorphism. This example shows that the two vector spaces $\mathcal{P}_2$ and $\mathbb{R}^3$, although described differently, have the same underlying vector space structure.

11.6 Problems

11.1 Suppose $\underline{u}, \underline{v} \in \mathbb{R}^n$ (where $n \in \mathbb{N}$). Simplify the following expressions.

a. $3(\underline{u} - \underline{v}) + \frac{1}{2}(\underline{v} - \underline{u})$
b. $\underline{u} - \frac{1}{2}(\underline{v} + 3\underline{u}) + \frac{3}{4}(\underline{v} - \underline{u})$

11.2 Solve to find the vector $\underline{x}$.

a. $\underline{x} + 2[\underline{x} + (2,2)] = 3[\underline{x} + (3,3)]$
b. $3[\underline{x} + (-1,3,4)] = (5,5,5) - 5\underline{x}$

11.3 Express the vector $\underline{w}$ in terms of $\underline{u}$ and $\underline{v}$.

a. $5(\underline{w} - \underline{u}) = \underline{v} + 3\underline{w}$
b. $-\frac{1}{2}\underline{w} + \underline{v} - \underline{u} = \frac{1}{3}(\underline{w} + \underline{u})$

11.4 Find real numbers a, b, c so that the following equation holds true:

$$a(2,4,-6) + b(1,-1,0) + c\left(\frac{2}{5}, 7, -2\right) = (2,-36,7).$$

11.5 The *length* of a vector $\underline{v} = (v_1, v_2, \ldots, v_n) \in \mathbb{R}^n$ is defined as

$$||\underline{v}|| = \sqrt{v_1^2 + v_2^2 + \cdots + v_n^2}.$$

a. Calculate the length of each of the following vectors.

$$\underline{v} = (2,-3) \qquad \underline{u} = (1,-7,8) \qquad \underline{w} = (1,-1,1,-1)$$

b. For an arbitrary $n \in \mathbb{N}$, calculate the length of the following vectors in $\mathbb{R}^n$. Your answer may depend on n.

$$\underline{x} = (n,n,\ldots,n) \quad \underline{y} = \left(\frac{1}{\sqrt{n}}, \frac{1}{\sqrt{n}}, \ldots, \frac{1}{\sqrt{n}}\right) \quad \underline{z} = (1,2,3,\ldots,n)$$

c. For every $t, s \in \mathbb{R}$ define the following vector in $\mathbb{R}^3$:

$$\underline{r} = (\cos t \sin s, \sin t \sin s, \cos s)\,.$$

Show that the length of $\underline{r}$ is independent of t and s.

d. Prove that for any $\underline{v} \in \mathbb{R}^n$ and $t \in \mathbb{R}$ we have $||t\underline{v}|| = |t| \cdot ||\underline{v}||$.

11.6 Two geometric vectors $\vec{AB}$ and $\vec{CD}$ in $\mathbb{R}^n$ are said to be *equivalent* if they are represented by the same algebraic vector $\underline{u}$. Verify that this is, as expected, an equivalence relation.

11.7 In Figure 11.13, E and F are the midpoints of AB and AC, respectively. Denote $\underline{u} = \vec{AB}$ and $\underline{v} = \vec{AC}$. Express $\vec{EF}$ in terms of $\underline{u}$ and $\underline{v}$. What is the geometrical interpretation of the result obtained?

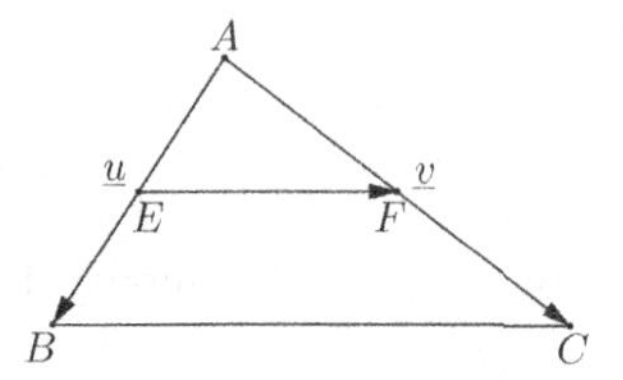

Figure 11.13 E and F are the midpoints of AB and AC, respectively.

11.8 In Figure 11.14, E and F are the midpoints of AB and DC, respectively. If $\underline{u} = \vec{AB}$ and $\underline{v} = \vec{BC}$, express the vector $\vec{EF}$ in terms of $\underline{u}$ and $\underline{v}$. (Hint: Express $\vec{EF}$ in two ways using the tip-to-tail rule.) What is the geometrical interpretation of the result obtained?

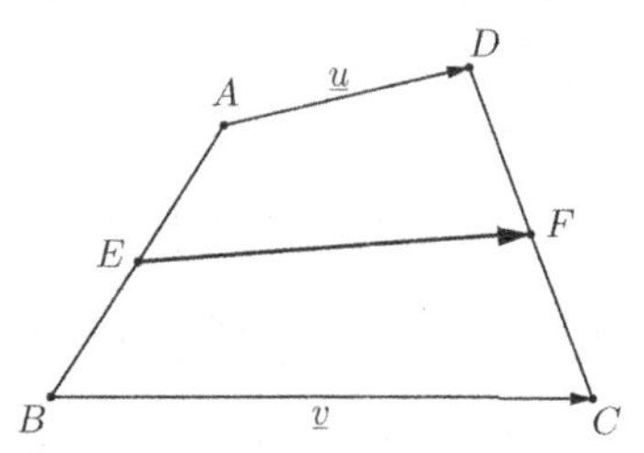

Figure 11.14 E and F are the midpoints of AB and DC, respectively.

11.9 Consider the cube in Figure 11.15. Denote $\underline{u} = \vec{AB}$, $\underline{v} = \vec{AD}$ and $\underline{w} = \vec{AA'}$. Express the vectors $\vec{D'C}$, $\vec{DB'}$ and $\vec{B'C}$ in terms of $\underline{u}$, $\underline{v}$ and $\underline{w}$.

11.10 Consider the pyramid in Figure 11.16. Denote $\underline{u} = \vec{DA}$, $\underline{v} = \vec{AB}$ and $\underline{w} = \vec{DC}$. Show that $\vec{AC} + \vec{BD} = \vec{AD} + \vec{BC}$ and express these equal quantities in terms of $\underline{u}$, $\underline{v}$ and $\underline{w}$.

11.11 Let V be a vector space over a field $\mathbb{F}$. Prove the following.

a. $-v = (-1)v$ and $-(-v) = v$ for all $v \in V$.
b. $-(cv) = (-c)v$ for all $c \in \mathbb{F}$ and $v \in V$.

11.12 Let V be a vector space over a field $\mathbb{F}$, $u, v, w \in V$ and $0 \neq c \in \mathbb{F}$. Prove the following statements.

a. If $u + v = w + v$ then $u = w$.
b. If $cu = cv$ then $u = v$.

11.13 Suppose we replace the vector space Axioms 3 and 4 (in Definition 11.3.1) by the following.

3′ $v + 0 = 0 + v = v$ for all $v \in V$.
4′ For every $v \in V$, there is a $u \in V$ such that $v + u = u + v = 0$.

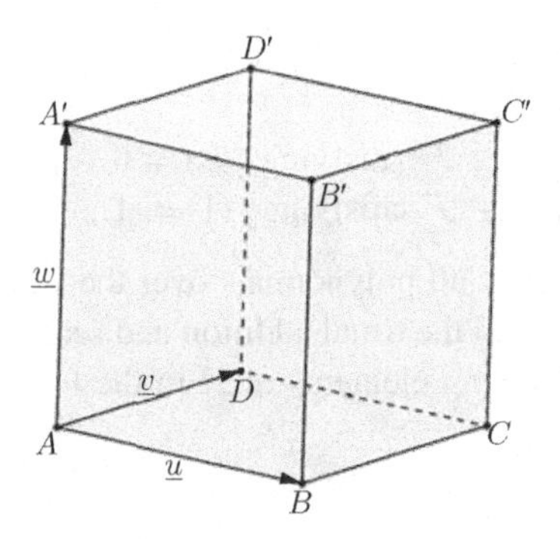

Figure 11.15 A cube.

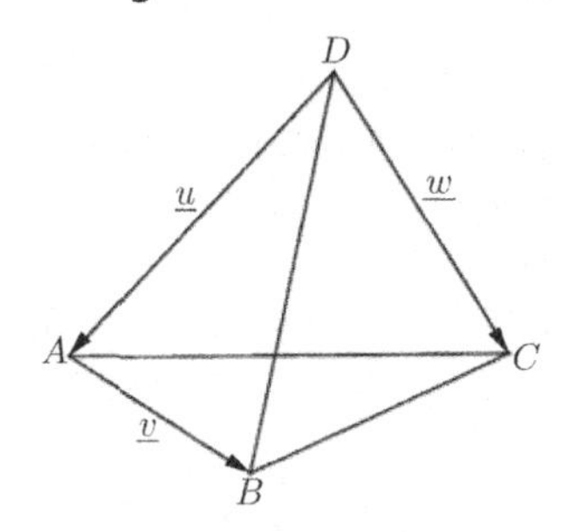

Figure 11.16 A triangular pyramid.

Prove that, in this case, commutativity of addition (Axiom 1) follows from the other vector space axioms.
(Hint: Expand the expression $(1+1)(u+v)$ in two different ways.)

11.14 Let U and V be two vector spaces over a field $\mathbb{F}$. On $U \times V$ we define addition and scalar multiplication as follows:

$$(u_1, v_1) + (u_2, v_2) = (u_1 + u_2, v_1 + v_2) \qquad \text{and} \qquad c(u, v) = (cu, cv)$$

(where $u_1, u_2, u \in U$, $v_1, v_2, v \in V$, and $c \in \mathbb{F}$). Show that $U \times V$ is a vector space over $\mathbb{F}$ with respect to these operations.

11.15 Let $\{U_\alpha\}_{\alpha \in J}$ be a collection of subspaces of a vector space V over a field $\mathbb{F}$. Prove that the intersection

$$\bigcap_{\alpha \in J} U_\alpha = \{v \in V : v \in U_\alpha \text{ for all } \alpha \in J\}$$

is a subspace of V.

11.16 Let $\mathcal{F}$ be the vector space (over $\mathbb{R}$) of all functions $f : \mathbb{R} \to \mathbb{R}$, with the usual addition and scalar multiplication. Which of the following subsets of $\mathcal{F}$ are subspaces? Justify your answer.

a. Even functions.

b. Odd functions.
c. Positive functions.
d. All functions $f \in \mathcal{F}$ satisfying $f(0) = 0$.
e. All functions $f \in \mathcal{F}$ satisfying $f(1) = 1$.

11.17 Let $\mathcal{P}$ be the set of all polynomials over the field with three elements $\mathbb{F}_3 = \{0, 1, \alpha\}$, with the usual addition and scalar multiplication. Write each of the following elements of $\mathcal{P}$ in the form $ax^2 + bx + c$ where $a, b, c \in \mathbb{F}_3$.

a. $(x^2 + 1) + (x^2 + x)$
b. $\alpha \cdot (\alpha x^2)$
c. $-(1 + \alpha x + x^2)$
d. $(-1)x - \alpha - x^2$

11.18 Let U, W be subspaces of a vector space V over a field $\mathbb{F}$. We say that V is the direct sum of U and V, and write $V = U \oplus W$, if the following two conditions hold.

- Each $v \in V$ can be expressed as $v = u + w$ for some $u \in U$ and $w \in W$.
- $U \cap W = \{0\}$.

Consider the following subsets of $\mathbb{R}^3$:

$$\begin{aligned} U &= \{(a, a, a) : a \in \mathbb{R}\} \\ W &= \{(0, 2b, 3c) : b, c \in \mathbb{R}\}. \end{aligned}$$

Show that U, W are subspaces of $\mathbb{R}^3$, and that $U \oplus W = \mathbb{R}^3$.

11.19 Let V be the vector space of all infinite real sequences, with component-wise addition and scalar multiplication. We say that a sequence $a = (a_1, a_2, \ldots)$ is *eventually zero* if there is an $n_0 \in \mathbb{N}$ such that $a_n = 0$ for all $n \geq n_0$. Show that the set of all sequences in V which are eventually zero form a subspace.

11.20 Let V and W be two vector spaces over a field $\mathbb{F}$. Show that a function $T: V \to W$ is a linear map if and only if $T(cv_1 + v_2) = cT(v_1) + T(v_2)$ for all $v_1, v_2 \in V$ and $c \in \mathbb{F}$.

11.21 Let $\mathcal{F}$ be the vector space of all functions $f: \mathbb{R} \to \mathbb{R}$, and let $c \in \mathbb{R}$. Show that the map $T: \mathcal{F} \to \mathbb{R}$ given by $T(f) = f(c)$ is a linear map.

11.22 Let $T: \mathbb{R}^2 \to \mathbb{R}$ be a linear map, and denote $\underline{e}_1 = (1, 0)$ and $\underline{e}_2 = (0, 1)$.

a. Show that every $\underline{v} \in \mathbb{R}^2$ can be written as $x\underline{e}_1 + y\underline{e}_2$ for some $x, y \in \mathbb{R}$.
b. Show that there are $a, b \in \mathbb{R}$ such that $T(x, y) = ax + by$ for all $x, y \in \mathbb{R}$.

This problem shows that every linear map $T\colon \mathbb{R}^2 \to \mathbb{R}$ has the form $T(x, y) = ax + by$ for some $a, b \in \mathbb{R}$.

11.23 Generalize the result of Problem 11.22. Show that every linear map $T\colon \mathbb{R}^n \to \mathbb{R}^m$ has the form

$$T(x_1, \ldots, x_n) = \underline{a}_1 x_1 + \cdots + \underline{a}_n x_n$$

for some fixed vectors $\underline{a}_1, \ldots, \underline{a}_n \in \mathbb{R}^m$.

11.24 Recall that two vector spaces V, W over the same field $\mathbb{F}$ are said to be isomorphic if there is an isomorphism T between them. Show that the relation of being isomorphic vector spaces is an equivalence relation.

11.25 Let U, V, W be vector spaces over a field $\mathbb{F}$ and $T\colon U \to V, S\colon V \to W$ be two linear maps. Show that the composition $S \circ T$ is also a linear map.

11.7 Solutions to Exercises

Solution to Exercise 11.1.4
Definition 11.1.2 implies that

$$(3\underline{u}) + (2\underline{v}) = (0, -12, 1.5) + (2, 0, -2) = (2, -12, -0.5).$$

Solution to Exercise 11.1.6
We prove the remaining properties. In our proof, c and d represent real numbers, and $\underline{u}, \underline{v}, \underline{w}$ represent vectors in $\mathbb{R}^n$ with coordinates u_i, v_i and w_i, respectively.

1. We have

$$\begin{aligned} \underline{v} + \underline{u} &= (v_1 + u_1, \ldots, v_n + u_n) \\ \underline{u} + \underline{v} &= (u_1 + v_1, \ldots, u_n + v_n). \end{aligned}$$

As $v_i + u_i = u_i + v_i$ for all $1 \le i \le n$, we conclude that $\underline{v} + \underline{u} = \underline{u} + \underline{v}$.

2. Again, computing each side separately gives:

$$\begin{aligned} (\underline{v} + \underline{u}) + \underline{w} &= ((v_1 + u_1) + w_1, \ldots, (v_n + u_n) + w_n) \\ \underline{v} + (\underline{u} + \underline{w}) &= (v_1 + (u_1 + w_1), \ldots, v_n + (u_n + w_n)). \end{aligned}$$

Associativity of addition in $\mathbb{R}$ implies that the two vectors obtained are equal.

3. This follows immediately as

$$\underline{v} + \underline{0} = (v_1 + 0, \ldots, v_n + 0) = (v_1, \ldots, v_n) = \underline{v}.$$

5. This is straightforward as well:

$$1 \cdot \underline{v} = (1 \cdot v_1, \ldots, 1 \cdot v_n) = (v_1, \ldots, v_n) = \underline{v}.$$

7. Finally, computing each side gives

$$c \cdot (\underline{v} + \underline{u}) = c(v_1 + u_1, \ldots, v_n + u_n) = (c(v_1 + u_1), \ldots, c(v_n + u_n))$$
$$c \cdot \underline{v} + c \cdot \underline{u} = (cv_1, \ldots, cv_n) + (cu_1, \ldots, cu_n) = (cv_1 + cu_1, \ldots, cv_n + cu_n),$$

and both sides are equal due to the distributive law for real numbers.

Solution to Exercise 11.1.7
Write $\underline{v} = (v_1, \ldots, v_n)$.

1. If $t \neq 0$, and $t\underline{v} = \underline{0}$, then for each $1 \leq i \leq n$ we get

$$tv_i = 0 \qquad \Rightarrow \qquad v_i = 0.$$

We conclude that $\underline{v} = (0, 0, \ldots, 0) = \underline{0}$, as needed.
2. $(-1)\underline{v} = ((-1)v_1, \ldots, (-1)v_n) = (-v_1, \ldots, -v_n) = -\underline{v}$
3. $-(t\underline{v}) = (-tv_1, \ldots, -tv_n) = (-t) \cdot (v_1, \ldots, v_n) = (-t)\underline{v}$

Solution to Exercise 11.2.4
If $A = (a_1, \ldots, a_n)$ and $B = (b_1, \ldots, b_n)$, then

$$\underline{u} = (b_1 - a_1, \ldots, b_n - a_n).$$

On the other hand, the vector $\vec{BA}$ is represented by

$$\vec{BA} = (a_1 - b_1, \ldots, a_n - b_n) = -(b_1 - a_1, \ldots, b_n - a_n) = -\underline{u}.$$

In conclusion, $\vec{BA}$ is represented by the algebraic vector $-\underline{u}$. That is, $\vec{BA} = -\underline{u}$.

Solution to Exercise 11.2.5
Suppose that two vectors $\vec{AB}$ and $\vec{CD}$ in $\mathbb{R}^2$ have the same length and direction. Then we can translate $\vec{AB}$ so that it coincides with $\vec{CD}$. Denote by Δx and Δy the horizontal and vertical shifts, and by a_i, b_i, c_i and d_i the coordinates of A, B, C and D.

The translation moves points A to C and B to D, and so we have:

$$\begin{aligned} C &= (c_1, c_2) = (a_1 + \Delta x, a_2 + \Delta y) \\ D &= (d_1, d_2) = (b_1 + \Delta x, b_2 + \Delta y). \end{aligned}$$

The corresponding algebraic vectors are:

$$\begin{aligned} \vec{AB} &= (b_1 - a_1, b_2 - a_2) \\ \vec{CD} &= (d_1 - c_1, d_2 - c_2) = ((b_1 + \Delta x) - (a_1 + \Delta x), (b_2 + \Delta y) - (a_2 + \Delta y)) \\ &= (b_1 - a_1, b_2 - a_2) \end{aligned}$$

and we see that $\vec{AB} = \vec{CD}$, as needed.

Solution to Exercise 11.2.8

If $\vec{OB}$ and $\vec{OD}$ lie on the same straight line, the two triangles are similar. Set

$$t = -\frac{DF}{BE} = -\frac{OF}{OE}$$

(the negative sign is needed to ensure that multiplication by t switches signs from negative to positive and vice versa). We get

$$OF = -t \cdot OE \qquad \Rightarrow \qquad d_1 = tb_1$$

and

$$DF = -t \cdot BE \qquad \Rightarrow \qquad d_2 = tb_2,$$

from which it follows that $\underline{v} = t\underline{u}$.

Conversely, if $\underline{v} = t\underline{u}$ for some $t \neq 0$, then $d_1 = tb_1$ and $d_2 = tb_2$, and we conclude that

$$\frac{OF}{OE} = |t| = \frac{DF}{BE}.$$

It follows that triangles ODF and OEB are similar, and hence the angles $\angle O_1$ and $\angle O_2$ are the same. We conclude that $\vec{OD}$ and $\vec{OB}$ lie on the same straight line, as needed.

Solution to Exercise 11.3.6

1. a. This set is a vector space. Denote by $\mathcal{P}$ all the polynomials with real coefficients of degree at most 3:

   $$\mathcal{P} = \{ax^3 + bx^2 + cx + d\colon a, b, c, d \in \mathbb{R}\}.$$

   The zero vector is simply the zero polynomial $f(x) = 0x^3 + 0x^2 + 0x + 0$, and the negative of an element $ax^3 + bx^2 + cx + d$ is $(-a)x^3 + (-b)x^2 + (-c)x + (-d)$. The vector space axioms can be easily verified. They all follow directly from the field axioms satisfied in $\mathbb{R}$.

   b. The set of all such functions is *not* a vector space, as negatives will not always exist. For example, $f(x) = e^x$ is a function satisfying $f(x) \geq -1$ for all $x \in \mathbb{R}$, but there is no such function $g(x)$ for which $f + g = 0$.

2. In a field with three elements, the following equalities hold (see Problem 2.50 on page 60):

   $$a^2 = 1, \qquad 1 + 1 = a, \qquad 1 + a = 0.$$

   We use these results and properties of fields to calculate the given expressions.

   a. $a \cdot (1, a) = (a \cdot 1, a \cdot a) = (a, a^2) = (a, 1)$.

   b. $(a, 1) + (0, 1) = (a + 0, 1 + 1) = (a, a)$.

c. $-(1,1)+a(0,a) = (-1)\cdot(1,1)+(a\cdot 0, a\cdot a) = a(1,1)+(0,a^2) = (a,a)+(0,1) = (a+0,a+1) = (a,0)$. Note how we used the fact that $-1 = a$ in $\mathbb{F}$.

Solution to Exercise 11.3.8

The scalar multiplication operation takes, as an input, a scalar and a vector, and provides, as an output, a vector. That is, in every equality of the form $cv = u$ in the theorem, c is a scalar and u, v are vectors. We restate the theorem, and underline each occurence of the zero vector. Zero scalars have no underline. Let V be a vector space. Then we have the following.

1. $0\cdot v = \underline{0}$ for all $v \in V$.
2. $c\cdot \underline{0} = \underline{0}$ for all $c \in \mathbb{F}$.
3. For all $c \in \mathbb{F}$ and $v \in V$, if $cv = \underline{0}$, then $c = 0$ or $v = \underline{0}$.

Solution to Exercise 11.3.9

We use a nearly identical argument to the proof of Part 1 to show that $c\cdot 0 = 0$:

$$\begin{aligned} c\cdot 0 &\overset{(3)}{=} c\cdot 0+0 \overset{(4)}{=} c\cdot 0+[c\cdot 0+(-c\cdot 0)] \overset{(2)}{=} [c\cdot 0+c\cdot 0]+(-c\cdot 0) \\ &\overset{(7)}{=} c\cdot(0+0)+(-c\cdot 0) \overset{(3)}{=} c\cdot 0+(-c\cdot 0) \overset{(4)}{=} 0. \end{aligned}$$

Solution to Exercise 11.4.8

There are many possible examples. For instance, the following are subspaces of $\mathbb{R}^2$ (check!):

$$\begin{aligned} U &= \{(x,0) : x \in \mathbb{R}\} \\ V &= \{(0,y) : y \in \mathbb{R}\}. \end{aligned}$$

However, $U \cup V$ is not a subspace, as it is not closed under addition. We have $(1,0) \in U$ and $(0,1) \in V$, but their sum $(1,1)$ is not in $U \cup V$. Therefore, the union is *not* a subspace of $\mathbb{R}^2$.

Solution to Exercise 11.5.3

We use Condition 2 from Definition 11.5.1 with $c = 0$ and $v = 0$, and get:

$$T(0\cdot 0) = 0\cdot T(0) \qquad \Rightarrow \qquad T(0) = 0.$$

Note that we applied the fact that $0v = 0$ for any $v \in V$ (Theorem 11.3.7).

INDEX

Made in the USA
Middletown, DE
11 January 2024

47590792R00199